ANNALS OF DISCRETE MATHEMATICS

annals of discrete mathematics

Managing Editor

Peter L. HAMMER, University of Waterloo, Ont., Canada

Advisory Editors

C. BERGE, Université de Paris, France
M.A. HARRISON, University of California, Berkeley, CA, U.S.A.
V. KLEE, University of Washington, Seattle, WA, U.S.A.
J.H. VAN LINT, California Institute of Technology, Pasadena, CA, U.S.A.
G.-C. ROTA, Massachusetts Institute of Technology, Cambridge, MA, U.S.A.

N·H
P C

NORTH-HOLLAND PUBLISHING COMPANY – AMSTERDAM • NEW YORK • OXFORD

ANNALS OF DISCRETE MATHEMATICS 3

ADVANCES IN GRAPH THEORY

Edited by

B. BOLLOBÁS, University of Cambridge, Cambridge CB2 1SB, England

1978

NORTH-HOLLAND PUBLISHING COMPANY – AMSTERDAM • NEW YORK • OXFORD

PRINTED IN THE NETHERLANDS

FOREWORD

The Cambridge Combinatorial Conference was held at Trinity College from 12 to 14 May 1977, under the auspices of the Department of Pure Mathematics and Mathematical Statistics. Twenty two of the participants, many from abroad, were invited to give talks. This volume consists of most of the papers they presented, together with two additional articles which are closely connected with the themes of the conference. The opportunity was taken, where necessary, to revise and amend the papers, each of which has been thoroughly refereed. It is a pleasure to acknowledge the rapid and efficient work of both referees and authors.

This volume is dedicated to Professor W.T. Tutte in acknowledgement of his great contributions to graph theory and combinatorics. Professor Tutte had spent two months in Cambridge, with the financial support of the Science Research Council, and the date of the conference was arranged to coincide with his sixtieth birthday. On Friday 13 May a celebration dinner was held in Trinity College. Professor P.W. Duff, Regius Professor of Civil Law Emeritus, who was Professor Tutte's tutor while he was a student at Trinity, proposed a most memorable toast which received an equally memorable reply.

Several of the papers were quickly and efficiently retyped by Mrs. J.E. Scutt. The editorial burden was greatly relieved by the excellent work of Mr. A.G. Thomason.

Béla Bollobás
Cambridge

3 August, 1977

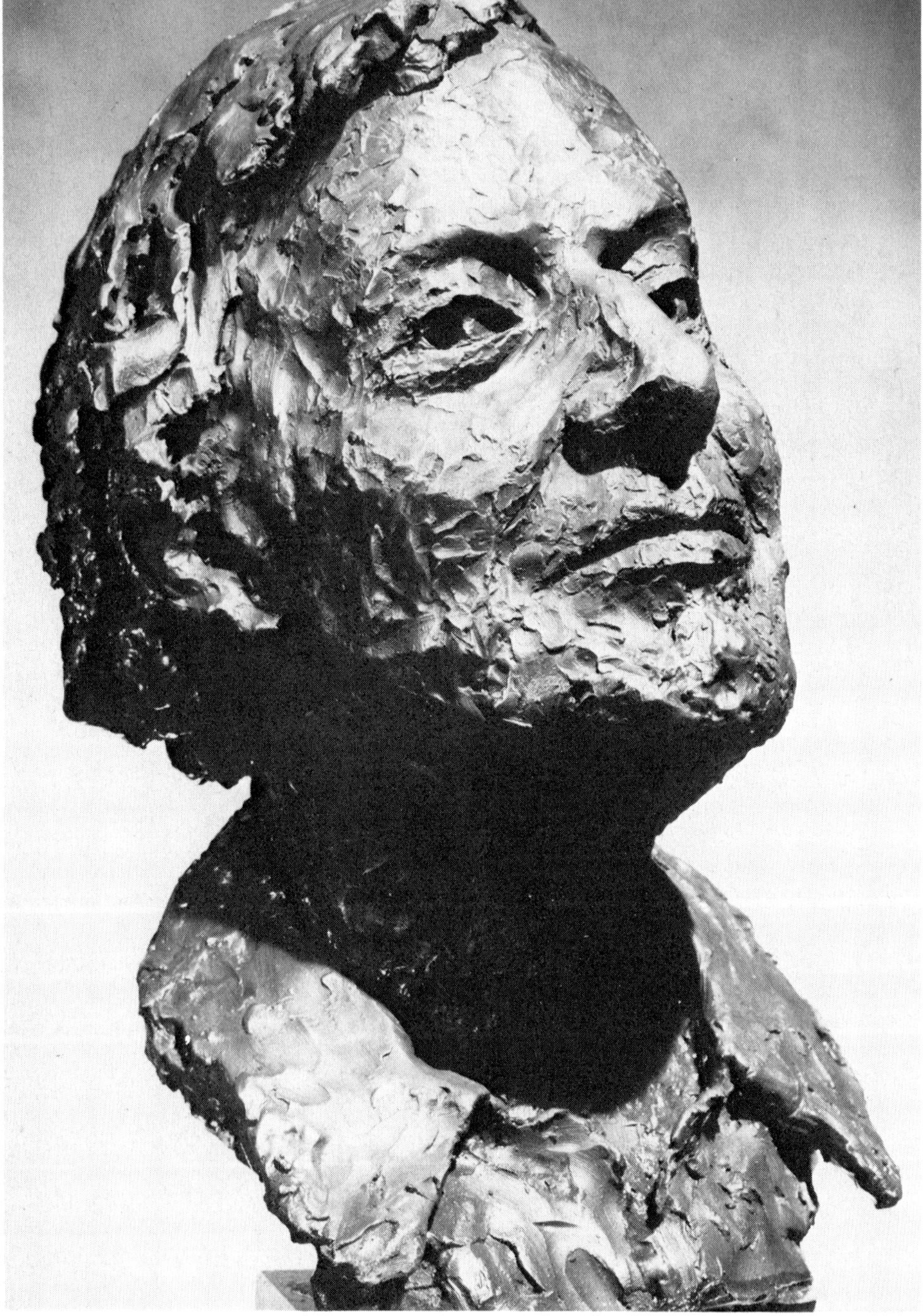

W.T. TUTTE

CONTENTS

Annals of Discrete Mathematics 3 (1978) 1–10.

LINEAR SEPARATION OF DOMINATING SETS IN GRAPHS*

C. BENZAKEN
Université Scientifique et Medicale de Grenoble, Mathématiques Appliquées et Informatique, 38041 Grenoble, France

P.L. HAMMER
University of Waterloo, Department of Combinatorics and Optimization, Waterloo, Ontario N2L 3G1, Canada

The class of finite undirected graphs G having the property that there exist real positive numbers associated to their vertices so that a set of vertices is dominating if and only if the sum of the corresponding weights exceeds a certain threshold θ is characterized: (a) by forbidden induced subgraphs; (b) by the linearity of a certain partial order on the vertices of G; (c) by the global structure of G. The class properly includes that of threshold graphs and is properly included in that of perfect graphs.

1. Introduction, notations, main results

We shall consider in this paper only finite, simple, loopless, undirected graphs $G=(V, E)$ (where V is the vertex set of G, and E is the edge set of G). The terminology follows that in [1] or [5].

For any $x \in V$, we shall denote by $N(x)$ the set of vertices adjacent to x and by $M(x)$ the set of vertices of G not belonging to $x \cup N(x)$ (for simplicity we shall usually put x instead of $\{x\}$).

The edgeless graph on k vertices will be denoted by I_k. The complete graph with k vertices will be denoted by K_k. The complement of the perfect matching of $2k$ vertices will be denoted by J_{2k}. (Note that $I_0=K_0=J_0=\emptyset$, $I_1=K_1$, $I_2=J_2$.)

Following Zykov's terminology [8], for two graphs $G_1=(V_1, E_1)$ and $G_2=(V_2, E_2)$, with $V_1 \cap V_2=\emptyset$, we shall define their direct sum G_1+G_2 as being $(V_1 \cup V_2,\ E_1 \cup E_2)$ and their direct product $G_1 \times G_2$ as being $(V_1 \cup V_2,\ E_1 \cup E_2 \cup E_{12})$, where E_{12} is the set of all edges linking points in V_1 to points in V_2.

A subset S of the vertex set V of a graph G is called a *dominating* set of G (in abbreviation S dom G) if any vertex $x \notin S$ is adjacent to at least one vertex $y \in S$. A vertex v is called *universal* (or *dominating*) if $\{v\}$ dom G. Every set containing a dominating set is dominating.

A subset S of V is called an *independent* set of G when the induced subgraph G_S is edgeless. Every subset of an independent set is independent.

*This research has been carried out at the University of Waterloo (December 1976) and completed at the University of Grenoble (March 1977).

A maximal independent set of G is a minimal dominating set of G. The converse is generally not true. A *domistable* graph is a graph such that every minimal dominating set is independent.

A *domishold* graph is a graph having the property that there exist positive real numbers associated to their vertices so that S is dominating if and only if the sum of the corresponding "weights" of vertices of S exceeds a certain threshold θ.

Examples and counterexamples. Both I_n and K_n are domishold and domistable graphs. Each weight is 1, and the thresholds θ are n (for I_n) and 1 (for K_n). For $p>1$, the graph J_{2p} is domishold (each weight is 1, and the threshold θ is 2), but not domistable.

Let $H_1 = K_2 + K_2$, let H_2 be the simple path on 4 vertices, and let $H_3 = I_3 \times I_3$, $H_4 = (I_1 + K_2) \times I_3$, $H_5 = (I_1 + K_2) \times (I_1 + K_2)$ (see Fig. 1). It is easy to notice that none of the graphs in Fig. 1 are domishold.

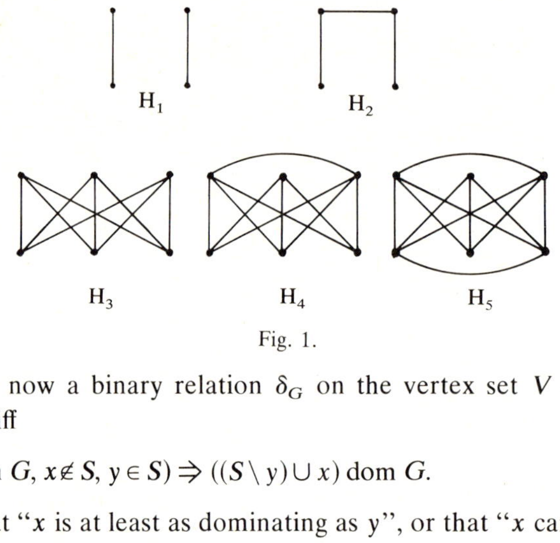

Fig. 1.

Let us define now a binary relation δ_G on the vertex set V of G, by putting $x\delta_G y$ $(x, y \in V)$ iff

$$(S \text{ dom } G, x \notin S, y \in S) \Rightarrow ((S \setminus y) \cup x) \text{ dom } G.$$

We shall say that "x is at least as dominating as y", or that "x can replace y".

Lemma 1.1. *δ_G is a reflexive and transitive relation (i.e. a preorder).*

Proof. The reflexivity is obvious. Assume $i\delta_G j$ and $j\delta_G k$ (i, j, k – distinct), and let S be a dominating set of G, containing k, but not i. If $j \notin S$, then $((S\setminus k)\cup j)$ dom G (because $j\delta_G k$) and does not contain i; therefore $((S\setminus k)\cup i)$ dom G because i can replace j. If $j \in S$ then $((S\setminus j)\cup i)$ dom G, contains k but not j. So $((((S\setminus j)\cup i)\setminus k)\cup j)$ dom G, i.e. $((S\setminus k)\cup i)$ dom G. In both cases $i\delta_G k$.

The main results of this paper are the following:

Theorem 1.2. *The following properties are equivalent:*

(i) *G is domishold.*
(ii) *The preorder δ_G is linear.*
(iii) *G has no induced subgraph isomorphic to H_1, H_2, H_3, H_4 or H_5.*

(iv) *G is built from the empty graph by the repeated application of* $G' \to G''$ *where*

$$G'' = (G' + I_p) \times K_q \times J_{2r} \qquad (p+q+r \neq 0).$$

Corollaries. (a) *Every induced subgraph of a domishold graph is domishold* (because of (iii)).

(b) *Every domishold graph is perfect* (follows from [6] where it is proved that *a* graph without any induced subgraphs isomorphic to H_2 is perfect).

Theorem 1.3. *G is a domishold graph iff, the vertex set V of G can be partitioned into three (possibly empty) subsets* V_1, V_2, V_3 ($|V_3|$ *being even*) *inducing respectively the graphs* $I_{|V_1|}$, $K_{|V_2|}$, $J_{|V_3|}$ *with the following properties*:

Any vertex of V_2 *is adjacent to any vertex of* V_3.

For any $i \in V_1$, $N(i) \cap V_3$ *induces the complement of a perfect matching* J_{2k} *with* $2k = |N(i) \cap V_3|$.

The elements of V_1 *can be indexed so that*

$$N(i_1) \supseteq N(i_2) \supseteq \cdots \supseteq N(i_{|V_1|}).$$

The proofs of these results are given in Section 2.

Section 3 deals with connections between threshold and domishold graphs. Consider an arbitrary threshold graph *G*, and let *L* be an arbitrary subset of vertices, inducing a complete subgraph in *G*. A one to one correspondence is established between the set of all pairs (*G*, *L*) (taken for all threshold graphs *G* and all their complete subsets *L*) and the set of all domishold graphs.

Section 4 deals with Boolean aspects of the previously obtained results and with algorithms for recognizing domishold graphs.

2. Proof of the main results

Proposition 2.1. *If G is domishold, then* δ_G *is a linear preorder.*

Proof. Indeed, if *G* is domishold and a_i are the weights associated to its vertices, then it is obvious that for any pair of vertices *j*, *k* one of the relations $j\delta_G k$ (if $a_j \geqslant a_k$) or $k\delta_G j$ (if $a_k \geqslant a_j$) must hold.

A vertex *m* of *G* is called *maximal* if it is maximal with respect to $\delta_G(m\delta_G i, \forall i \in V_G)$.

Remarks. (1) Any dominating vertex is maximal.

(2) If a graph has a dominating vertex, then every maximal vertex is dominating.

Lemma 2.2. *Let G be a graph such that the corresponding preorder* δ_G *is linear and let m be a maximal vertex of it.*

If m is neither an isolated nor a dominating vertex then every pair $\{x, y\}$ with $x \in N(m)$, $y \in M(m)$ is a dominating set of G and every vertex $y \in M(m)$ is dominating in $G_{V \setminus m}$.

Proof. Let S be a maximal independent set of G not containing m (its existence is guaranteed by the fact that m is not isolated). Thus S dom G. Let i be any element of S. Since $m \notin S$ and $m \delta_G i$ we must have $(S \setminus i) \cup m$ dom G; i is not adjacent to any vertex of $S \setminus i$, hence it must be adjacent to m. So $S \subseteq N(m)$.

Now, if $\{x, y\}$ is such that $x \in N(m)$, $y \in M(m)$ then x is adjacent to y (otherwise $\{x, y\}$ is included in a maximal independent set of G not containing m and not included in $N(m)$). This means that $\{x, y\}$ dom G (every $x' \in N(m)$ is adjacent to y, every $y' \in M(m)$ is adjacent to x and m is adjacent to x).

But m can replace x and $\{m, y\}$ dom G. Hence every y' in $M(m)$ is adjacent to y, proving the Lemma.

Lemma 2.3. *If δ_G is linear, m is a maximal vertex of G and G_m the subgraph induced by $V \setminus m$, then the preorder δ_{G_m} is also linear.*

Proof. Let $i, j \in V \setminus m$ and assume $i \delta_G j$. Let S be a dominating set of G_m containing j but not i. Assume first that m is isolated in G. Then $S \cup m$ dom G contains j but not i. Hence $((S \cup m) \setminus j) \cup i$ dom G and by deleting m, $(S \setminus j) \cup i$ dom G_m showing that $i \delta_{G_m} j$. If m is a dominating vertex of G then S dom G and $(S \setminus j) \cup i$ dom G_m, showing that $i \delta_{G_m} j$. Finally let us consider the case where m is neither isolated nor dominating. If i or j belongs to $M(m)$ then $i \delta_{G_m} j$ (or $j \delta_{G_m} i$) because by Lemma 2.2 i(resp. j) is dominating and so maximal in G_m. If i and j belong to $N(m)$ then S dom G and $(S \setminus j) \cup i$ dom G does not contain m so that $(S \setminus j) \cup i$ dom G_m. Hence $i \delta_{G_m} j$.

Lemma 2.4. *If δ_G is linear, m is a nonisolated vertex of it, and $i, j \in V \setminus m$ such that $i \in M(m) \cap M(j)$, then m is not a maximal vertex of G.*

Proof. Otherwise (by Lemma 2.2) $i \in M(m)$ must be dominating in $G_{V \setminus m}$, which is impossible since i is not adjacent to j ($j \neq m$).

Lemma 2.5. *If δ_G is linear, $m \in V$, i and j are adjacent vertices in $M(m)$ and if $h, k, l \in N(m)$ are such that $l \in M(h) \cap M(k)$, then m is not a maximal vertex of G.*

Proof. Assume m is maximal. Since it is neither isolated nor dominating, it follows from Lemma 2.2 that $\{h, i\}$ and $\{k, j\}$ are dominating sets of G. However $\{i, j\}$ and $\{k, h\}$ are not dominating (because $m \in M(i) \cap M(j)$ and $l \in M(h) \cap M(k)$). Hence neither $j \delta_G h$ nor $h \delta_G j$ hold, in contradiction with the assumed linearity of δ_G.

Proposition 2.6. *A graph G having the property that the preorder δ_G is linear, cannot have any induced subgraph isomorphic to H_1, H_2, H_3, H_4 or H_5.*

Proof. Assume that G with linear preorder δ_G has an induced subgraph H isomorphic to an H_t $(t=1,\ldots,5)$.

By removing a maximal vertex $m \notin H$ (if possible) and continuing this process as many times as possible, we shall eventually arrive (by Lemma 2.3) to a graph G' (with linear preorder) having a maximal vertex m in its induced subgraph H.

If $t=1,2$ or 3 then we can find two vertices n $(\neq m)$ and p such that $p \in M(m) \cap M(n)$. By Lemma 2.4, m is not maximal (a contradiction). If $t=4$ or 5 then $H=(I_1+K_2)\times H'$ (where $H'=I_3(t=4)$ or $H'=I_1+K_2(t=5)$).

By the same argument as above $m \notin K_2$. Similarly, if $H'=I_3$, $m \notin I_3$. So we may suppose $m=I_1$. Then $K_2 \subseteq M(m)$ while H' is a subset of $N(m)$. It follows now, from Lemma 2.5, that m is not maximal. In any case, we have a contradiction.

Lemma 2.7. (Wolk [7].) *If G is a connected graph without a dominating vertex, then the complementary graph $\bar{G}$ contains an induced subgraph isomorphic to H_1 or H_2.*

Lemma 2.8. *If G has no isolated or dominating vertex and no induced subgraph isomorphic to H_t $(t=1,2,\ldots,5)$ then its complement $\bar{G}$ has an isolated edge (i.e. an edge which is not adjacent to any other edge).*

Proof.[1] $\bar{G}$ has no dominating vertex. If $\bar{G}$ is connected then by Lemma 2.7, G contains a subgraph isomorphic to H_1 or H_2 (a contradiction). If $\bar{G}$ is not connected then every connected component has at least two vertices (G has no dominating vertex). If one component has exactly two vertices the lemma is proved. Otherwise each component contains a subgraph isomorphic to one of the following

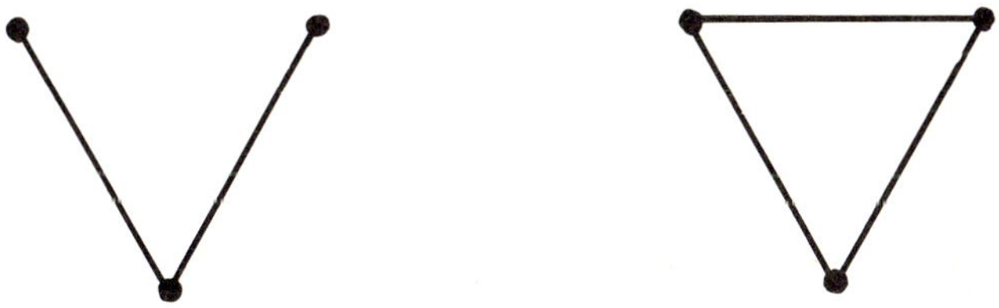

Hence, G contains a subgraph isomorphic to $L_1 \times L_2$ where the L_i $(i=1,2)$ are I_3 or I_1+K_2. Thus G contains a subgraph isomorphic to H_t $(t=3$ or 4 or 5).

Lemma 2.9. *If G is not empty and has no induced subgraph isomorphic to H_t $(t=1,2,\ldots,5)$ then G has one of the forms*

$$G=G'+I_1,$$

$$G=G'\times K_1,$$

$$G=G'\times J_2,$$

where G' has no induced subgraph isomorphic to any H_t $(t=1,\ldots,5)$.

[1] The use of Wolk's result in the present proof was recommended by Ch.Payan and has produced a substantial simplification over our original proof.

Proof. The decomposition follows from Lemma 2.8; the fact that G' has no induced subgraph isomorphic to H_t is obvious.

Proposition 2.10. *If G has no induced subgraph isomorphic to H_t $(t=1,2,\ldots,5)$ then G is built from the empty graph by the repeated application of $G' \to G''$ where*

$$G''=(G'+I_p)\times K_q\times J_{2l} \quad \text{with} \quad p+q+l\neq 0.$$

Proof. Obvious from the repeated application of Lemma 2.9, from the associativity and commutativity of + and ×, and from the following relations:

$$I_p=I_1+I_1+I_1+\cdots+I_1 \quad (p \text{ times}),$$
$$K_q=K_1\times K_1\times K_1\times\cdots\times K_1 \quad (q \text{ times}),$$
$$J_{2l}=J_2\times J_2\times J_2\times\cdots\times J_2 \quad (l \text{ times}).$$

Proposition 2.11. *If G is built from the empty graph by the repeated application of $G'\to G''$ defined above then G is domishold.*

Proof. The empty graph is obviously domishold. Assume now that $G=(G'+I_p)\times K_q\times J_{2r}$, and that G' is domishold. Let w_l represent the weight of the vertices $l\in V_{G'}$, and w_0 the threshold for G'. Let $w^*=\frac{1}{2}\min_l w_l$. We can always assume that $2w^*\leqslant w_0$, since otherwise $G'=K_q$, and we could take all weights w_l $(l\in V_{G'})$, as well as w_0, equal to 1 (in which case again $2w^*\leqslant w_0$). Let us also put $W=1+\sum_{l\in V_{G'}} w_l$ and let us define $\hat{w}_0=w_0+pW$ and

$$\hat{w}_i=\begin{cases} w_i & i\in V_G{}', \\ W & i\in I_p, \\ w_0+pW & i\in K_q, \\ w_0+pW-w^* & i\in J_{2r}. \end{cases}$$

The "weights" $\hat{w}_i$ and the "threshold" $\hat{w}_0$ of G characterize the dominating sets of G. Indeed, any minimal dominating set D of G is of one of the following three types: (i) $D=\{k\}$, $k\in K_q$; (ii) $D=\{j,e\}$, with $j\in J_{2r}$, $j\neq e$, and $e\in J_{2r}\cup I_p\cup V_{G'}$; (iii) $D=D'\cup I_p$, where D' is a minimal dominating set of G'.

Proof of Theorem 1.2. Follows from Propositions 2.1, 2.6, 2.10, 2.11.

Proof of Theorem 1.3. *Necessity.* From property (iv) of Theorem 1.2, we can define $G_0, G_1,\ldots, G_t$ with

$$G_0=\emptyset \qquad G_t=G$$

and

$$G_{i+1}=(G_i+I_{p_i})\times K_{q_i}\times J_{2r_i} \qquad (i=0,1,\ldots,t-1).$$

Putting

$$V_1 = \bigcup_i I_{p_i} \qquad V_2 = \bigcup_i K_{q_i} \qquad V_3 = \bigcup_i J_{2r_i},$$

it is clear by induction that this partition has the desired properties.
Sufficiency. By induction. If G has the prescribed properties it is obvious that if $V_1 \neq \emptyset$ then $i_{|V_1|}$ is either isolated (and after its elimination we get a graph G' with the same properties), or is adjacent to a vertex k in V_2 (k is dominating and its elimination leads to a graph G' with the same properties), or is adjacent to a non-adjacent pair $\{j, j'\}$ of V_3 so that $G = G' \times J_2$ with G' having the same properties.

If $V_1 = \emptyset$ then every $k \in V_2$ (in case of $V_2 \neq \emptyset$) is dominating and G_{V-k} has the same properties.

If $V_2 = \emptyset$ then $G = J_{2r}$ is domishold.

3. Threshold and domishold graphs

We shall recall that, as in [2], by a *threshold graph* we shall mean a graph such that real non negative numbers can be associated to its vertices so that two vertices are adjacent iff the sum of their weights exceeds a certain threshold. Alternatively, a graph is threshold iff there exist real numbers associated to its vertices so that the sum of these numbers associated to vertices belonging to an independent set (a dependent set) is $<$ (is $\geq$) than a certain threshold. Several characterizations of such graphs can be found in [2].

We recall also that, as in [3], by a *split graph* we shall mean a graph whose vertex set V can be partitioned in two (possibly empty) subsets V_1, V_2 such that V_1 induces $I_{|V_1|}$ and V_2 induces $K_{|V_2|}$.

Theorem 3.1. *Every threshold graph is domishold and has all the following properties:*

(a) *It has no induced square* $[I_2 \times I_2]$.
(b) *It is split.*
(c) *It is domistable.*
(d) *It is an interval graph.*

Conversely a domishold graph having any one of the mentioned properties is threshold.

Proof. It has been proved in [2] that a threshold graph is characterized by the absence of induced subgraphs isomorphic to H_1, H_2 and $I_2 \times I_2$. From this, it follows that it has no subgraph isomorphic to H_t ($t = 1, 2, \ldots, 5$). Hence it is domishold and satisfies (a).

In [2] it is proved that a threshold graph is split (b). Moreover if i and j are two adjacent vertices of a threshold graph, we have:

$$N(i) \subseteq N(j) \cup j \quad \text{or} \quad N(j) \subseteq N(i) \cup i.$$

Therefore a minimal dominating set will never contain both i and j, and hence it is independent, proving (c).

Finally it has been proved [2] that in a threshold graph (having split structure $V_1 \cup V_2$) one can index the elements of V_1 in such a way that $N(i_1) \subseteq N(i_2) \subseteq \cdots \subseteq N(i_{|V_1|})$. Associate to each element $i_\alpha \in V_1$ the interval $[\alpha, \alpha]$ and to each element $k \in V_2$ the interval $[m_k, |V_1|+1]$ where m_k is the least integer such that $k \in N(i_{m_k})$ (if any), and otherwise $m_k = |V_1|+1$. It is easy to see that the corresponding interval graph is isomorphic to the original one, proving (d).

Conversely a domishold graph without an induced square $I_2 \times I_2$ (and of course without H_1, H_2) is threshold. A split graph has no square and if it is domishold, it is threshold.

If a graph G is domishold and domistable then a maximal vertex m (for δ_G) is either dominating or isolated. Otherwise by Lemma 2.2 any set $\{x, y\}$ with $x \in N(m)$, $y \in M(m)$ is minimal dominating (the minimality follows from the fact that y is obviously not dominating, and neither is x, otherwise m should be dominating). Hence by removing m, we get again a domistable and domishold graph. By induction it follows that the original graph is threshold.

Finally if an interval graph is domishold then it does not contain a square. Indeed assume there exists a square and $[a, b]$, $[c, d]$ are the corresponding intervals associated to two opposite vertices of this square. We have $[a, b] \cap [c, d] = \emptyset$. Obviously, the two intervals $[e, f]$, $[g, h]$ associated to the other two vertices of the square must intersect both $[a, b]$ and $[c, d]$ and therefore intersect each other (a contradiction).

Definition. Let i be a vertex of a graph G. The *i-duplication of G* is the graph G' obtained by adding a new vertex i' to V_G with $N(i') = N(i)$. Conversely, we shall say that G is the *(i, i')-fusion* of G'.

We can extend this definition to *W-duplication* of G ($W \subseteq V_G$) by duplicating sequentially each vertex of W (this operation does not depend on the order of duplications).

Also if $U \subseteq V_G$, induces the complement of a perfect matching ($J_{|U|}$) and if every pair (i, i') of non adjacent vertices in U have the same neighbourhood ($N(i) = N(i')$), by the *U-fusion* of G, we mean the graph obtained by the sequential repetition of all the (i, i')-fusions of G.

Theorem 3.2. *If G is a domsihold graph and S a maximal subset of V_G inducing a subgraph $J_{|S|}$ then the S-fusion of G is threshold.*

Conversely if G is threshold and L a subset of a maximal subset of V_G inducing a clique of G then the L-duplication of G is a domishold graph.

Proof. A direct consequence of Theorem 1.3.

Remark. Despite the fact that the class of domishold graphs includes properly that one of threshold graphs, this theorem seems to point to a (to us) surprising similarity between threshold and domishold graphs.

4. Boolean aspects and algorithms

4.1. *Recognizing domishold graphs*

It is easy now to construct a procedure for the recognition of domishold graphs; the time needed by this procedure will be polynomial in the number n of vertices. The procedure can start by searching for isolated vertices and eliminating them. When no more isolated vertices can be found, the procedure could search for dominating vertices. After repeating the above two steps as many times as possible we shall obtain a graph without isolated or dominating vertices; in this graph we shall look for two non-adjacent vertices, both of which are linked to every other vertex. The graph is domishold if and only if the above three steps can be repeated until the total exhaustion of the vertex set.

4.2. *Recognizing the linear separator of a domishold graph*

A linear inequality

$$\sum_{i=1}^{n} w_i x_i \geqslant w_0, \qquad x_i \in \{0, 1\} \qquad (i = 1, \ldots, n)$$

is called *domigraphic* if there exists a domishold graph of n vertices such that the w_i's are the weights of the vertices, and w_0 is the threshold. In other words, the inequality holds if and only if $(x_1, \ldots, x_n)$ is the characteristic vector of a dominating set. We can obviously assume that $w_1 \geqslant \cdots \geqslant w_n$.

The condition

$$\sum_{i=1}^{n} w_i \geqslant w_0$$

is obviously necessary for a linear inequality to be domigraphic. In the case $n = 1$, it is sufficient too. For $n = 2$, this condition along with $(w_2 < w_0) \Rightarrow (w_1 < w_0)$ are again sufficient.

Theorem 4.1. *The inequality $\sum_{i=1}^{n} w_i x_i \geqslant w_0$ $(n \geqslant 3)$ is domigraphic if and only if one of the following conditions hold*:

(i) $w_1 \geqslant w_0$ *and* $\sum_{i=2}^{n} w_i x_i \geqslant w_0$ *is domigraphic*;
(ii) $w_1 < w_0$, $\sum_{i=2}^{n} w_i < w_0$ *and* $\sum_{i=2}^{n} w_i x_i \geqslant w_0 - w_1$ *is domigraphic*;
(iii) $w_1 < w_0$, $w_2 + w_n \geqslant w_0$ *and* $\sum_{i=3}^{n} w_i x_i \geqslant w_0$ *is domigraphic.*

The proof is based on Theorem 1.2 and is omitted here.

4.3. *Boolean functions associated to domishold graphs*

A monotone Boolean function $f(x_1, \ldots, x_n)$ is called *domigraphic* if there exists a graph with n vertices such that the set of characteristic vectors X of all its dominating sets is the same as the set of solutions to the equation $f(x_1, \ldots, x_n) = 1$.

Theorem 4.2. *$f(x_1, \ldots, x_n)$ is a domigraphic Boolean function if and only if it has one of the forms*

$$f(x_1, x_2, \ldots, x_n) = x_1 \vee g(x_2, \ldots, x_n)$$

$$f(x_1, x_2, \ldots, x_n) = x_1 \cdot g(x_2, \ldots, x_n)$$

$$f(x_1, x_2, \ldots, x_n) = x_1 x_2 \vee (x_1 \vee x_2)\left(\bigvee_{i=3}^{n} x_i\right) \vee g(x_3, \ldots, x_n),$$

where g is a domigraphic Boolean function.

Noticing that the preorder δ_G introduced in Section 1 is the same as the Winder-type preorder defined on the set of variables of a Boolean function, Theorem 1.2 will have the following.

Corollary. *A domigraphic Boolean function is threshold if and only if it is 2-monotonic.*

Acknowledgement

Partial support through a Canada Council Cultural Exchange Grant and through the National Research Council's Grant A-8552 are gratefully acknowledged.

References

[1] C. Berge, Graphes et Hypergraphes (Dunod, Paris, 1970).
[2] V. Chivatal and P.L. Hammer, Aggregation of inequalities in integer programming, Ann. Discr. Math. 1 (1977) 145–162.
[3] S. Foldes and P.L. Hammer, Split graphs, Proc. Eighth Southeastern Conference on Combinatorics, Graph Theory and Computing (1977).
[4] M.C. Golumbic, Comparability graphs and a new matroid, J. Combinatorial Theory B-22 (1977) 68–90.
[5] F. Harary, Graph Theory (Addison-Wesley, Reading, MA, 1969).
[6] D. Seinsche, On a property of the class of n-colorable graphs, J. Combinatorial Theory B-16 (1974) 191–193.
[7] E.S. Wolk, A note on the comparability graph of a tree, Proc. of A.M.S. 16 (1965) 17–20.
[8] A.A. Zykov, On some properties of linear complexes, Translations of A.M.S., (Series One) 17; Algebraic Topology (1962) 418–449.

Annals of Discrete Mathematics 3 (1978) 11–19.

REGULARISABLE GRAPHS

Claude BERGE

University of Paris 4 Pl. Jussieu, 75230 Paris, Cédex 06, France

1. Introduction

A graph G is said to be *regularisable* if a regular multigraph can be obtained from G by adding edges parallel to the edges of G. In this paper, we give several characterisations of regularisable graphs; in particular, a connected non-bipartite graph has a unique optimal fractional transversal (with all coordinates equal to $\frac{1}{2}$) if and only if it is regularisable.

This class of graphs contains the edge-critical graphs with no isolated vertex and the line-graphs of graphs with no pendent edge. The Fulkerson–Hoffman theorem (Corollary 2.5 to Theorem 2.2) and a property of the edge-critical graphs due to Hajnal (Corollary 4.1 to Theorem 3.1) follow immediately from the main results.

2. Optimal k-transversals of a graph

In this paper, G always denotes a simple graph (with no loops and no multiple edges), but the results and concepts are also valid for a multigraph with no loops. The vertex-set of G is denoted by X, and the edge-set by E.

Let $G=(X, E)$ be a graph. For $x \in X$ write $\Gamma_G x$ or Γx for the set of neighbours of x. For $A \subset X$ put $\Gamma A = \bigcup_{x \in A} \Gamma x$. Also, for $A \subset X$, $B \subset X$ write $m_G(A, B)$ or $m(A, B)$ for the number of edges of G having one endvertex in A and the other in B.

A *fractional transversal* of G is a non-negative function $p(x)$, defined for $x \in X$, such that

$$[x,y] \in E \Rightarrow p(x)+p(y) \geq 1.$$

A *k-transversal* of G is a function $p(x)$ on X such that:

(i) $p(x) \in \{0, 1, 2, \ldots, k\}$,

(ii) $p(x)+p(y) \geq k$ for every edge xy.

$\tau_k(G)$ denotes the minimum of $\sum_{x \in X} p(x)$ when p ranges over all the k-transversals of G. Thus $\tau_1(G)$ is the usual transversal number $\tau(G)$, i.e. the minimum cardinality of a set $T \subset X$ which meets all the edges.

Let kG denote the multigraph obtained from G by multiplying each edge by k.

A partial H of kG is called a *k-matching* if at each vertex x, the degree $d_H(x)$ does not exceed k.

Denote by $\nu_k(G)$ the maximum number of edges in a k-matching. Thus $\nu_1(G)=\nu(G)$ is the usual matching number.

It is well-known that $\max(\nu_k(G)/k)=\min(\tau_k(G)/k)$. The common value, called the *fractional transversal number* of G and denoted by $\tau^*(G)$, satisfies the following inequalities (see Berge, and Simonovits [5], Lovász [10]):

Lemma 2.1. *For every hypergraph G we have*

$$\nu(G)=\min_k \frac{\nu_k(G)}{k} \leqslant \max_{G'\subset G} \frac{m(G')}{\Delta(G')} \leqslant \frac{\nu_k(G)}{k} = \tau^*(G)$$

$$= \min_k \frac{\tau_k(G)}{k} \leqslant \max_k \frac{\tau_k(G)}{k} = \tau(G).$$

Theorem 2.2. *Let G be a connected graph. Then there exists an optimal 2-matching H of G such that each connected component of H is either a single vertex or a pair of parallel edges ("double edge") or an odd cycle.*

For every 2-matching, there exists and optimal 2-transversal p(x) with values as follows: 0 *if x belongs to a singleton of H*, (0, 2) *or* (1, 1) *for the two vertices belonging to a double edge of H, and* 1 *for each vertex belonging to an odd cycle of H.*

Proof. Let $H\subset 2G$ be a maximum 2-matching. Every connected component of H which is a path or a cycle of even length can be replaced by a set of pairwise disjoint double edges without changing $m(H)$. No component of H is an odd path (i.e. a path of odd length) since $m(H)$ is maximum. Thus H is now of the type described by the theorem.

We shall label each vertex with 0, 1 or 2, by an iterative procedure described by the following rules:

(1) Label with 0 each vertex which is a singleton of H.

(2) Label with 2 each vertex which is adjacent in G to a vertex previously labelled with 0.

(3) Label with 0 every vertex which is adjacent in H to a vertex previously labelled with 2.

(4) Every vertex which cannot be labelled by the iterative procedure described by rules 1, 2, and 3, will be labelled 1.

No odd chain, starting from a singleton of H and consisting alternately of edges of G–H and of double edges of H, ends with a singleton because such a chain would constitute a connected component of a 2-matching H' with $m(H')>m(H)$. Similarly no odd chain of that kind can end in an odd cycle of H. No odd chain of that kind can cross itself at a vertex labelled with 0 (because there would be a better 2-matching having as connected components an odd cycle and a set of double edges).

Thus a unique label $t(x)$ can be given to a vertex x by the above rules,

$t(x)=0$ if x is a singleton of H,

$t(x)=2$ and $t(y)=0$ (or vice versa) if xy is a double edge connectable to a singleton (otherwise $t(x)=t(y)=1$), $t(x)=1$ if x belongs to an odd cycle of H.

The rules show also that $t(x)$ is a 2-transversal of G. Furthermore, we have

$$\tfrac{1}{2}m(H)=\tfrac{1}{2}\nu_2(G)\leq\tau^*(G)\leq\tfrac{1}{2}\tau_2(G)\leq\sum_{x\in X}t(x)=\tfrac{1}{2}m(H).$$

So these inequalities hold as equalities; hence $t(x)$ is an optimal 2-transversal.

Corollary 2.3. [11, 15]. *For a multigraph G,*

$$\tau^*(G)=\tfrac{1}{2}\nu_2(G)=\tfrac{1}{2}\tau_2(G).$$

Corollary 2.4. (*König*). *For a bipartite graph $G=(X, Y, E)$, the maximum number of edges in a matching equals*

$$\min_{A\subseteq X}(|X-A|+|\Gamma_G(A)|).$$

Proof. Let $M=\min_{A\subset X}(|A-A|+|\Gamma_G(A)|)$ and let N be the maximum number of edges in a matching. Clearly $M\geq N$. Let H be the maximum 2-matching given by Theorem 2.2, and let $\bar{X}\subset X$ be the set of singletons of H in X. Further let $X_i\subset X-\bar{X}$ be the set of remaining elements of X labelled i, $0\leq i\leq 2$. Since the double edges of H define a matching of G, and as H has no odd cycles, we have $N\geq\sum_{i=0}^{2}|X_i|$. However the only vertices of X labelled with 2 are in $\Gamma_H(X_0)$, and so $\Gamma_G(X_0\cup\bar{X})=\Gamma(X_0)$. Hence

$$\begin{aligned} M&\leq|X-(X_0\cup\bar{X})|+|\Gamma_G(X_0\cup\bar{X})|\\ &=|X_1|+|X_2|+|\Gamma_H(X_0)|\\ &=|X_1|+|X_2|+|X_0|\leq N, \end{aligned}$$

and the corollary is proved.

Corollary 2.5. (*Fulkerson et al.* [7]). *Let G be a regular connected graph such that every pair of vertex-disjoint odd cycles is joined by an edge ("semi-bipartite graph", like the Peterson graph). Then there exists a matching which has at most one unsaturated vertex.*

Proof. As above, the 2-matching H has no singleton. If two connected components are odd cycles, they can be replaced by double edges. If only one odd cycle remains, it can be replaced by one singleton and a set of double edges, so that the double edges constitute a matching with at most one unsaturated vertex.

Remark. Theorem 2.2 shows that an optimal fractional tranversal of a graph G can easily be obtained; we first construct a maximum 2-matching (as a maximum flow in a bipartite transportation network), and then we apply the algorithm described in Theorem 2.2 to get a minimum 2-transversal. This shows also the fact, quoted by several authors, that the linear programming which describes the maximum stable sets of a graph always has a solution with coordinates 0, 1 or $\frac{1}{2}$.

3. Regularisable graphs

From Lemma 2.1, it follows immediately that a regular graph G satisfies $\tau^*(G)=\frac{1}{2}n$, and therefore, the vector $\mathbf{1}=(1,1,\ldots,1)$ is an optimal 2-transversal. In this section, we shall first consider the problem of the uniqueness of the optimal 2-transversal; more precisely, for which graphs is $\mathbf{1}$ the unique optimal 2-transversal?

Let $G=(X,Y,E)$ be a bipartite graph; then, from Theorem 2.2, we have $\tau_2(G)=n$ if and only if G possesses a perfect matching. In this case, there exist at least three optimal 2-transversals: one with all weights equal to 1, and two with weights equal to 0 in one vertex-class, and to 2 in the other vertex class (since $|X|=|Y|$). So, *if a graph G has* $\mathbf{1}$ *as an unique optimal 2-transversal, no connected component of G is a bipartite graph.*

Theorem 3.1. *Let G be a connected graph which is not bipartite. Then the following conditions are equivalent*:

(1) *G is regularisable,*
(2) *$t(x)\equiv 1$ is the only optimal 2-transversal,*
(3) *for every non-empty stable ("independent") set S of vertices, $|\Gamma S|>|S|$.*

Proof. (1) *implies* (2). Let G be a graph and let H be a regular multigraph obtained from G by edge-multiplication. Then

$$\tau_2(G)=2\tau^*(G)=2\tau^*(H)=2\frac{n(H)}{\Delta(H)}=2\frac{n(H)}{2}=n.$$

Thus, $t(x)\equiv 1$ is an optimal 2-transversal for G.

Now, assume that there exists another optimal 2-transversal $t'(x)$, and for $s=0,1,2$, put

$$A_s=\{x\colon x\in X,\ t'(x)=s\}.$$

Then $|A_0|=|A_2|\neq 0$. The set A_0 is stable (otherwise $t'(x)$ would not be a 2-transversal), and $\Gamma A_0\subseteq A_2$. We have $\Gamma A_0=A_2$ (otherwise, $t'(x)$ would not be optimal; a better 2-transversal can be obtained from $t'(x)$ by replacing a 2-value by a 1-value).

Since H is regular,

$$\Delta(H)\,|A_0| = m_H(A_0, A_2) \leqslant \sum_{x \in A} m_H(x, A_0) \leqslant |A_2|\,\Delta(H) = \Delta(H)\,|A_0|.$$

Hence $m_H(x, A_0) = \Delta(H)$, and no edge goes out of $A_0 \cup A_2$. Since G is connected, its vertex set is $A_0 \cup A_2$ and G is a bipartite graph having two vertex classes with the same cardinality. This contradicts the hypothesis.

(2) *implies* (3). Let $S \neq \emptyset$ be a stable set, and let $H \subset 2G$ be an optimal 2-matching as described in Theorem 2.2.

Since $t(x) \equiv 1$ is an optimal 2-transversal, we have $\tau_2(G) = n$, so the connected components of H are either double edges or odd cycles. Hence

$$|\Gamma_G S| \geqslant |\Gamma_H S| \geqslant |S|.$$

If $|\Gamma_G S| = |S|$, it would follow that all the components of H meeting S are double edges. We can then define a 2-transversal $t'(x)$ by putting

$$t'(x) = \begin{cases} 0 & \text{if } x \in S, \\ 2 & \text{if } x \in \Gamma S, \\ 1 & \text{if } x \in X - (S \cup \Gamma S). \end{cases}$$

Since $t'(x)$ would also be an optimal 2-transversal of G, this contradicts the uniqueness of the optimal 2-transversal. Thus $|\Gamma S| > |S|$.

(3) *implies* (1). Now assume that $|\Gamma S| > |S|$ for every non-empty stable set S of G. Let H be a bipartite graph whose vertex-classes are two copies X and $\bar{X}$ of the vertex set of G, the vertices $x \in X$ and $\bar{y} \in \bar{X}$ being joined by an edge in H if and only if x and y are adjacent in G.

Let $B \subset X$, $B \neq \emptyset$, $B \neq X$, be a set such that the subgraph G_B has no isolated vertex. Then $\Gamma_H(B) \supset \bar{B}$. Now let $S \subset X$ be a set such that G_S has only isolated vertices. Then S is a stable set of G, and by (3),

$$|\Gamma_H S| = |\Gamma_G S| > |S|.$$

So, for every set $A = B \cup S \subset X$, $A \neq \emptyset$, $A \neq X$, we have

$$|\Gamma_H A| > |A|,$$

noting that $\Gamma_H S \cap \bar{B} = \emptyset$ if there are no edges between B and S.

First, we shall show that each edge $a\bar{b}$ of H belongs to at least one perfect matching, that is the subgraph H' of H induced by $(X \cup \bar{X}) - \{a, \bar{b}\}$ has a perfect matching. For every $A \subset X - \{a\}$,

$$|\Gamma_{H'} A| = |\Gamma_H A - \{\bar{b}\}| \geqslant |\Gamma_H A| - 1 \geqslant |A|.$$

Thus, by König's theorem, H has such a matching.

Consequently, for each edge ab of G, there exists a 2-matching which saturates all the vertices and which uses the edge ab. The union of all these possible

2-matchings defines a regular multigraph, which arises from G by edge-multiplications. Thus G is regularisable.

Corollary 3.2. *Let G be a connected regularisable graph which is not bipartite: then the graph G' obtained from G by adding a new edge is also regularisable.*

Proof. Let S' be a stable set of $G'=(X, \Gamma')$. Since S' is also stable for $G=(X, \Gamma)$, $|\Gamma S'| \geq |S'|+1$, and

$$|\Gamma' S'| \geq |\Gamma S'| \geq |S'|+1.$$

Hence G' is regularisable.

(Note that Corollary 3.2 does not hold for bipartite graphs; the quadrilateral C_4 is regularisable, but the graph obtained by adding a diagonal is not regularisable.)

Corollary 3.3. *If, for every vertex x of a graph $G=(X, \Gamma)$, there exists a maximum stable set T_x such that $x \notin T_x$, $\Gamma x \not\subset T_x$, then G is regularisable.*

Proof. Clearly, we may assume that G is connected without loss of generality; we shall show, by induction on $|S|$, that $|\Gamma S| > |S|$ for every stable set S.

First, let $S=\{x\}$ be a singleton. Then x is not an isolated vertex (because $T_x \cup \{x\}$ would be a stable set larger than T_x). Also, x is not incident to only one edge, say $[x, y]$, because $x \notin T_x$, hence $y \in T_x$, hence $\Gamma x \subset T_x$, which is a contradiction.

Thus, $|\Gamma S| > |S|$.

Now, assume that every stable set S with cardinality $\leq p-1$ satisfies $|\Gamma S| > |S|$, and consider a stable set S_0 with cardinality p. Let $a \in S_0$; we have

$$|\Gamma S_0 \cap T_a| \geq |S_0 - T_a| \tag{1}$$

Otherwise, $|\Gamma S_0 \cap T_a| < |S_0 - T_a|$, and $T_a - (\Gamma S_0 \cap T_a) \cup (S_0 - T_a)$ would be a stable set larger than T_a, which is a contradiction.

Case 1: $S_o \cap T_a = \emptyset$. Then, by eq. (1),

$$|\Gamma S_0| \geq |\Gamma S_0 \cap T_a| + |\Gamma a - T_a| > |\Gamma S_0 \cap T_a| \geq |S_0 - T_a| = |S_0|.$$

Case 2: $S_0 \cap T_a \neq \emptyset$. Then $S_0 \cap T_a$ is a stable set with cardinality $\leq p-1$, and by the induction hypothesis, $|\Gamma(S_0 \cap T_a)| > |S_0 \cap T_a|$. Hence

$$|\Gamma S_0| \geq |\Gamma(S_0 \cap T_a)| + |\Gamma S_0 \cap T_a| > |S_0 \cap T_a| + |S_0 - T_a| = |S_0|.$$

Thus, in each case, $|\Gamma S_0| > |S_0|$, and this is true for every stable set S_0. Hence G is not bipartite, and, by Theorem 3.1, G is regularisable.

4. Application to α-edge-critical graphs

Let G be a graph with stability number (or "independence number") $\alpha(G)=k$. An edge e of G is said to be *α-critical* if the subgraph $G-e$ satisfies $\alpha(G-e) > k$.

The graph G is *α-edge-critical* (for short, α-critical or edge-critical) if every edge is α-critical. Edge-critical graphs have been extensively studied, in particular by Plummer [13], Erdös and Gallai [6], Hajnal [9], Berge [3], Wessel [16], George [8], Andrásfai [1], Surányi [14], Lovász [12], and Zykov [17].

Generalizing a result of Hajnal, we can show the following as an immediate application of Corollary 3.3.

Corollary 4.1. *Every edge-critical graph with no isolated vertex is regularisable.*

Proof. We may assume without loss of generality that the edge-critical graph G is of order larger than 2 (because K_2 is regularisable). It is easy to see that G has no pendent vertex. Every vertex x is incident to an edge, say $[x, y]$, and the removal of this edge creates a stable set $S_{x,y}$ with cardinality $\alpha(G)+1$. Thus, $S_{x,y}-\{x\}=T_x$ fulfils the conditions of Corollary 3.3, and consequently G is regularisable.

5. Application to line-graphs

Jeager and Payan [12] have shown that the line-graph $L(H)$ of a connected graph H with no pendent edge is regularisable. We prove here the following result:

Theorem 5.1. *If H is an r-uniform hypergraph with no vertex of degree one, such that each edge meets at least r other edges, then $L(H)$ is regularisable.*

Let E be the edge-set of H, or the vertex set of $L(H)$; let $F\subseteq E$ be a matching of H, or a stable set of $L(H)$. Let G be the bipartite graph obtained by removing from $L(H)$ the edges which are not incident to F. The degree in G of a vertex $e\in E$ is denoted by $d_G(e)$.

For $e\in E-F$, we have $d_G(e)\leq r$, because in H the edge e has only r elements. For $f\in F$, we have $d_G(f)\geq r$, because in H the edge f meets at least r other edges. Hence

$$r\,|F|\leq \sum_{f\in F} d_G(f)=m_G(F, E-F)=\sum_{e\in \Gamma_G F} d_G(e)\leq |\Gamma_G F|.$$

Thus, $|\Gamma_G F|\geq |F|$. If the equality never occurs, then the graph $L(H)$ is regularisable by Theorem 3.1. If the equality holds for some stable set F_0, then

$$d_G(f)=r, \qquad (f\in F_0.)$$

$$d_G(e)=r, \qquad (e\in \Gamma_G F_0.)$$

It follows that G is a regular bipartite graph, and $E-F_0=\Gamma_G F_0$ (because in H an edge $e\in\Gamma_G F_0$ is covered by r edges of the matching F_0 so any edge in $E-F_0$ which meets e meets also an edge of F_0 and consequently belongs to $\Gamma_G F_0$). Thus $|E-F_0|=|F_0|$. If H has no vertex of degree 1, $\bigcup_{f\in F_0} f=\bigcup_{e\in E-F_0} e$, and $L(H)=G$ which is a regular graph. This completes the proof.

6. Index of regularisability

We shall now consider the index of regularisability $k(G)$; this is a number equal to $+\infty$ if G is not regularisable, and otherwise equal to the least k for which a k-regular multigraph can be obtained from G by edge-multiplication.

Theorem 6.1. *Let G be a simple graph, and denote by Q the set of all pairs (S, T), where*

S is a non-empty stable set,
T is a stable set (possibly empty) disjoint from S and satisfying $\Gamma T \subseteq S, |T| \leqslant |S|$.
Put

$$\mu(G) = \max_{(S,T)\in Q} \frac{m(S, X-T)}{|S|-|T|}.$$

(This maximum takes into account the value $p/0 = +\infty$, but not the undetermined value $0/0$.)
Then $k(G) \geqslant \mu(G)$.

Proof. Clearly, $Q \neq \emptyset$, because $(\{x\}, \emptyset) \in Q$. If $k(G) = +\infty$ then $k(G) \geqslant \mu(G)$ (trivially). If $k(G) < +\infty$, then there exists a k-regular multigraph H minimum degree obtained from G by edge-multiplication. Let $(S, T) \in Q$. Then

$$k|S| - k\,|T| = m_H(S, X-S) - m_H(S, T) = m_H(S, X-T) \geqslant m_G(S, X-T).$$

Since this is true for all $(S, T) \in Q$, we have $k(G) = k \geqslant \mu(G)$.

For bipartite graphs, we have a more precise result, which is a straightforward application of the Theorem of Berge and Hoffman [4] for unimodular hypergraphs:

Theorem 6.2. (reformulation of [4, Theorem 3.2]). *Let $G = (X, Y, E)$ be a bipartite graph; then $k(G) = [\mu(G)]^*$. In other words, G is regularisable if and only if $\mu(G) < +\infty$; in this case, the least degree for a regular multigraph H is the least integer $\geqslant \mu(G)$.*

References

[1] B. Andrásfai, On critical graphs, Théorie des Graphes (Rome I.C.C., Paris, 1967) 9–19.
[2] C. Berge, Graphes et Hypergraphes (Dunod, Paris, 1970).
[3] C. Berge, Une propriété des graphes k-stables critiques, Combinatorial Structures (Gordon and Breach, New York, 1970) 7–11.
[4] C. Berge and A.J. Hoffman, Multicolorations dans les hypergraphes unimodulaires, Proc. Paris Conf. July 1976, C.N.R.S. Publ. (to appear).
[5] C. Berge and M. Simonovits, The coloring numbers of the direct product of two hypergraphs, Hypergraph Seminar 1972, Lecture Notes 411 (Springer, Berlin, 1974) 21–33.

[6] P. Erdös and T. Gallai, On the minimal number of vertices representing the edges of a graph, Publ. Math. Inst. Hung. Acad. Sci. 6 (1961) 181–203.
[7] D.R. Fulkerson, A.J. Hoffman and M.H. McAndrew, Some properties of graphs with multiple edges, Can. J. Math. 17 (1965) 166–177.
[8] A. George, On line-critical graphs, Thesis, Vanderbilt Univ., Nashville, TN (1971).
[9] A. Hajnal, A theorem on k-saturated graphs, Can. J. Math. 17 (1965) 720–772.
[10] L. Lovász, Minimax theorems for hypergraphs, Hypergraph Seminar 1972, Lecture Notes 411 (Springer, Berlin, 1974) 111–126.
[11] L. Lovász, 2-matchings and 2-covers of hypergraphs, Acta Math. Acad. Sci. Hungar 26 (1975) 433–444.
[12] F. Jaeger and C. Payan, A class of regularisable graphs, Annals of Discrete Mathematics 3 (1978) 125–127.
[13] M.D. Plummer, On a family of line critical graphs, Monatsh. Math. 71 (1967) 40–48.
[14] L. Surányi, On line-critical graphs, Infinite and finite sets, (North-Holland, Amsterdam, 1975) 1411–1444.
[15] W.T. Tutte, The 1-factors of oriented graphs, Proc. Am. Math. Soc. (1953) 922–931.
[16] W. Wessel, Kanten-kritische Graphen, Manuscripts Math. 2 (1970) 309–334.
[17] A.A. Zykov, On some properties of linear complexes, Math. USSR Sb. 24 (1949) 163–188.

Annals of Discrete Mathematics 3 (1978) 21–28.

HAMILTONIAN DECOMPOSITIONS OF GRAPHS, DIRECTED GRAPHS AND HYPERGRAPHS

J.-C. BERMOND
C.M.S., 54 Bd. Raspail, 75006 Paris, Cédex 06, France

0. Definitions

Definitions not given here can be found in [3].

We will say that a *graph G* (undirected or directed) can be *decomposed into Hamiltonian cycles or paths* if we can partition its edges (arcs in the directed case) into hamiltonian cycles or paths (directed cycles or directed paths in the directed case). Our notation is as follows:

K_n — the complete graph on n vertices;

K_n^* — the complete symmetric directed graph on n vertices;

$K_{r\times n}$ — the complete r-partite graph whose vertex set is the disjoint union of r sets of n elements, two vertices being joined iff they belong to two different sets;

C_r (resp. $\vec{C}_r$) — a cycle (resp. directed cycle) of length r;

S_n — an independent set of n vertices;

$G_1 \times G_2$ — the cartesian sum (also called product) of two graphs $G_1 = (X_1, E_1)$ and $G_2 = (X_2, E_2)$ is the graph with vertex set $X_1 \times X_2$ in which (x_1, x_2) is joined to (y_1, y_2) whenever $x_1 = y_1$ and x_2 is joined to y_2 in G_2, or $x_2 = y_2$ and x_1 is joined to y_1 in G_1;

$G_1 \otimes G_2$ — the lexicographic product (also called composition) of two graphs G_1 and G_2 is the graph with vertex set $X_1 \times X_2$ in which (x_1, x_2) is joined to (y_1, y_2) whenever x_1 is joined to y_1 in G_1 or $x_1 = y_1$ and x_2 is joined to y_2 in G_2:

$G_1 \cdot G_2$ — the cartesian product (also called conjunction) of two graphs G_1 and G_2 is the graph with vertex set $X_1 \times X_2$ in which (x_1, x_2) is joined to (y_1, y_2) whenever x_1 is joined to y_1 in G_1 and x_2 is joined to y_2 in G_2.

1. Hamiltonian decompositions of graphs

The first two results are folklore.

1.1. Theorem. *K_{2n} can be decomposed into n hamiltonian paths.*

1.2. Theorem. *K_{2n+1} can be decomposed into n hamiltonian cycles and K_{2n+2} can be decomposed into n hamiltonian cycles and a perfect matching (or 1-factor).*

1.3. Theorem (Auerbach and Laskar [2]). *$K_{r\times n}$ can be decomposed into hamiltonian cycles iff $n(r-1)$ is even. If $n(r-1)$ is odd $K_{r\times n}$ can be decomposed into hamiltonian cycles and a perfect matching.*

We will now give a survey of hamiltonian decompositions of the three products defined in Section 0 and pose some problems concerning them.

1.4. Theorem (Kotzig [12]). *$C_r \times C_n$ can be decomposed into 2 hamiltonian cycles.*

1.5. Remark. The case $r=n$ is also proved in Myers [15].

1.6. Corollary. *If each of G_1 and G_2 can be decomposed into p hamiltonian cycles, then $G_1 \times G_2$ can be decomposed into 2p hamiltonian cycles.*

1.7. Theorem (Myers [15]). *$K_n \times K_n$ can be decomposed into $(n-1)$ hamiltonian cycles.*

1.8. Conjecture (Kotzig [12]). *$C_r \times C_n \times C_m$ can be decomposed into hamiltonian cycles.*

1.9. Remarks on Conjecture 1.8. G. Koester (personal communication, 1977) has proved that $C_4 \times C_4 \times C_4$ can be decomposed into hamiltonian cycles. In fact he informed me that the problem of the existence of a decomposition of $C_4 \times C_4 \times \cdots \times C_4$ (n times) was posed by Ringel [16, Problem 2] as the existence of a decomposition of the $2n$-cube ($2n$-dimensional Würfel) into hamiltonian cycles. Ringel [16] proved this conjecture for n a power of 2; this also follows from Theorem 1.4 with $r=n=4$ and Corollary 1.6.

Very recently I learned that the existence of a decomposition of $C_3 \times C_3 \times C_3$ into hamiltonian cycles was proved by M. Foregger (personal communication of R. Brualdi, 1977).

1.10. Conjecture. *$K_m \times K_n$ can be decomposed into $\frac{1}{2}(n+m-2)$ hamiltonian cycles iff $n+m$ is even and into $\frac{1}{2}(n+m-3)$ hamiltonian cycles and a perfect matching if $n+m$ is odd.*

1.11. Conjecture. *If G_1 can be decomposed into p_1 hamiltonian cycles and if G_2 can be decomposed into p_2 hamiltonian cycles, then $G_1 \times G_2$ can be decomposed into p_1+p_2 hamiltonian cycles.*

1.12. Theorem (Laskar [13]). *$C_r \otimes S_n$ can be decomposed into n hamiltonian cycles.*

1.13. Remark. Theorem 1.12 can be used to give a short proof of Theorem 1.3 since

$K_{r\times n}=K_r\otimes S_n$. If r is odd, then by Theorem 1.2 K_r can be decomposed into hamiltonian cycles, that is $K_r=\bigcup_i C_r^{(i)}$ (where $\bigcup$ means the edge-disjoint union and $1\leq i\leq r$) and $K_{r\times n}=\bigcup_i C_t^{(i)}\otimes S_n=\bigcup_{i,j} C_{rn}^{(i,j)}$ by Theorem 1.12. If r is even, then $K_{\frac{1}{2}r\times 2}$ is the graph obtained from K_r by deleting a perfect matching and so can be decomposed into hamiltonian cycles by Theorem 1.2. Finally $K_{r\times 2p}=(K_{r\times 2})\otimes S_p=\bigcup_i C_{2r}^{(i)}\otimes S_p=\bigcup_{i,j} C_{2rp}^{(i,j)}$. And thus if $n(r-1)$ is even $K_{r\times n}$ can be decomposed into hamiltonian cycles.

1.14. Theorem (Laskar [13]). *$C_r\otimes C_n$ can be decomposed into $n+1$ hamiltonian cycles if n is odd or r is even.*

1.15. Conjecture. *$C_r\otimes C_n$ can always be decomposed into $n+1$ hamiltonian cycles.*

1.16. Conjecture. *If G_1 can be decomposed into p_1 hamiltonian cycles and if G_2 can be decomposed into p_2 hamiltonian cycles, then $G_1\otimes G_2$ can be decomposed into $p_1n_2+p_2$ hamiltonian cycles (where n_2 is the number of vertices of G_2).*

1.17. Remark. I can prove that Conjecture 1.15 implies the truth of Conjecture 1.16 for $p_1\geq p_2$.

1.18. Theorem. *$C_r\cdot C_n$ can be decomposed into 2 hamiltonian cycles.*

Proof. We can suppose $r\geq n$; let the vertex set be $\mathbb{Z}_r\times\mathbb{Z}_n$ (where $\mathbb{Z}_n$ denotes the additive group of residues mod n). Then two hamiltonian cycles are $x_0, x_1, \ldots, x_{rn-1}$ and $y_0, y_1, \ldots, y_{rn-1}$ where $x_{in+j}=(i+j, j)$ and $y_{in+j}=(-i-j, j)$, $0\leq i\leq r-1$, $0\leq j\leq n-1$. □

1.19. Corollary. *If G_1 and G_2 can be decomposed into hamiltonian cycles, then $G_1\cdot G_2$ also can be decomposed into hamiltonian cycles.*

Proof. This follows from the distributivity of the product · with respect to the edge disjoint union of graphs (a property not holding for the cartesian sum and the lexicographic product). □

1.20. Many other problems similar to those above can be considered; in particular we can consider decompositions into cycles of given length (see [6]). We mention also that Huang and Rosa [9] have considered "orthogonal" hamiltonian decompositions, and finally we give the following conjecture of Kotzig [11].

1.21. Conjecture (Kotzig [11]). *K_{2n} can be decomposed into perfect matchings, i.e. has a 1 factorisation, in such a manner that the union of any two perfect matchings is a hamiltonian cycle.*

Partial results have been obtained on this problem, see for example Anderson [1].

2. Hamiltonian decompositions of directed graphs

The same problems can be asked for the directed graphs, but they are more difficult. Many of the known results are obtained in the following easy way; associate with a hamiltonian decomposition of G a hamiltonian decomposition of the directed graph G^* (obtained from G by associating to each edge of G two opposite arcs) by associating with each hamiltonian cycle two opposite directed hamiltonian cycles. For example Theorem 1.2 gives the following.

2.1. Theorem. *K^*_{2n+1} can be decomposed into $2n$ directed hamiltonian cycles.*

The problem of the existence of a hamiltonian decomposition of K^*_n has been solved only recently.

2.2. Theorem (Tillson [17]). *If $2n \geq 8$, then K^*_{2n} can be decomposed into $2n-1$ directed hamiltonian cycles.*

2.3. For $2n=4$ and $2n=6$, such a decomposition is impossible. The problem seems to have been asked first by Strauss for hamiltonian paths (see Mendelsohn [14]). In [14] Mendelsohn showed how the existence of sequenceable groups implies the existence of a hamiltonian decomposition of K^*_{2n} and that gives the result for $2n=22$ [14], 28 [10], 40, 56, 58 [18]. By computer the existence of a hamiltonian decomposition of K^*_{2n} for $8 \leq 2n \leq 18$ was obtained (see [5]). With Faber we proposed Theorem 2.2 as conjecture in [5] and [4]. After that A. Bouchet (personal communication, 1976) showed that if K^*_{2n} can be decomposed into directed hamiltonian cycles, then so can K^*_{4n-2}.

2.4. Conjecture. *One can easily ask many other problems, for example the directed versions of the results or problems of Section 1. But there are also problems peculiar to the directed case like Kelly's conjecture that every regular tournament can be decomposed into directed hamiltonian cycles.*

3. Hamiltonian decompositions of hypergraphs

3.1. For hypergraphs the number of problems grows quickly, because one can give different definitions of a hamiltonian cycle. I will restrict myself to a definition and a problem considered in [7]. If H is a hypergraph with n vertices then a hamiltonian cycle is a sequence $x_1E_1x_2 \cdots x_iE_ix_{i+1} \cdots x_nE_nx_1$ such that

(i) the n vertices x_i are all different (and thus are the n vertices of the hypergraph),

(ii) the n edges E_i are all different,

(iii) $\{x_i, x_{i+1}\} \subset E_i (1 \leq i \leq n-1)$ and $\{x_1, x_n\} \subset E_n$.

Let K_n^h denote the complete h-uniform hypergraph; its edges are all the h-subsets of a set X of cardinality n.

In [7] we conjectured that K_n^h can be decomposed into hamiltonian cycles if and only if $\binom{n}{h}/n$ is an integer and proved this conjecture for n a prime.

Here I want to prove two theorems concerning the case $h=3$.

3.2. Theorem. *If K_n^3 can be decomposed into hamiltonian cycles then K_{2n}^3 also can be decomposed into hamiltonian cycles.*

Proof. (This proof was obtained with D. Sotteau.) In order to shorten the writing, I will write a hamiltonian cycle as $E_1, \ldots E_i, \ldots, E_n$ where $E_i = (x_i, y_i, x_{i+1})$. Let the vertex set of K_{2n}^3 be $X \cup X'$ with $|X| = |X'| = n$. With each hamiltonian cycle of the decomposition of K_n^3 we associate 4 hamiltonian cycles of K_{2n}^3 in the following manner:

(i) if n is even, we associate with $(x_1y_1x_2)(x_2y_2x_3), \ldots, (x_ny_nx_1)$ the following

$$(x_1y_1x_2)(x_2y_2x_3)\cdots(x_{n-1}y_{n-1}x_n)(x_ny_nx_1')$$
$$(x_1'y_1x_2')\cdots(x_{n-1}'y_{n-1}x_n')(x_n'y_nx_1),$$

$$(x_1y_1x_2')(x_2'y_2x_3)(x_3y_3x_4')\cdots(x_{n-1}y_{n-1}x_n')(x_n'y_nx_1')$$
$$(x_1'y_1x_2)(x_2y_2x_3')\cdots(x_{n-1}'y_{n-1}x_n)(x_ny_nx_1),$$

and the two cycles obtained by exchanging the vertices of X and those of X';

(ii) if n is odd we associate with $(x_1y_1x_2)(x_2y_2x_3), \ldots, (x_ny_nx_1)$ the following:

$$(x_1y_1x_2)(x_2y_2x_3)\cdots(x_{n-2}y_{n-2}x_{n-1})(x_{n-1}y_{n-1}'x_n')(x_n'y_nx_1')$$
$$(x_1'y_1x_2')(x_2'y_2x_3')\cdots(x_{n-2}'y_{n-2}x_{n-1}')(x_{n-1}'y_{n-1}'x_n)(x_ny_nx_1),$$

$$(x_1'y_1'x_2')(x_2'y_2'x_3')\cdots(x_{n-2}'y_{n-2}'x_{n-1}')(x_{n-1}'y_{n-1}'x_n')(x_n'y_n'x_1)$$
$$(x_1y_1'x_2)(x_2y_2'x_3)\cdots(x_{n-2}y_{n-2}'x_{n-1})(x_{n-1}y_{n-1}'x_n)(x_ny_n'x_1'),$$

$$(x_1y_1x_2')(x_2'y_2x_3)\cdots(x_{n-2}y_{n-2}x_{n-1}')(x_{n-1}'y_{n-1}x_n)(x_ny_nx_1')$$
$$(x_1'y_1x_2)(x_2y_2x_3')\cdots(x_{n-2}'y_{n-2}x_{n-1})(x_{n-1}y_{n-1}x_n')(x_n'y_nx_1),$$

and

$$(x_1'y_1'x_2)(x_2y_2'x_3')\cdots(x_{n-2}'y_{n-2}'x_{n-1})(x_{n-1}y_{n-1}x_n)(x_ny_n'x_1)$$
$$(x_1y_1'x_2')(x_2'y_2'x_3)\cdots(x_{n-2}y_{n-2}'x_{n-1}')(x_{n-1}'y_{n-1}x_n')(x_n'y_n'x_1').$$

Thus we have obtained a decomposition of the edges of K_{2n}^3 not of the form (x, x', y) or (x, x', y'). We will use Theorems 2.1 and 2.2 to decompose these remaining triples. Indeed with the directed hamiltonian cycle $x_1, \ldots, x_n$ of a decomposition of K_n^* we associate the following hamiltonian cycle of K_n^3:

$$(x_1x_1'x_2)(x_2x_2'x_3)\cdots(x_{n-1}x_{n-1}'x_n)(x_nx_n'x_1')$$
$$(x_1'x_1x_2')\cdots(x_{n-1}'x_{n-1}x_n')(x_n'x_nx_1).$$

Thus the proof is complete. One can check that we have found $[\frac{4}{6}(n-1)(n-2)]+n-1=\frac{1}{6}(2n-1)(2n-2)$ hamiltonian cycles in K_{2n}^3. □

3.3. Theorem. *If* $n\equiv 2 \bmod (3)$ K_n^3 *can be decomposed into hamiltonian cycles.*

We are grateful to A.E. Brouwer (personal communication, 1976) for the following idea on which the proof is based.

3.4. *A choice design of order n* is a system of representatives of the triples of K_n^3 such that:

(i) each point is chosen equally often as a representative;

(ii) among the $n-2$ triples containing a given pair $\{a, b\}$, a is chosen $\frac{1}{3}(n-2)$ times and b $\frac{1}{3}(n-2)$ times also.

For example, when $n=5$, we have underlined the element chosen.

$$0\underline{1}2;\ 01\underline{3};\ \underline{0}14;\ \underline{0}23;\ 0\underline{2}4;\ 03\underline{4};\ 1\underline{2}3;\ 12\underline{4};\ \underline{1}34;\ 2\underline{3}4.$$

3.5. Theorem. *A choice design of order n exists if and only if* $n\equiv 2 (\bmod 3)$.

Proof. The necessary condition is obvious as $\frac{1}{3}(n-2)$ must be an integer. We will prove that the condition is sufficient by induction. Suppose there exists a choice design of order n. Let the elements of K_{n+3}^3 be $\{1, 2, \ldots n\}\cup\{\alpha, \beta, \gamma\}$. For a triple of elements of $\{1, 2, \ldots, n\}$ we choose the element defined by the choice design of order n. For the triples we choose

$$\begin{array}{cccl}
(i, j, \alpha) & (i, j, \beta) & (i, j, \gamma) & \text{with } i<j, \\
i & j & \gamma & \text{if } i+j\equiv 0 \ (\text{mod } 3), \\
j & \beta & i & \text{if } i+j\equiv 1 \ (\text{mod } 3), \\
\alpha & i & j & \text{if } i+j\equiv 2 \ (\text{mod } 3).
\end{array}$$

For the triples we choose

$$\begin{array}{cccl}
(i, \alpha, \beta) & (i, \alpha, \gamma) & (i, \beta, \gamma) & \\
i & \gamma & \gamma & \text{if } i\equiv 0 \ (\text{mod } 3), \\
\beta & \alpha & i & \text{if } i\equiv 1 \ (\text{mod } 3), \\
\alpha & i & \beta & \text{if } i\equiv 2 \ (\text{mod } 3).
\end{array}$$

For the triple (α, β, γ) we choose γ.

We leave to the reader the care of checking that we obtain a choice design of order $n+3$; the only non-immediate part is to check property (ii) for the triples containing a pair (i, α) or (i, β) or (i, γ). □

3.6. Proof of Theorem 3.3. If n is odd, there exists a decomposition of K_n into $\frac{1}{2}(n-1)$ hamiltonian cycles. To each of these cycles $(x_1, x_2, \ldots, x_n)$ we associate the following $\frac{1}{3}(n-2)$ hamiltonian cycles of $K_n^3: (x_1, y_1^i, x_2)(x_2, y_2^i, x_3) \cdots (x_{n-1}, y_{n-1}^i, x_n)(x_n, y_n^i, x_1)$, where $i = 1, 2, \ldots, \frac{1}{3}(n-2)$, $y_j^i \neq y_j^k$ for $i \neq k$ and where the y_j^i are defined according to the existence of a choice design of order n (by Theorem 3.5). The set $\{y_j^i : i = 1, 2, \ldots, \frac{1}{3}(n-2)\}$ consists of the $\frac{1}{3}(n-2)$ elements representatives of the $\frac{1}{3}(n-2)$ triples (x_j, x_{j+1}, y) containing the pair $\{x_j, x_{j+1}\}$ and where neither x_j nor x_{j+1} has been chosen. Thus we have constructed $\frac{1}{6}(n-1)(n-2)$ hamiltonian cycles of K_n^3 and it suffices to verify that no triple (edge) appears twice, but that follows from the definition of a choice design of order n.

3.7. Example. Let (0, 1, 2, 3, 4) and (0, 2, 4, 1, 3) be two hamiltonian cycles of K_5; by using the choice design of the example we obtain the two hamiltonian cycles of K_5^3:

$$(0, 3, 1)(1, 4, 2)(2, 0, 3)(3, 1, 4)(4, 2, 0),$$
$$(0, 1, 2)(2, 3, 4)(4, 0, 1)(1, 2, 3)(3, 4, 0).$$

If n is even the proof is similar. We use a decomposition of K_n^* into $n-1$ directed hamiltonian cycles (Theorem 2.2). To each of these directed cycles $(x_1, x_2, \ldots, x_n)$ we associate $\frac{1}{6}(n-2)$ hamiltonian cycles of K_n^3:

$$(x_1, y_1^i, x_2)(x_2, y_2^i, x_3) \cdots (x_{n-1}, y_{n-1}^i, x_n)(x_n, y_n^i, x_1),$$

where $i = 1, 2, \ldots, \frac{1}{6}(n-2)$; $y_j^i \neq y_j^k$ for $i \neq k$ and where the set $\{y_j^i : i = 1, 2, \ldots, \frac{1}{6}(n-2)\}$ is determined as follows. Consider the $\frac{1}{3}(n-2)$ elements representative of the triples (x_j, x_{j+1}, y) containing the pair $\{x_j, x_{j+1}\}$ and where neither x_j nor x_{j+1} has been chosen. Then split these elements into two sets of cardinality $\frac{1}{6}(n-2)$: $Y_{\{x_j, x_{j+1}\}}$ and $Y'_{\{x_j, x_{j+1}\}}$. Then the set $\{y_j^i : i = 1, 2, \ldots, \frac{1}{6}(n-2)\}$ is either the set $Y_{\{x_j, x_{j+1}\}}$ or $Y'_{\{x_j, x_{j+1}\}}$ according as the arc (x_j, x_{j+1}) or the arc (x_{j+1}, x_j) appears in the directed hamiltonian cycle. □

Note added in proof

M.F. Foregger has proved Conjecture 1.8 (Hamiltonian Decompositions of Product of Cycles).

References

[1] B.A. Anderson, A perfectly arranged room square, Proc. 4th Southeastern Conference on Combinatorics, Graph Theory, and Computing, Utilitas Math. (1973) 141–150.

[2] B. Auerbach and R. Lasker. On decompositions of r-partite graphs into edge — disjoint hamiltonian circuits, Discrete Math. 14 (1976) 265–268.

[3] C. Berge, Graphs and Hypergraphs (North-Holland, Amsterdam, 1973).

[4] J.-C. Bermond and V. Faber, in: Problems of the 5th British Combinatorial Conference, Aberdeen 1975, Utilitas Math. Congressus Numerantium XV (1975). 695.
[5] J.-C. Bermond and V. Faber, Decomposition of the complete directed graph into k-circuits, J. Combinatorial Theory, 21(B) (1976) 146–155.
[6] J.-C. Bermond and D. Sotteau, Graph Decompositions and G-designs, Proc. 5th Combinatorial Conference, Aberdeen 1975, Utilitas Math. Congressus Numerantium XV (1975) 53–72.
[7] J.-C. Bermond, A. Germa, M.-C. Heydemann and D. Sotteau, Hypergraphes hamiltoniens, in: Coll. Int. C.N.R.S., Problèmes Combinatoires et Théorie des Graphes, Orsay, 1976.
[8] J. Denes and A.D. Keedwell, Latin squares and their applications, (Akademiai Kiado, Budapest and English University Press, London, 1974).
[9] C. Huang and A. Rosa, On sets of orthogonal hamiltonian circuits, Proc. 2nd Manitoba Conference on Numerical Mathematics, Utilitas Math. Congressus Numerantium VII (1972), 327–332.
[10] A.D. Keedwell, Some problems concerning complete latin squares, Combinatorics, Proc. British Combinatorial Conference, 1973, L.M.S. Notes 13, Cambridge University (1974) 89–96.
[11] A. Kotzig, Problem, Theory of Graphs and Its Applications (Academic Press, New York, 1964) 162 and 63–82.
[12] A. Kotzig, Every cartesian product of two circuits is decomposable into two hamiltonian circuits, Centre de Recherches Mathematiques, Montreal (1973).
[13] R. Laskar, Decomposition of some composite graphs into hamiltonian cycles, Proc. 5th Hungarian Colloquium, Keszthely, 1976, (to appear).
[14] N.S. Mendelsohn, Hamiltonian decomposition of the complete directed n-graph, Theory of Graphs (Akademiai Kiado, Budapest, 1968) 237–241.
[15] B.R. Myers, Hamiltonian factorization of the product of a complete graph with itself, Networks 2 (1972) 1–9.
[16] G. Ringel, Über drei kombinatorische Probleme am n-dimensionalen Würfel und Würfelgitter, Abh. Math. Sem. Hamburg 20 (1954) 10–19.
[17] T. Tillson, A hamiltonian decomposition of K^*_{2m}, $2m \geq 8$, J. Combinatorial Theory (B) (to appear).
[18] L.L. Wang, A test for sequencing a class of finite groups with two generators, Notices Am. Math. Soc. 20 (1973) A632.

Annals of Discrete Mathematics 3 (1978) 29–41.

EXTREMAL GRAPHS WITHOUT LARGE FORBIDDEN SUBGRAPHS

B. BOLLOBÁS

Department of Pure Mathematics and Mathematical Statistics, Cambridge CB2 1SB, England

P. ERDÖS

Hungarian Academy of Sciences, Budapest, Hungary

M. SIMONOVITS

Department of Mathematics, Eötvös University, Budapest VIII, Hungary

E. SZEMERÉDI

Mathematical Institute, Hungarian Academy of Sciences, Budapest V, Hungary

The theory of extremal graphs without a fixed set of forbidden subgraphs is well developed. However, rather little is known about extremal graphs without forbidden subgraphs whose orders tend to ∞ with the order of the graph. In this note we deal with three problems of this latter type. Let L be a fixed bipartite graph and let $L+E^m$ be the join of L with the empty graph of order m. As our first problem we investigate the maximum of the size $e(G^n)$ of a graph G^n (i.e. a graph of order n) provided $G^n \not\supset L+E^{[cn]}$, where $c>0$ is a constant. In our second problem we study the maximum of $e(G^n)$ if $G^n \not\supset K_2(r, cn)$ and $G^n \not\supset K^3$. The third problem is of a slightly different nature. Let $C^k(t)$ be obtained from a cycle C^k by multiplying each vertex by t. We shall prove that if $c>0$ then there exists a constant $l(c)$ such that if $G^n \not\supset C^k(t)$ for $k=3, 5, \ldots, 2l(c)+1$, then one can omit $[cn^2]$ edges from G^n so that the obtained graph is bipartite, provided $n>n_0(c, t)$.

Our notation is that of [1]. Thus G^n is an arbitrary graph of order n, K^p is a complete graph of order p, E^p is a null graph of order p (that is one with no edges), C^m is a cycle of length m, $G_r(n_1, \ldots, n_r)$ is an r-partite graph with n_i vertices in the ith class, $K_r(n_1, \ldots, n_r)$ is a complete r-partite graph. $C^m(t)$ is a graph obtained from C^m by multiplying it by t, that is by replacing each vertex by t independent vertices. We use H^m, S^m, T^m, U^m to denote graphs of order m with properties specified in the text. We write $|A|$ for the cardinality of a set A, $|G|$ for the order of a graph G and $e(G)$ for the number of edges (the *size*) of G. The set of neighbours of a vertex x is denoted by $\Gamma(x)$ and $d(x)=|\Gamma(x)|$ is the degree of x. The minimum degree in G is $\delta(G)$.

Let $\mathscr{F}$ be a family of graphs, called the family of *forbidden* graphs. Denote by $\mathrm{EX}(n, \mathscr{F})$ the set of graphs of order n with the *maximal* number of edges that does not contain any member of $\mathscr{F}$. The graphs in $\mathrm{EX}(n, \mathscr{F})$ are the *extremal graphs of order n for* $\mathscr{F}$. Write $\mathrm{ex}(n, \mathscr{F})$ for the size of the extremal graphs: $\mathrm{ex}(n, \mathscr{F})=e(H)$, where $H\in \mathrm{EX}(n, \mathscr{F})$. The problem of determining $\mathrm{ex}(n, \mathscr{F})$ or $\mathrm{EX}(n, \mathscr{F})$ may be called a Turán type extremal problem. We shall prove some Turán type extremal results in which the forbidden graphs depend on n. The first deep theorem of this kind was proved by Erdös and Stone [8] in 1946. This theorem is the basis of the theory of extremal graphs without forbidden subgraphs (see [1, Ch. VI]). Considerable extensions of it were proved by Bollobás and Erdös [2] and by Bollobás, Erdös and Simonovits [3].

For fixed r and t the extremal graphs $\mathrm{EX}(n, K_3(2, r, t))$ were studied by Erdös and Simonovits [7]. Our first aim in this note is to describe $\mathrm{EX}(n, K_3(2, r, cn))$, where $r \geq 2$ and $c > 0$. In fact, we prove the following somewhat more general result.

Theorem 1. *Let L be a bipartite graph. Put*

$$q(n, L) = \max\{n_1 n_2 + \mathrm{ex}(n_1, L) + \mathrm{ex}(n_2, L) : n_1 + n_2 = n\}. \tag{1}$$

There exist $c > 0$ and n_0 such that if $n > n_0$ and

$$e(G^n) > q(n, L), \tag{2}$$

then G^n contains an $L + E^{[cn]}$. If in addition for every m there exists an extremal graph $S^m \in \mathrm{EX}(m, L)$ with maximum degree $< \frac{1}{2} cm$, then

$$\mathrm{ex}(n, L + E^{[cn]}) = q(n, L) \tag{3}$$

and every extremal graph $U^n \in \mathrm{EX}(n, L + E^{[cn]})$ can be obtained from an $S^m \in \mathrm{EX}(m, L)$ and an $S^{n-m} \in \mathrm{EX}(n-m, L)$ as $S^m + S^{n-m}$.

Remarks. (i) If $L = K_2(2, r)$, then the maximum degree of any $S^m \in \mathrm{EX}(m, L)$ is $o(m)$ and the same holds if L is not a tree, but there exists a vertex $v \in L$ for which $L - v$ is a tree. Thus Theorem 1 gives

$$\mathrm{ex}(n, K_3(2, r, [cn])) = q(n, K_2(2, r)).$$

It also gives information on the structure of the extremal graphs.

(ii) Theorem 1 states that $q(n, L)$ is an upper bound for $\mathrm{ex}(n, L + E^{[cn]})$. A lower bound for $\mathrm{ex}(n, L + E^{[cn]})$ can be obtained by observing that if $S^m \in \mathrm{EX}(n, L)$, then $S^m + E^{n-m} \not\supset L + E^{[cn]}$, so

$$\mathrm{ex}(n, L + E^{[cn]}) \geq \max\{n_1 n_2 + \mathrm{ex}(n_1, L) : n_1 + n_2 = n\}. \tag{4}$$

In some cases, for instance if L consists of independent edges, (4) is sharp.

(iii) The essential part of Theorem 1 states that a graph G^n not containing an $L + E^{[cn]}$ can not have more edges than $S^p + S^{n-p}$, where $S^p \in \mathrm{EX}(p, L)$, $S^{n-p} \in \mathrm{EX}(n-p, L)$ and p is suitably chosen. It is unfortunate that $S^p + S^{n-p}$ may contain an $L + E^{[cn]}$ and we need an additional condition to exclude this possibility.

The proof of Theorem 1 is based on five lemmas.

Lemma 2. $q(n+1, L) - q(n, L) \geq n/2$.

Proof. Let $q(n, L) = n_1 n_2 + \mathrm{ex}(n_1, L) + \mathrm{ex}(n_2, L)$, where $n_1 \leqslant n_2$. Then

$$q(n+1, L) \geqslant (n_1+1)n_2 + \mathrm{ex}(n_1+1, L) + \mathrm{ex}(n_2, L) \geqslant q(n, L) + n/2.$$

The next lemma is an immediate consequence of Lemma 2 and the straightforward Lemma V.3.2 of [1].

Lemma 3. *Given $c_1 > 0$ there exists $c_2 > 0$ such that if $e(G^n) > q(n, L)$ then G^n contains a subgraph G^p satisfying $p \geqslant c_2 n$, $e(G^p) > q(p, L)$ and $\delta(G^p) > (\frac{1}{2} - c_1)p$.*

Lemma 4. *There exists a constant $c_L > 0$ such that if $\delta(G^n) \geqslant (\frac{1}{2} - \frac{1}{100})n$ and $K = K_3(9r, 9r, 9r) \subset G^n$, where $r = |L|$, then G^n contains an $L + E^t$ with $t \geqslant c_L n$.*

Proof. Put $H = G^n - K$. Since at least $27r \cdot \frac{49}{100}n - (27\,r)^2$ edges join K to H, at least $\frac{1}{20}n$ vertices of H are joined to at least $11r$ vertices of K. Let $c_L = \frac{1}{20} \cdot 2^{-27r}$. Then H contains $t \geqslant c_L n$ vertices that are joined to the same set of at least $11r$ vertices of K. The subgraph of $K = K_3(9r, 9r, 9r)$ spanned by this set of vertices contains a $K_2(r, r)$ so $K_2(r, r) + E^t \subset G^n$. Since $L \subset K_2(r, r)$ we have $L + E^t \subset G^n$. □

The first part of the next lemma is a weak form of Theorem V.2.2 in [1], the second part is an immediate consequence of the first part.

Lemma 5. (i) *If $G = G_2(m, n)$ does not contain a $K_2(s, t)$ whose first class is in the first class of G then*

$$e(G) < t^{1/s} m n^{1-1/s} + sn.$$

(ii) *Given d and R, there exist $\varepsilon > 0$ and n_0 such that if $n \geqslant n_0$ and if in $G = G_d(n, n, \ldots, n)$ at least $(1-\varepsilon)n^2$ edges join any two classes then G contains a $K_d(R, R, \ldots, R)$.*

The last lemma needed in the proof of Theorem 1 is a slight extension of some results proved by Erdös and Simonovits [5, 6, 10].

Lemma 6. *Given c, $0 < c < 1$, and natural numbers d and R, there exist $M = M(c, d, R)$, $\delta = \delta(c, d, R) > 0$, and $n_0 = n(c, d, R)$ such that if $n > n_0$, $e(G^n) > (1 - 1/d - \delta)\frac{1}{2}n^2$ and $K_{d+1}(R, \ldots, R) \not\subset G^n$, then the vertices of G^n can be divided into d classes, say $A_1, A_2, \ldots, A_d$, such that the following conditions are satisfied.*

(i) $|n_i - n/d| < cn$, *where $n_i = |A_i|$.*

(ii) *The subgraph $G_i = G^n[A_i]$ of G^n, spanned by A_i, satisfies*

$$e(G_i) < cn^2$$

(iii) *Call a pair $\{x, y\}$ of vertices a missing edge if x and y do not belong to the same class A_i and xy is not an edge of G^n. The number of missing edges is less than cn^2.*

(iv) *Let* B_i *be the set of vertices in* A_i *joined to at least* cn *vertices of the same class* A_i. *Then* $|B_i|<M$.

Proof. Let $M_0=R$ and choose natural numbers $M_1<M_2<\ldots<M_d$ such that $M_i/M_{i+1}<\frac{1}{2}$. Put $M=M_d$. Pick η such that $0<\eta<(\frac{1}{2}c)^M$.

By Lemma 5 (ii) we can choose ε, $0<\varepsilon<c$, and n_1 such that if $N=[\eta n]$, $n\geq n_1$ and in $H=G_d(N, N, \ldots, N)$ at most εn^2 edges are missing between any two classes then H contains a $K_d(R, R, \ldots, R)$.

The above mentioned theorem of Erdös and Simonovits (see Theorem V.4.2 in [1]) implies that there exist $n_0\geq n_1$ and $\delta>0$ with the following properties. If G^n is as in our lemma and $A_1, A_2, \ldots, A_d$ is a partition with the minimal number of missing edges (cf. condition (iii) of the lemma) then (i), (ii) and (iii) hold.

Suppose (iv) fails, say $|B_1|\geq M$. Then by the minimality of the partition each vertex of B_1 is joined to at least cn vertices in each A_i. Since

$$M_{i+1}cn>(\eta n)^{1/M_i}M_{i+1}n^{1-1/M_i}+M_i n,$$

repeated applications of Lemma 5 (i) imply that there are sets $\tilde{B}\subset B_1$, $\tilde{A}_i\subset A_i$, $i=1,2,\ldots,d$, such that $|\tilde{B}|=R$, $|\tilde{A}_i|=N$ and each vertex of B is joined to each vertex of $A=\bigcup_1^d \tilde{A}_i$. Now it follows from (iii) and the choice of ε that $G[A]$ contains a $K_d(R, R, \ldots, R)$. Hence $G[A\cup B]$ contains a $K_{d+1}(R, R, \ldots, R)$. □

Proof of Theorem 1. It is easy to see that if G, H are graphs containing no L and no $K_2(1, cm/2)$, then $G+H$ contains no $L+E^{[cm]}$. Hence the second assertion of Theorem 1 is trivial. To prove the first assertion assume indirectly that G^n contains no $L+E^{[cn]}$ and $e(G^n)>q(n, L)$. We shall show that this is impossible if $c>0$ is sufficiently small. By Lemma 3 and Lemma 4 we may and will assume that $\delta(G^n)\geq(\frac{1}{2}-\frac{1}{10}r^{-1})n$, $R=gr$, and $G^n\not\supset K_3(9r, 9r, 9r)$. Applying Lemma 6 with $d=2$, we obtain a partition (A_1, A_2), satisfying (i)–(iv) of Lemma 6. For the sake of convenience in the sequel a subset H of the vertices of G^n and the corresponding spanned subgraph may be denoted by the same letter. Clearly, if m is the number of missing edges, then

$$e(G^n)=e(A_1)+e(A_2)+n_1n_2-m, \tag{5}$$

where $n_i=|A_i|$. Trivially, if neither A_1 nor A_2 contain L, then

$$e(G^n)\leq \mathrm{ex}(n_1, L)+\mathrm{ex}(n_2, L)+n_1n_2\leq q(n, L).$$

Thus $L\subset A_1$ may be assumed.

Let us assume that A_1-B_1 contains a subgraph L_0 isomorphic to L. To each $x\in L_0$ we find $\frac{1}{2}n(1-\frac{1}{2}r^{-1})$ or more vertices in A_2-B_2 joined to this x: since x is joined to cn or less vertices of A_1, it is joined to at least $(\frac{1}{2}-\frac{1}{10}r^{-1})n-\frac{1}{10}nr^{-1}$ vertices of A_2. Thus at least $n_2-r\cdot\frac{1}{4}nr^{-1}>\frac{1}{5}n$ vertices of A_2-B_2 are completely joined to L_0, yielding an $L+E^t$ for $t=[\frac{1}{5}n]$. This proves that $L\not\subset A_i-B_i$. Hence

$$e(A_i)\leq e(A_i-B_i)+|B_i|\,n_i\leq \mathrm{ex}(n_i, L)+|B_1|\,n_i.$$

Now we fix some constants and give the basic ideas of the proof. The details are given afterwards.

We fix a constant T such that $|B_i| \leq T\,(i=1,2)$. Lemma 6 guarantees the existence of such a T. A constant $c_L > 0$ is fixed so that for $a_i = r^{2T-i}c_L$ we have $a_1 < (10T)^{-2}$. Given a set $W \subset A_i$, denote by $F(W)$ the set of vertices of A_{3-i} not joined to at least one vertex of W. Observe that if W has at least $c_L n$ vertices, then $F(W)$ represents each $L \subset A_{3-i}$, for otherwise there would be an L completely joined to W and therefore $L + E^{c_L n} \subset G^n$ and we are home. Thus we may assume that $F(W)$ represents all the L's in A_{3-i}. Let

$$k_i = \frac{1}{n_i}(e(A_i) - \mathrm{ex}\,(n_i, L)).$$

Clearly, to represent all the L's in A_i we need vertices, the omission of which diminishes $e(A_i)$ by at least $e(A_i) - \mathrm{ex}\,(n_i, L)$, hence we have to omit at least k_i vertices: $F(W)$ has at least k_i vertices. This is the basic idea of the proof, but this in itself will not be enough. We shall prove the existence of a set Q_i of O(1) vertices in A_i such that the number of missing edges incident with this Q_i is at least $k_i n_i + \frac{1}{25} n T^{-1}$ if $L \subset A_i$. We have already checked the case, when no L occurs in A_1 and A_2. Let us consider the case, when $A_1 \supset L$ but $A_2 \not\supset L$. By (5) we have

$$e(G^n) \leq \mathrm{ex}\,(n_1, L) + k_1 n_1 + \mathrm{ex}\,(n_2, L) + n_1 n_2 - \left(k_1 n_1 + \frac{n}{25T}\right) < q(n, L).$$

If $A_1 \supset L$, $A_2 \supset L$, then the number of missing edges is estimated by the sum of the missing edges incident with Q_1 and Q_2 minus the number of missing edges between Q_1 and Q_2, which is only O(1). Hence

$$e(G^n) \leq \sum_i \left(\mathrm{ex}\,(n_i, L) + k_i n_i - \left(k_i n_i + \frac{n}{25T}\right)\right) + \mathrm{O}(1) < q(n, L).$$

This completes the sketch of the proof.

Let us see now how the argument above can be made precise. Recall that $L \subset A_1$. Let $L_1, \ldots, L_p, \ldots$ be subgraphs of A_1 isomorphic to L. For any $W = W_1 \subset A_2$ and $\tilde{W} \subset W_1$, $|W_1| \geq a_1 n$, $|\tilde{W}| \geq c_L n$, $F(\tilde{W})$ represents all the L_p's, among them L_1, hence for at least $a_1 n - c_L n$ vertices of W_1 there exists a vertex in L_1 not joined to it. Hence there exists an $x_1 \in L_1$ and a $W_2 \subset W_1$, $|W_2| \geq a_2 n$, such that x_1 is not joined to W_2 at all. If x_1 does not represent all the L_p's, we may assume that $x_1 \notin L_2$. Iterating this argument we find an $x_2 \in L_2$ not joined to a $W_3 \subset W_2$ at all, where $|W_3| \geq a_3 n$, and if x_1, x_2 do not represent all the L_p's, we define x_3 and W_4 in the same way.

Generally, if x_p and W_{p+1} are already defined, we check whether the set $X_p = \{x_1, \ldots, x_p\}$ represents all the $L \subset A_1$. If it does or if $p = 2T$, the procedure stops, otherwise we find an L_{p+1} and an x_{p+1} in it and a

$$W_{p+2} \subset W_{p+1}, \quad |W_{p+2}| \geq a_{p+2} n$$

so that x_{p+1} is not joined to W_{p+2} at all. At the end of the procedure we have an $X=X_p$ and a $W'=W_{p+1}$ not joined at all to each other.

Let $B'\subset B_1$ be the class of vertices of degree at least $\frac{1}{2}n-\frac{1}{10}nT^{-1}$ in A_1. Let D be the set of vertices of A_2 joined to B' completely. D is relatively large. Indeed, by the minimum property of the partition (A_1, A_2) any $x\in B'$ is joined to at least $n(\frac{1}{2}-\frac{1}{10}T^{-1})$ vertices of A_2, hence at least $n(\frac{1}{2}+\varepsilon)-Tn(\frac{1}{10}T^{-1}+\varepsilon)$ vertices of A_i are joined to B' completely. Thus $|D|\geqq\frac{1}{3}n$ if $\varepsilon\leqq\frac{1}{20}T^{-1}$.

Now we define another procedure, in each step of which the above procedure is applied to a set $W_j\subset A_2$ yielding a pair of sets W'_j and X_j not joined to each other at all: $|W_j|\geqq a_1 n$, $|W'_j|\geqq c_L n$, $|X_j|\leqq 2T$. Let

$$W_1=D\quad(|D|\geqq a_1 n),$$

$$W_j=D-\bigcup_{i<j} W'_i\quad\text{until}\quad |W_j|<a_1 n,$$

then

$$W_j=A_2-\bigcup_{i<j} W'_i.$$

The corresponding sets in A_1 are $X_1,\dots,X_j$. The procedure stops if for $W_j=A_2-\bigcup_{i<j} W'_i$ we have $|W_j|<a_1 n$. By $|W'_1|\geqq c_L n$ this will happen for some $j\leqq c_L{}^{-1}$. Let $X=\bigcup X_j$. Clearly, $|X|\leqq 2T/c_L=\mathrm{O}(1)$. We shall show that there exist at least $k_1 n_1+\frac{1}{25}nT^{-1}$ missing edges joining X to A_2. This will complete the proof.

We need a lower bound for the number of missing edges joining a W'_i to X: this lower bound is $|X_i|$. By the definition of D, if $W_i\subset D$, then each vertex of $X_i\subset A_1-B'$ has degree $\leqq\frac{1}{2}n-\frac{1}{10}nT^{-1}$ in A_1. These vertices represent all the L's in A_1, hence they represent at least $k_1 n_1$ edges:

$$|X_i|\geqq\frac{k_1 n_1}{\frac{1}{2}n-\dfrac{n}{10T}}\geqq k_1+\frac{1}{6T},$$

if n is sufficiently large, ε sufficiently small and $k_1\geqq\frac{5}{6}$. If $k_1\leqq\frac{5}{6}$, we use $|X_i|\geqq 1$, $|T|\geqq 1$ (which can be assumed). Thus

$$|X_i|\geqq k_1+\frac{1}{6T}$$

again. In the other case, when $W_i\not\subset D$, we use a weaker lower bound. Since $|W'_i|\geqq c_L n$ and no $x\in X_i$ is joined to W'_i, such an x is joined to at most $n_2-c_L n$ vertices of A_2, and consequently, to at most $n_2-c_L n=\frac{1}{2}n-c_L n+\varepsilon n$ vertices of A_1, we obtain now that

$$|X_i|\geqq k_1+k_1 c_L\geqq k_1.$$

Thus the number of missing edges incident to X can be estimated from below by

$$\frac{n}{3}\left(k_1+\frac{1}{6T}\right)+\frac{n}{6}\cdot k_1-(a_1+\varepsilon)n\left(k_1+\frac{1}{6T}\right).$$

Here the third term stands for the vertices not belonging to any W_i' and for the difference between n_2 and $\frac{1}{2}n$. If ε is sufficiently small, T large, then by $k_1 \leq T$ and $a_1 \leq (10T)^{-2}$ we obtain that at least $k_1 n_1 + \frac{1}{25}nT^{-1}$ missing edges are between X and A_2. Thus the proof is complete. □

Remark. Theorem 1 can be generalized to higher chromatic numbers, that is, an analogous theorem holds for $L + K_{d-1}(r, \ldots, r, cn)$. The proof of this generalization is essentially the same as for the particular case considered above.

Our second theorem concerns ex $(n, K^3, K_2(r, [cn]))$ and, more generally, ex $(n, C^{2j+1}(t), K_2(r, [cn]))$. An interesting feature of the result is that the value does not really depend on j and t.

Theorem 7. *Let j, r, t be natural numbers, let $k = 2j+1$ and let $c > 0$. If $e(G^n) \geq cn^2$ and G^n does not contain a $C^k(t)$, then G^n contains a $K_2(r, m)$, where*

$$m = 2^{2^r-1}c^r n + o(n).$$

Proof. We shall show first that if instead of a $C^k(t)$ (and so a fortiori a C^k) we prohibit *all* odd cycles, then G^n contains a $K_2(r, m)$ with

$$m = 2^{2^r-1}c^r n + o(n),$$

but if $\varepsilon > 0$ then G^n need not contain a $K_2(r, m')$ with

$$m' = (2^{2^r-1}c^r + \varepsilon)n + o(n).$$

(This will show that the value of m given in the theorem is as large as possible and that the main thrust of the theorem is that the condition "G^n is bipartite" can be replaced by the much weaker condition "G^n does not contain a $C^k(t)$" without decreasing the value of m we can guarantee.)

The first assertion is an immediate consequence of Lemma 5. Instead of the second we prove the following stronger assertion.

Let n be even and let G^n be a random subgraph of $K_2(\frac{1}{2}n, \frac{1}{2}n)$ obtained by taking an edge of $K_2(\frac{1}{2}n, \frac{1}{2}n)$ with probability $4c$. Then, with probability tending to 1, G^n has $cn^2 + o(n^2)$ edges and if $t = t(G^n)$ is the maximal number for which G^n contains a $K_2(r, t)$ then, again with probability tending to 1, we have

$$t = 2^{2^r-1}c^r n + o(n).$$

In order to prove this assertion, we denote by A and B the two classes of $K_2(\frac{1}{2}n, \frac{1}{2}n)$. We say that a vertex $x \in B$ forms a *cap* with a set U if $U \subset A$, $|U| = r$

and x is joined to every vertex in U. The expected number of vertices forming a cap with a given r-set in A is $2^{2r-1}c^r n$ and the variance of the event that an $x \in B$ forms a cap with U is $d_0^2 = 4c(1-4c)$. By the well known Bernstein inequality for binomial distributions (see p. 387 in [9]) the probability that U is joined to more than $2^{2r-1}c^r n + n^{2/3}$ or to less than $2^{2r-1}c^r n - n^{2/3}$ vertices $x \in B$ completely is $O(\exp(-c_1 n^{1/3}))$. Hence with probability tending to 1 on each U there is a $K_2(r, t)$ for $t = 2^{2r-1}c^r n - n^{2/3}$ but on no U for $t = 2^{2r-1}c^r n + n^{2/3}$, since

$$\binom{n}{r} \cdot O(\exp(-c_1 n^{1/3})) = o(1).$$

A similar application of Bernstein's inequality yields that $|e(G^n) - cn^2| \leq n^{5/3}$ with probability tending to 1.

Exactly the same argument gives that if G^n is a random subgraph of K_n of size $[cn]^2$ (or is obtained from K_n by choosing each edge with probability $2c$) then G^n will contain a $K_2(r, t)$ for $t = 2^r c^r n - n^{2/3}$ with probability tending to one, but for $t = 2^r c^r n + n^{2/3}$ only with probability tending to 0. This shows that prohibiting the odd cycles results in an increase of the constant from $2^r c^r$ to $2^{2r-1}c^r$ and that the main point of our theorem is that the same result can be obtained by prohibiting just one odd cycle.

The proof of our theorem is based on the following result of Szemerédi [11].

Lemma 8 (Uniform Density Lemma). *Given two subsets U, V of the vertex set of a graph G^n, denote by $e(U, V)$ the number of edges joining U to V and put*

$$d(U, V) = \frac{e(U, V)}{|U|\,|V|}.$$

There exists for a given constant $\beta > 0$ an integer $M(\beta)$ such that for any G^n the vertices of G^n can be divided into disjoint classes $V_0, \ldots, V_k$ for some $k < M(\beta)$ so that $|V_i| = |V_j|$ if $i \neq 0$, $j \neq 0$, $|V_i| \leq \beta n$ if $i = 0, 1, \ldots, k$ and for all but βk^2 pairs (i, j) the following condition holds.

(*) *Whenever $U_i \subset V_i$, $U_j \subset V_j$ and $|U_i| > \beta |V_i|$, $|U_j| > \beta |V_j|$, then*

$$|d(U_i, U_j) - d(V_i, V_j)| < \beta^2.$$

Let us turn now to the main body of the proof of Theorem 7.

(A) Let $e(G^n) = cn^2$ and let $\beta > 0$ be an arbitrarily small constant, much smaller than c. Applying the Uniform Density Lemma to G^n we obtain the classes $V_0, V_1, \ldots, V_k$. Let $m = |V_i|$ $(i = 1, \ldots, k)$. Instead of G^n we consider a graph G' of $n - |V_0|$ vertices, obtained from $G^n - V_0$ by omitting all the edges

(i) joining vertices from the same V_i $(i = 1, \ldots, k)$;
(ii) joining a V_i to a V_j for an "exceptional pair", that is, (*) does not hold;
(iii) joining a V_i to a V_j, when $d(V_i, V_j) < \beta^{1/2}$.

Clearly,

$$n-\beta n\leq |G'|\leq n,$$

and

$$0\leq e(G^n)-e(G')\leq 2\beta^{1/2}n^2$$

If β is sufficiently small. Therefore instead of proving Theorem 7 for G^n it is sufficient to prove it for G'. Hence we may and shall assume that $G'=G^n$.

(B) Let R^k be the graph whose vertices are the classes V_i $(i=1,\ldots,k)$ and V_i is joined to V_j in R^k if there exists an edge (u, v) in G^n joining V_i to V_j. We prove that R^k does not contain a triangle (K^3). Let us assume that V_1, V_2 and V_3 from a triangle in R^k. Put

$$U^+=\{x\in V_3\colon\ d(x, V_1)\leq\beta\},$$
$$U^{++}=\{x\in V_3\colon\ d(x, V_2)\leq\beta\}.$$

For any $x\in U=V_3-U^+-U^{++}$ there exist a $U_{1,x}$ and a $U_{2,x}$ in V_1 and V_2 respectively, joined to x completely, where $|U_{i,x}|\geq\beta m$. Hence the number of edges joining $U_{1,x}$ to $U_{2,x}$ is at least

$$(\beta m)^2(\beta^{1/2}-\beta)>\beta^3m^2.$$

This is a lower bound on the number of triangles on x, with the other two vertices in V_1 and V_2. Hence the total number of triangles (K^3's) of form (x, y, z), $x\in V_3$, $y\in V_1$, $z\in V_2$ is at least $(1-2\beta)\beta^3m^3$: by (*) $|U^+|\leq\beta m$, $|U^{++}|\leq\beta m$. A theorem of Erdös [4] asserts, that if in an r-uniform hypergraph H of n vertices there are at least $cn^{r-(r-1)/t}$ hyperedges, then H contains a subgraph of the following form: $C_1,\ldots,C_r$ are vertex-disjoint t-tuples and we take all the r-tuples (= hyperedges) of form $(x_1,\ldots,x_r)$, $x_i\in C_i$ for $i=1,\ldots,r$. Applying this theorem to the system of K^3's obtained above we get a $C_i\subset V_i$ $(i=1,2,3)$ with $|C_i|=t$ and such that each K^3 of the form (x, y, z), $x\in C_3$, $y\in C_1$, $z\in C_2$ belongs to G^n. Thus $K_3(t,t,t)=C^3(t)\subset G^n$. This contradiction proves the assertion of (B) for $k=3$. In the general case we apply the theorem with kt instead of t and observe that $K_3(kt, kt, kt)\supset C^k(t)$, again completing the proof of (B).

(C) Now we fix a $c_1\in(0,c)$ and assume indirectly that

$$e(G^n)=cn^2,\qquad G^n\not\supset C^k(t)\quad\text{and}\quad G^n\not\supset K_2(2^{2r-1}c_1^rn, r).$$

Let $d_i=d(V_i, V-V_i)$, where V is the vertex set of G^n. We may assume that $d_1=\max d_i=d$. Let us permute the indices of V_i so that $V_2,\ldots,V_{s+1}$ are the classes joined to V_1, the others are independent of it. Clearly, $V_2,\ldots,V_{s+1}$ form a set of ms independent vertices. Hence

$$e(G^n)\leq\sum_{i=s+2}^{k}(d_in)m+(d_1n)m\leq(dn)(n-a). \tag{6}$$

where

$$a = \left| \bigcup_{1 \le i \le s+1} V_i \right| = (s+1)m.$$

To obtain an upper bound of d in terms of a, we apply Lemma 5 to the bipartite graph determined by the classes $\bigcup_{2 \le i \le s+1} V_i$ (=first class) and V_1 (=second class). We find that

$$G^n \supset K_2(r, t) \quad \text{with } t = (1-o(1))d^r n^r a^{-(r-1)}. \tag{7}$$

By the assumption $G^n \not\supset K_2(r, 2^{2r-1} c_1^r n)$ and by (7)

$$d^r n^{r-1} a^{-(r-1)} \le (1+o(1)) 2^{2r-1} c_1^r. \tag{8}$$

Let us assume that $d > 2c_1$ (this will be shown later). From (8) and $c_1^r < \frac{1}{2} d c_1^{r-1}$ we obtain $d < (1+o(1))\, 4c_1(a/n)$. This and (6) yield

$$cn^2 \le e(G^n) \le dn(n-a) \le (1+o(1))n^2 \cdot 4c_1 \frac{a}{n}\left(1 - \frac{a}{n}\right) \le c_1 n^2,$$

which is a contradiction.

To prove $d > 2c_1$ observe that "essentially, dn is the maximum degree":

$$cn^2 \le e(G^n) = \frac{1}{2} \sum_i d(V_i, V - V_i) m(n-m) \le km \cdot d(n-m) = dn(n-m). \tag{9}$$

Until now β and c_1 were independent, now we may agree that β is chosen depending on c_1 and it is so small that $1 - \beta > (c_1/c)$. This, $(m/n) < \beta$ and (9) yield the desired inequality $d > 2c_1$. □

Remark. The method used to prove Lemma 6 and the method used to prove that K^3 does not occur in the graph R^k are equivalent: both can be used in both cases. The proof becomes slightly shorter if we consider only the case $t = 1$.

Theorem 9. *Let t be a natural number and let $c > 0$. Then there exists an n_0 such that if $n > n_0$ and $G^n \not\subset C^m(t)$ for $m = 3, 5, \dots, 2l(c)+1$, where $2l(c)+1 > c^{-1}$, then G^n can be made bipartite by the omission of not more than cn^2 edges.*

Remark. Theorem 9 is sharp, apart from the value of $l(c)$ which is probably $O(c^{-1/2})$. This $l(c) = O(c^{-1/2})$ would be sharp if true. To see this put $n = (2l+3)$, and $G^n = C^{2l+3}(m)$. If $c = (2l+4)^{-2}$, then more than cn^2 edges must be omitted to turn G^n into a bipartite graph and C^n does not contain C^k if k is odd and smaller than $2l+3$.

Proof. Our proof consists of two parts. We shall give two versions of the second part.

Part I. Let $c'<c$ be fixed. We shall say that the edges are regularly distributed, if for every partition $V(G^n)=A\cup B$ we have $d(A,B)\geq 2c'$. If we have an arbitrary G^n, we shall find a G^m in it, in which the edges are regularly distributed and $h(G^m)\geq cm^2$, $m>n_0'$ also hold, where $h(G)$ denotes the minimum number of edges one has to omit to change G into a bipartite graph. Therefore it will be enough to prove the theorem for the case, when the edges are regularly distributed and this will be just Part II. Let us assume that the edges are not regularly distributed in G^n; $V(G^n)=A\cup B$ and $d(A,B)<2c'$. Clearly,

$$h(G^n)<h(G[A])+h(G[B])+d(A,B)\,|A\,\|B|,$$

therefore we may assume that

$$h(G[A])>c\,|A|^2+(c-c')\,|A\,\|\,B|. \tag{10}$$

Hence

$$\tfrac{1}{4}|A|^2>(c-c')\,|A\,\|\,B|,$$

that is, $|A|>c''\,|B|$ for $c''=4(c-c')$. This also shows that $|A|>(1+c'')^{-1}n$. Furthermore, by (10),

$$h(G[A])>c_1\,|A|^2 \quad \text{for } c_1=c+(c-c')\frac{|B|}{|A|}.$$

Put $G_0=G^n$, $c_0=c$, $G_1=G[A]$ and repeat the step above until either we arrive at a G_j in which the edges are regularly distributed or to a G_j with $\sqrt{n}$ or less vertices (and use always $c_j'=c_j-(c-c')$). It is easy to show that if n is sufficiently large, then G_j cannot go below $\sqrt{n}$, otherwise $c_j>1$ would occur. Hence the procedure will always stop with a graph G_j in which the edges are regularly distributed. This was to be proved.

Part II (First version). (A) We start with a graph G^n for which $h(G^n)\geq cn^2$, fix a $c''<c$ and then a $c'\in(c'',c)$ and a $\beta>0$, which is much smaller than c''. Using the first part we may assume that the edges are regularly distributed. We may repeat part (A) of the proof of Theorem 7 replacing $e(\ )$ by $h(\ )$ and c by c'. Then we may assume that $G'=G^n$, but have to decrease c': replace the original condition by condition $h(G^n)\geq c''n^2$. How we define the graph R^k as in the beginning of (B) of the proof of Theorem 7.

(B) We prove that if n is sufficiently large and $R^k\supset C^i$, then $G^n\supset C^i(t)$, where t is fixed, but arbitrarily large. Exactly as in the proof of Theorem 7, we can prove that G^n contains at least c_1n^j cycles C^j, where $c_1>0$ is a constant. Applying the theorem of Erdös on hypergraphs [4], we obtain j sets $X_1,\ldots,X_j$ with $|X_j|=T\to\infty$, such that if $x_1\in X_1,\ldots,x_j\in X_j$, then some permutation $(x_{i_1},\ldots,x_{i_j})$ is a cycle of G^n (we consider here the hypergraph whose hyperedges are the j-sets of vertices of j-cycles in G^n). Unfortunately the cycles will not determine a $C^j(T)$, since the permutation $i_1,\ldots,i_j$ may differ from j-tuple to j-tuple. However, let us apply the Erdös theorem again, now to the hypergraph whose vertices are in $X_1 \cup X_2 \cup \ldots \cup X_j$ and the hyperedges of which are some cycles of G^n of form $(x_{i_1}, \ldots, x_{i_j})$, $x_s \in X_s$, where we choose only one permutation $i_1, \ldots, i_j$, for which the number of cycles is at least $T^j/j!$. If T is large enough, we obtain j subsets $Y_i \subset X_i$ such that whenever $x_i \in Y_i$, then $(x_{i_1}, \ldots, x_{i_j})$ defines a cycle in G^n and $|Y_i| = t$. Thus we obtained a $C^j(t) \subset G^n$.

(C) Clearly, the only thing to prove is, that $R^k \supset C^{2s+1}$ for some $2s+1 \leq (c')^{-1}$. If e.g. $V_1, \ldots, V_j$ define a shortest odd cycle in R^k, by the assumption that the edges are regularly distributed in G^n, there must be a V_q, $q > j$, which is joined to at least $2c'j$ of the classes $V_1, \ldots, V_j$. If V_q is joined to a V_i and $V_{i'}$ for some i' farther from i than 2, then the arc $V_i V_q V_{i'}$ will create a shorter odd cycle. Hence either $C^3 \subset R^k$ or $2c''j \leq 2$, and, consequently, $j \leq (c')^{-1}$.

Part II (Second version). The difference between the two proofs is above all, that here we shall not use the Uniform Density Lemma.

(A) By the first part we may assume that the edges are regularly distributed. Let A_1 be an arbitrary set of $\sqrt{n}$ vertices. By $d(A_1, V - A_1) \geq c'$ (where V is the vertex set of G^n) and by Lemma 5 we can find a $B_1 \subset A_1$, for which $|B_1| = T = t^2$, and a set $B_2 \subset V - A_1$ for which $|B_2| \geq bn$ with $b = (c')^T$, so that B_1 and B_2 are completely joined. B_j is recursively defined:

$$\tilde{B}_j = \left\{ x\colon x \notin \bigcup_{i<j} B_i \quad \text{and} \quad d\left(x, \bigcup_{i<j} B_i\right) > c' \right\},$$

$$B_j = \tilde{B}_j - \bigcup_{i<j} B_i.$$

Clearly,

$$|B_j| \geq c' \left| V - \bigcup_{i<j} B_i \right|.$$

Hence for any fixed $\beta > 0$ we can find a $l_0 = l_0(c, \beta)$ such that

$$\left| V - \bigcup_{i<j} B_i \right| < \beta n \quad \text{if} \quad j \geq l_0.$$

Omitting all the edges between $\bigcup_{i \leq l} B_1$, and the rest of the graph we omit at most βn^2 edges. If now we omit all the edges (x, y) for which $x \in B_i$, $y \in B_{i+2p}$ for some i and p then we change the graph into a bipartite one. Hence there exists a pair (i, p) for which at least $(c'n^2 - \beta n^2)/l_0^2$ edges were omitted between B_i and B_{i+2p}. Hence there exists a $K_2(T, T)$ joining B_i to B_{i+2p} in the sense that the first (second) class of it is contained in B_i (B_{i+2p}). Let these classes be denoted by D_i and E_{i+2p}, respectively. If D_j is already defined, D_{j-1} can also be defined as follows: $|D_j| = T$, we find t vertices in D_j and $2T$ vertices in B_{j-1} joined to each other completely. By Lemma 5 this can be done if n is sufficiently large. The class

D_{j-1} contains T of these $2T$ vertices, E_{j-1} is obtained from E_j in the same way, but here we have to choose the T vertices outside of D_{j-1}. Finally we obtain a $C^{2i+2p-1}(t)$ in G^n, whose classes are E_0 in $B_1, E_2, \ldots, E_{i+2p}, D_i, D_{i-1}, \ldots, D_2$ in this cyclic order. This proves the theorem, except for the upper bound on the length of the cycle, which is very similar to that of the first version. We only sketch it here; if we already know the existence of a $C^{2s+1}(t)$ for any t and $s \le l_1$, then we take a $C^{2s+1}(t^2)$ for some very large t and find t vertices outside joined to the same $c'(2s+1)t^2$ vertices of this subgraph. If t is sufficiently large, at least $c'(2s+1)$ classes are joined to each of the considered t vertices by t or more edges. Thus we can find a shorter $C^{2s'+1}(t)$ if $2s+1>(c')^{-1}$. □

Remark. With essentially the same effort we could prove the existence of a $C^{2s+1}(t, cn, t, cn, t, cn, \ldots, t, cn, t)$ instead of the existence of a $C^{2s+1}(t)$, where $C^k(m_1, \ldots, m_k)$ is the graph obtained from the cycle C^k by replacing its ith vertex by m_i new independent vertices. In other words, we can guarantee that every second class of our graph contains cn vertices.

References

[1] B. Bollobás, Extremal Graph Theory (Academic Press, London, 1978).
[2] B. Bollobás and P. Erdös, On the structure of edge graphs, Bull. London Math. Soc. 5 (1973) 317–321.
[3] B. Bollobás, P. Erdös and M. Simonovits, On the structure of edge graphs II, J. London Math. Soc. 12 (2) (1976) 219–224.
[4] P. Erdös, On extremal problems of graphs and generalized graphs, Israel J. Math. 2 (1965) 183–190.
[5] P. Erdös, Some recent results on extremal graph problems in graph theory, in: Theory of Graphs, Int. Symp. Rome (1966) 118–123.
[6] P. Erdös, On some new inequalities concerning extremal properties of graphs, in: Theory of Graphs, Proc. Coll. Tihany, Hungary (1966) 77–81.
[7] P. Erdös and M. Simonovits, An extremal graph Problem, Acta Math. Acad. Sci. Hung. 22 (3–4) (1971) 275–282.
[8] P. Erdös and A.H. Stone, On the structure of linear graphs, Bull. Amer. Math. Soc. 52 (1946) 1087–1091.
[9] A. Rényi, Probability Theory (North-Holland, Amsterdam, 1970).
[10] M. Simonovits, A method for solving extremal problems in graph theory, stability problems, in: Theory of Graphs, Proc. Coll. Tihany, Hungary (1966) 279–319.
[11] E. Szemerédi, Regular partitions of graphs (to appear).

Annals of Discrete Mathematics 3 (1978) 43–48.

HAMILTONIAN CYCLES IN REGULAR GRAPHS

Béla BOLLOBÁS

Department of Pure Mathematics and Mathematical Statistics, Cambridge CB2 1SB, England

Arthur HOBBS

Texas A. and M. University, College Station, TX, U.S.A.

Dirac [4] proved over 20 years ago that if in a graph of order n every vertex has degree at least $\frac{1}{2}n$ then the graph contains a Hamiltonian cycle. This theorem of Dirac was the first in a long line of results (see [2, 3, 5, 9, 11, 12], etc.) concerning *forcibly Hamiltonian* degree sequences, that is degree sequences all whose realizations are Hamiltonian. The conditions given in these results are such that if a sequence $(d_k)_1^n$ satisfies them then so does every sequence $(d'_k)_1^n$ majorizing $(d_k)_1^n$, that is satisfying $d'_k \geq d_k$, $k = 1, 2, \ldots, n$. In fact, Chvátal [3] proved the best possible result of this kind.

Very little is known about graphic sequences that together with some other restrictions on the graph force the graph to be Hamiltonian, and which are such that not every sequence majorizing them has that property. Szekeres raised the question whether a two-connected $(m-k)$-regular graph G of order $2m$ is Hamiltonian if k (≥ 1) is sufficiently small. It is clear that if instead of regularity we ask only that the minimal degree is $m-1$ then the answer is negative. Similarly one can not discard the condition that the graph is two-connected. Erdös and Hobbs [7, 8] proved that the answer to the question of Szekeres is in the affirmative if $k = 1, 2$ or $k < c\sqrt{m}$, where c is a positive constant. On the other hand, if $m = 3k-4$ and $k \geq 3$ then G need not contain a Hamiltonian cycle. An example showing this can be obtained by omitting some edges from $K^{m-k} \cup 2K^{m-k+1}$ and suitably joining these three components to two vertices (see Fig. 1).

The aim of this paper is to show that the order of k in the example above is best possible: the graph has to be Hamiltonian if $k < c_1 m - c_2$ for some positive constants c_1, c_2. It seems very likely that the best value of c_1 is $\frac{1}{3}$, but we can not prove this.

Throughout the paper we use the terminology and notation of [1].

Theorem 1. *Let k and m be natural numbers satisfying $m \geq 9k$. Let G be a 2-connected $(m-k)$-regular graph of order $2m - \varepsilon$, $\varepsilon \in \{0, 1\}$. Then G is Hamiltonian.*

Proof. In order to reduce the number of symbols floating around, we shall take $\varepsilon = 0$. The case $\varepsilon = 1$ can be treated in exactly the same way.

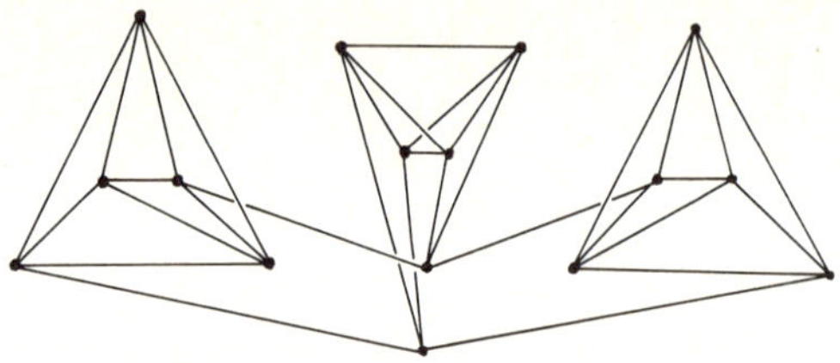

Fig. 1. A two-connected 4-regular non-Hamiltonian graph of order 16, $k=4$, $m=8$.

Let us assume that G does not contain a Hamiltonian cycle. Our aim is to arrive at a contradiction. We prepare the ground by proving five lemmas about the structure of G; the first three lemmas were proved in [8].

Let L be a longest cycle in G. Give L an orientation. Put $R=V(G)-V(L)$ and $r=|R|$. By a theorem of Dirac [4] (see also [1; Theorem III.4.10]) we have $|L|\geq 2(m-k)$ so $1\leq r\leq 2k$.

Let $v\in R$. Let C be the set of vertices of L adjacent to v. Denote by B the set of vertices of L immediately preceding vertices in C ("before") and denote by A the set of vertices of L immediately following vertices in C ("after"). As L is a longest cycle, we have $B\cap C=C\cap A=\emptyset$.

Lemma 2. *$A\cup\{v\}$ and $B\cup\{v\}$ are independent sets. Furthermore, if $w\in R-\{v\}$ then w is joined to at most one vertex of A and at most one vertex of B.*

Proof. If one of the assertions of the lemma failed to hold, the graph G would contain a cycle longer than L, as shown in Fig. 2.

Lemma 3. *The set R consists of independent vertices.*

Proof. Let P be a longest path in $G[R]$ and suppose $p=|P|\geq 2$. Let a_1 and a_2 be the endvertices of P. Then each a_i $(i=1,2)$ is joined to at most $p-1$ vertices of R and so at least $2(m-k-p+1)$ edges join $\{a_1,a_2\}$ to L.

If x_1 and x_2 are vertices of L at distance d on L and $1\leq d\leq p$, then either x_1 is not adjacent to a_1 or x_2 is not adjacent to a_2, since otherwise there is a cycle longer than L, as shown in Fig. 3. Hence at most two edges join any set of $p+1$ consecutive vertices of L to the set $\{a_1,a_2\}$. Consequently at most $2/(p+1)\times(2m-r)$ edges join $\{a_1,a_2\}$ to L so

$$2(m-k-p+1)\leq 2(2m-r)/(p+1).$$

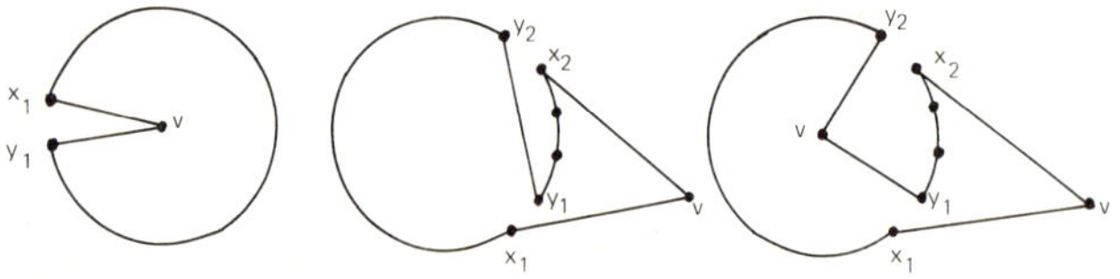

Fig. 2. The vertices of L are on a circle, $x_i\in C$, $y_i\in A$ and y_i follows x_i, $i=1,2$. The cycles in the thick line are longer than L.

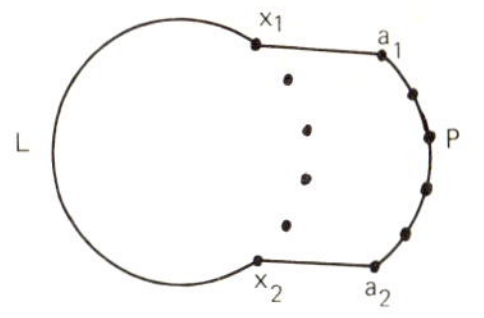

Fig. 3. The case $p=5$, $d=5$.

Noting that $p \leqslant r \leqslant 2k$, we see that

$$m \leqslant k(p+1)/(p-1)+p-1/(p-1),$$

contradicting the assumption on the relation between m and k. □

Remark 4. The last inequality shows that the lemma holds if instead of $m \geqslant 9k$ we require only that $m \geqslant 3k+2$ (or $m \geqslant 3k+1$ if $\varepsilon=1$).

Let $D=V(G)-A \cup B \cup C \cup R$ and put $s=|A-B|=|B-A|$. Since $|A|=|B|=|C|=m-k$ and $|A \cup B \cup C|=2(m-k)+s=2m-r-|D|$, we have $|D|=2k-r-s$. In particular, $r+s \leqslant 2k$. Let D_0 be the subset of D whose elements are adjacent to no vertices in $A \cap B$.

Lemma 5. $|D_0| \geqslant k-s-\frac{1}{2}r$.

Proof. We may assume that $r+2s \leqslant 2k-1$ since otherwise there is nothing to prove. Suppose $s=0$. Then $A \cap B=A$ so $A \cup C=V(L)$ and the vertices of A and C alternate around L. However, this is impossible since then $r=2m-2(m-k)=2k$. Thus $s \geqslant 1$.

For each vertex $b_1 \in B-A$ there is an interval on L consisting of vertices b_1, c_1, f_1, c_2, $f_2, \ldots, f_{l-1}$, c_l, a_l, where $c_i \in C$, $f_i \in A \cap B$ and $a_l \in A\ B$. There are s such intervals and so the vertices of D also form S intervals, some of which may be empty. Let l_i be the length of the ith interval. Then $\sum_{i=0}^{s} l_i=|D|=2k-r-s$. The set $D-D_0$ does not contain adjacent vertices since otherwise G contains a longer cycle than L, as shown in Fig. 4. Consequently

$$|D_0| \geqslant \sum_{i=1}^{s} \lfloor \tfrac{1}{2} l_i \rfloor \geqslant \frac{1}{2} \sum_{i=1}^{s} (l_i-1)=k-s-\tfrac{1}{2}r. \quad \square$$

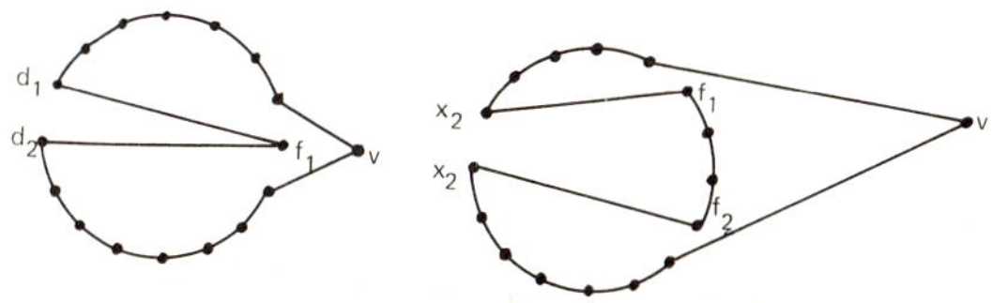

Fig. 4. $d_i \in D-D_0$ and $f_i \in A \cap B$. The cycle in the thick line is longer than L.

Let $W=V(G)-A$. By Lemma 3 we have $|A|=m-k$ and so $|W|=m+k$. Put $G_1=G[W]$. Order the vertices in W as $w_1, w_2, \ldots$, so that $\deg_W(w_i) \geq \deg_W(w_j)$ whenever $i<j$. Put $M=\{w_i: \deg_W(w_i) \geq m-3k\}$.

Lemma 6. $|M| \geq k$, *that is* $\deg_W(w_k) \geq m-3k$.

Proof. By Lemma 2 every vertex of R has degree at least $m-k-1$ in W. Furthermore no vertex of $(B-A) \cup D_0$ is joined to any vertex of $A \cap B$. Hence each vertex of $(B-A) \cup D_0$ has degree at least $m-k-s \geq m-k-(2k-r) \geq m-3k+r$ in W. Thus $R \cup (B-A) \cup D_0 \subset M$ and, by Lemma 5, $|R \cup (B-A) \cup D_0| \geq k$. □

Let now $p=\max\{t: \deg_W(w_{k+t}) \geq m-3k$ and $G[w_1, w_2, \ldots, w_{k+t}]$ contains $2t$ independent edges$\}$. Lemma 6 implies that $p \geq 0$. Since $2p$ independent edges have $4p$ vertices, we have $4p \leq k+p$ so $0 \leq p \leq \frac{1}{3}k$. Put $W_0=\{w_1, w_2, \ldots, w_{k+p}\}$.

Lemma 7. *There is an* x_1-x_2 *path* P *of length* $2k$ *in* G_1 *such that* $W_0 \subset V(P)$ *and* $x_1y_1, x_2y_2 \in E(G)$ *for some* $y_1, y_2 \in A$, $y_1 \neq y_2$.

Proof. Choose a set S of $2p$ independent edges in $G[W_0]$. Let T be the set of vertices w_i, $1 \leq I \leq K+P$, not incident with edges in S. Since $\deg_W(w_{k+p}) \leq m-3k$, for $1 \leq i<j \leq k+p$ we have

$$|\Gamma_W(w_i) \cap \Gamma_W(w_j)| \geq 2(m-3k-1)-(m+k-2)=m-7k \geq 2k.$$

This implies that we can connect the edges in S and the vertices in T in any order to form a path P', using a vertex of $W-W_0$ between each two elements of $S \cup T$. The path P' has length $2k-2$ and its endvertices have degrees at least $m-3k$ in W. Extend P' by one edge at each end to a path P so that the new endvertices x_1, x_2 have as small degrees in W as possible. Since W does not contain $m-4k$ vertices of degree at least $m-k-1$ in W, each endvertex x_i of P is adjacent to at least two vertices of A. Therefore we can select $y_1, y_2 \in A$, $y_1 \neq y_2$, such that x_i is adjacent to y_i, $i=1,2$. □

Armed with our lemmas, we shall show now that G *does contain* a Hamiltonian cycle.

Let P be the x_1-x_2 path whose existence is guaranteed by Lemma 7. Omit $V(P)$ from G together with every edge joining vertices in W and add a new vertex x' to the remainder. Join x' to y_1 and y_2. Denote by G' the graph constructed in this way. By construction G' is a bipartite graph with vertex classes A and $Z=(W-V(P)) \cup \{x'\}$. We shall show that G' contains a Hamiltonian cycle. Since x_1 is adjacent to y_1 and x_2 to y_2, a Hamiltonian cycle of G' can be pulled back to a Hamiltonian cycle of G: all we have to do is replace the path $x_1x'x_2$ by x_1Px_2.

In order to show that G' is Hamiltonian we make use of a result of Moon and Moser [10] (see also [1, Corollary III.4.7]). Somewhat unnaturally we state this result as an assertion about the graph G' at hand. Let $d_1 \leqslant d_2 \leqslant \cdots d_{m-k}$ be the degree sequence of the vertices in Z and let $d'_1 \leqslant d'_2 \leqslant \cdots \leqslant d'_{m-k}$ be the degree sequence of the vertices in A. *If* $d_l \leqslant l < m-k$ *implies* $d'_{m-k-l+1} \geqslant m-k-l+1$ *then* G' *is Hamiltonian.*

Suppose A contains $m-k+l+1$ vertices of degree at most $m-k-l$ in G'. Then the set of these vertices is joined to $\{w_1, w_2, \ldots, w_{k+p}\}$ by at least $(l-(k+1))(m-k-l+1)$ edges. Since each w_i, $1 \leqslant i \leqslant k+p$, is joined to at most $2k$ vertices of A,

$$(l-k+p-1)(m-k-l+1) \leqslant 2k(k+p).$$

This in equality implies

$$l < \tfrac{4}{3}k+1.$$

Hence it suffices to check that the condition of Moon and Moser is satisfied for *every* l *less than* $\frac{4}{3}k+1$.

Suppose $d_l \leqslant l$ for some l, $1 \leqslant l < \frac{4}{3}k+1$. In G' every vertex has degree at least 2, so $l \geqslant 2$. Put $W_1 = \{w_1, w_2, \ldots, w_{k+p+l-1}\}$. Then, by the choice of P and l, we have

$$\deg_W(w_i) \geqslant m-k-l \quad \text{for every} \quad w_i \in W_1. \tag{1}$$

In particular, each vertex of W_1 has degree at least $m-3k$ in W. Hence the definition of p implies that $G[W_0^+]$ has at most $2p+1$ independent edges where $W_0^+ = W_0 \cup \{w_{k+p+1}\} \subset W_1$. In turn, a theorem of Erdös and Gallai [6] (see also [1, Corollary II.1.10]) gives

$$e(G[W_0^+]) \leqslant \max\left\{\binom{k+p+1}{2} - \binom{k-p}{2}, \binom{4p+3}{2}\right\}.$$

Consequently

$$e(G[W_1]) \leqslant \binom{k+p+l-1}{2} - \min\left\{\binom{k-p}{2}, \binom{k+p+1}{2} - \binom{4p+3}{2}\right\}. \tag{2}$$

On the other hand, eq. (1) implies that some of the $k(m-k)$ edges of $G[W]$ must join vertices of W_1. More precisely,

$$e(G[W_1]) \geqslant (k+p+l-1)(m-k-l) - k(m-k). \tag{3}$$

Putting (2) and (3) together, we see that

$$\min\left\{\binom{k-p}{2}, \binom{k+p+1}{2} - \binom{4p+3}{2}\right\}$$
$$\leqslant \binom{k+p+l-1}{2} - (k+p+l-1)(m-k-l) + k(m-k). \tag{4}$$

To complete the proof of the theorem we shall show that (4) can not hold.

(a) Assume first that

$$\binom{k-p}{2} \leq \binom{k+p+l-1}{2} - (k+p+l-1)(m-k-l) + k(m-k).$$

Rearranging it we obtain

$$2(p+l-1)m \leq l(6k+3l-5) + p(6k+4l-4) - 4k + 2.$$

Since $6k+4l-3<m$, this inequality has to hold when we put $p=0$. In turn, one sees it has to hold for the minimal value of l, for $l=2$. However, then the inequality becomes

$$2m \leq 8k+4$$

and this does not hold.

(b) Assume now that

$$\binom{k+p+1}{2} - \binom{4p+3}{2} \leq \binom{k+p+l-1}{2} - (k+p+l-1)(m-k-l) + k(m-k).$$

Rearranging it we obtain

$$2(p+l-1)m \leq l(6k+4p+3l-5) + p(2k+16p+16) - 6k + 8.$$

Since $l<\frac{4}{3}k+1$ and $p \leq \lfloor \frac{1}{2}k \rfloor$ we see that the inequality must hold with $l=2$ and $p=0$. However, in that case we get the inequality

$$2m \leq 6k+10$$

which contradicts the assumption of the theorem. We have proved that (4) does not hold and so the proof of the theorem is complete. □

References

[1] B. Bollobás, Extremal Graph Theory (Academic Press, London, 1978).

[2] J.A. Bondy, Properties of graphs with constraints on the degrees, Studia Sci. Math. Hungar. 4 (1969) 473–475.

[3] V. Chvátal, On Hamilton's ideals, J. Combinatorial Theory 12(B) (1972) 163–168.

[4] G.A. Dirac, Some theorems on abstract graphs, Pro. London Math. Soc. 2 (3) (1952) 69–81.

[5] G.A. Dirac, On Hamilton Circuits and Hamilton Paths, Math. Ann. 197 (1972) 57–70.

[6] P. Erdös and T. Gallai, On the minimal number of vertices representing the edges of a graph, Publ. Math. Inst. Hungar. Acad. Sci. 6 (1961) 181–203.

[7] P. Erdös and A.M. Hobbs, A class of Hamiltonian regular graphs, J. Combinatorial Theory (B), to appear.

[8] P. Erdös and A.M. Hobbs, Hamiltonian cycles in regular graphs of moderate degree, to appear.

[9] M. Las Vergnas, Sur une proprieté des arbres maximaux dans un graphe, Comp. Rend. Acad. Sci. Paris Sér. A–B 272 (1971) 1297–1300.

[10] J.W. Moon and L. Moser, On a problem of Turán, Publ. Math. Inst. Hungar. Acad. Sci. 7 (1962) 283–286.

[11] O. Ore, Note on Hamiltonian circuits, Am. Math. Monthly 67 (1960) 55.

[12] L. Pósa, A theorem concerning Hamiltonian lines, Publ. Math. Inst. Hungar. Acad. Sci. 7 (1962) 225–226.

Annals of Discrete Mathematics 3 (1978) 49–53.

THE CHROMATIC INDEX OF THE GRAPH OF THE ASSIGNMENT POLYTOPE*

Richard A. BRUALDI
Department of Mathematics, University of Wisconsin, Madison, WI 53706, U.S.A.

1. Introduction

Let n be a positive integer, and let S_n denote the set of permutations of $\{1, \ldots, n\}$. We define a graph G_n as follows. The set of vertices of G_n is S_n. Two vertices $\sigma, \tau \in S_n$ are joined by an edge in G_n if and only if the permutation $\sigma^{-1}\tau$ has exactly one non-trivial cycle (that is, a cycle of length at least two). Let Ω_n denote the nth assignment polytope. Thus Ω_n is the $(n-1)^2$-dimensional polytope in R^{n^2} consisting of all $n \times n$ non-negative doubly stochastic matrices. It is proved in [1] and [3] that the vertex-edge graph of Ω_n is isomorphic to G_n. It is readily verified that G_n is the complete graph $K_{n!}$ for $n = 1, 2, 3$. As shown in [3], the graph G_4 is the complete 6-partite graph $K_{4,4,4,4,4,4}$. The following are some basic properties of G_n.

Property 1.1. *G_n has $n!$ vertices.*

Property 1.2. *G_n is a vertex-transitive graph.*

Property 1.3. *G_n is a regular graph of degree Δ_n where*

$$\Delta_n = \sum_{k=2}^{n} \binom{n}{k}(k-1)!.$$

Property 1.4 [3]. *For $n \geq 3$, the girth of G_n is 3. Indeed each edge is an edge of a cycle of length 3.*

Property 1.5 [1, 3]. *The diameter of G_n is 2 for $n \geq 4$. (It is, of course, 1 for $n = 1$, 2, 3.)*

Property 1.6 [3]. *G_n is a hamilton-connected graph.*

* Research performed while the author was visiting the Université de Paris VI and was partially supported by a grant from the Wisconsin Alumni Research Foundation and a grant from the National Science Foundation.

Since a hamilton-connected graph with at least three vertices has a hamilton cycle, it follows from Property 1.6 that G_n has a hamilton cycle for $n \geq 3$. As was pointed out by Balinski and Russakoff [1], the fact that G_n has a hamilton cycle is an immediate consequence of a well-known algorithm for generating the permutations in S_n.

Property 1.7 [4]. *For $n \geq 2$ the connectivity of G_n is Δ_n.*

Property 1.8 [3]. *For $n \geq 3$ the vertices of G_n can be partitioned into $n!/6$ cliques (maximal complete subgraphs) each having six vertices.*

For more information on these and other theorems concerning G_n one can consult [3].

2. The chromatic index of G_n.

Let G be a graph and let Δ be the largest degree of a vertex of G. A *matching* of G is a subset F of the edges of G such that no two edges of G have a common vertex. If the matching F has the property that each vertex of G meets an edge of F, then F is called a *perfect matching* or 1-*factor* of G. The *chromatic index* $q(G)$ of G is the smallest integer t such that the edges of G can be partitioned into t matchings. By a theorem of Vizing the chromatic index of the graph G is either Δ or $\Delta+1$. If G is a regular graph of degree Δ, then $q(G)=\Delta$ if and only if the edges of G can be partitioned into 1-factors. A partitioning of the edges of G into 1-factors is called a 1-*factorization* of G.

The graphs G_1, G_2, and G_3 being the complete graphs K_1, K_2, and K_6 respectively, have 1-factorizations. The graph G_4 is the 6-partite graph $K_{4,4,4,4,4,4}$, and a 1-factorization is readily found. One is naturally lead to conjecture that the graph G_n has a 1-factorization for each $n \geq 2$. The purpose of this note is to prove that $q(G_n)=\Delta_n$ $(n \geq 2)$. The following two lemmas are well known (see e.g. [2, p. 249]).

Lemma 2.1. *The complete graph K_n has a 1-factorization for each positive even integer n.*

Using the property that K_n is an induced subgraph of K_{n+1}, one quickly obtains the following.

Lemma 2.2. *For each positive odd integer n, the edges of the complete graph K_n can be partitioned into n sets each being a matching with $(n-1)/2$ edges.*

Let H be a subgroup of the symmetric group S_n with index t. Let $H_1 = H, H_2, \ldots, H_t$ be an enumeration of the right cosets of S_n with respect to H. Let

$G_n(H_i)$ denote the subgraph of G_n induced by the vertices in H_i $(1 \leq i \leq t)$. For distinct integers i and j with $1 \leq i, j \leq t$, let $G_n(H_i, H_j)$ denote the partial subgraph of G_n, where the set of vertices is $H_i \cup H_j$ and where two vertices are joined by an edge if and only if one is in H_i, the other is in H_j, and they are joined by an edge in G_n. In particular $G_n(H_i, H_j)$ is a bipartite graph with bipartition H_i, H_j.

Lemma 2.3. *The graphs $G_n(H_i)$, $i = 1, \ldots, t$, are isomorphic regular graphs.*

Proof. Let $H_i = H\sigma_i$ $(1 \leq i \leq t)$. Then it is readily verified that the correspondence between H and H_i defined by $\tau \to \tau\sigma_i$ $(\tau \in H)$ is an isomorphism between $G_n(H)$ and $G_n(H_i)$. Let σ be an element of H. Then the correspondence between H and itself defined by $\rho \to \sigma\rho$ $(\rho \in H)$ is an automorphism of $G_n(H)$ which takes the identity ε to ρ. The conclusions now follow.

Lemma 2.4. *$G_n(H_i, H_j)$ is a regular graph $(1 \leq i, j \leq t, i \neq j)$*

Proof. Let $\sigma, \tau \in H_i$ and let $\rho \in H_j$. Then $\tau\sigma^{-1} \in H$ and $\tau\sigma^{-1}\rho \in H_j$. Since $\tau^{-1}(\tau\sigma^{-1}\rho) = \sigma^{-1}\rho$, there is an edge joining σ and ρ if and only if there is an edge joining τ and $\tau\sigma^{-1}\rho$. Hence the correspondence $\rho \to \tau\sigma^{-1}\rho$ $(\rho \in H_j)$ shows that the degree of σ in $G_n(H_i, H_j)$ equals the degree of τ in $G_n(H_i, H_j)$, and we conclude that all vertices of H_i have the same degree in $G_n(H_i, H_j)$. Similarly, all vertices of H_j have the same degree in $G_n(H_i, H_j)$. Since $|H_i| = |H_j|$, $G_n(G_i, H_j)$ is a regular graph.

The collection of permutations σ of S_n such that $\sigma(n) = n$ form a subgroup of S_n which is isomorphic to S_{n-1}, and we identify this subgroup with S_{n-1}. Let

$$S_{n-1}^n = S_{n-1}, \qquad S_{n-1}^{n-1} = S_{n-1}(n-1, n), \ldots, \qquad S_{n-1}^1 = S_{n-1}(1, n)$$

be an enumeration of the right cosets of S_n with respect to S_{n-1}.

Lemma 2.5. *The graph $G_n(S_{n-1}^i, S_{n-1}^j)$ is a regular graph of degree γ_{n-1} where*

$$\gamma_{n-1} = \frac{\Delta_n - \Delta_{n-1}}{n-1} = \sum_{t=0}^{n-2} \binom{n-2}{t} t! \quad (1 \leq i, j \leq n, i \neq j).$$

Proof. According to Lemma 2.4 the graphs in question are regular graphs. Let i be an integer with $1 \leq i \leq n-1$ and consider the graph $G_n(S_{n-1}, S_{n-1}(i, n))$. Let $\sigma \in S_{n-1}$. Then there is an edge of this graph joining the identity ε and $\sigma(i, n)$ if and only if $\sigma(i, n)$ has exactly one non-trivial cycle. Since $\sigma(n) = n$, the latter is true if and only if $\sigma = \varepsilon$, or σ has exactly one non-trivial cycle and $\sigma(i) \neq i$. Hence the degree of regularity of $G_n(S_{n-1}, S_{n-1}(i, n))$ is the number of cycles of S_{n-1} of the form $(i, k_1, \ldots, k_t)$ where $0 \leq t \leq n-2$. Hence the degree of regularity is

$$\sum_{t=0}^{n-2} \binom{n-2}{t} t!.$$

Now let i and j be integers with $1 \leqslant i, j \leqslant n-1$ and $i \neq j$, and consider the graph $G_n(S^i_{n-1}, S^j_{n-1})$. Let $\sigma \in S_{n-1}$. Then there is an edge joining (i, n) and $\sigma(j, n)$ in this graph if and only if $(i, n)\sigma(j, n)$ has exactly one non-trivial cycle. Since σ varies over all permutations in S_{n-1}, it follows that the graphs $G_n(S^i_{n-1}, S^j_{n-1})$, $(1 \leqslant i, j \leqslant n-1, i \neq j)$ have the same degree of regularity γ_{n-1}. Since the degree of regularity of each of the graphs $G_n(S^i_{n-1})(1 \leqslant i \leqslant n)$ is Δ_{n-1}, it now follows that

$$\Delta_n - \Delta_{n-1} = (n-1) \sum_{t=0}^{n-2} \binom{n-2}{t} t!$$

and

$$\Delta_n - \Delta_{n-1} = \sum_{t=0}^{n-2} \binom{n-2}{t} t! + (n-2)\gamma_{n-1}$$

Hence

$$\gamma_{n-1} = \sum_{t=0}^{n-2} \binom{n-2}{t} t!,$$

and each of the graphs $G_n(S^i_{n-1}, S^j_{n-1})$ $(1 \leqslant i, j \leqslant n, i \neq j)$ is a regular graph of degree $\gamma_{n-1} = (\Delta_n - \Delta_{n-1})/n-1$. The lemma now follows.

Lemma 2.6. *$\Delta_n \geqslant \gamma_n$ for $n \geqslant 3$.*

Proof. By definition,

$$\Delta_n = \sum_{k=2}^{n} \binom{n}{k}(k-1)!, \qquad \gamma_n = \sum_{k=1}^{n} \binom{n-1}{k-1}(k-1)!.$$

Direct calculation shows that

$$\binom{n}{k}(k-1)! \geqslant \binom{n-1}{k-1}(k-1)! \quad (k = 2, \ldots, n),$$

with equality if and only if $k = n$. Since the first term $(k = 1)$ in the summation for γ_n is 1 and all terms in both summations are integers, it follows that for $n \geqslant 3$, $\Delta_n \geqslant \gamma_n$.

We are now ready to state and prove the main result.

Theorem 2.7. *The graph G_n admits a 1-factorization and consequently*

$$q(G_n) = \Delta_n \quad (n \geqslant 2).$$

Proof. We prove the theorem by induction on n. We have already observed that G_n has a 1-factorization for $n = 2$ and 3. Let $n \geqslant 4$. It follows from Lemma 2.3 that the graphs $G_n(S^i_{n-1})(i = 1, \ldots, n)$ are isomorphic to G_{n-1} and hence by the inductive hypothesis each of these graphs has a 1-factorization. Since $G_n(S^i_{n-1}, S^j_{n-1})$ is a regular bipartite graph of degree γ_{n-1} by Lemma 2.5 ($1 \leq i, j \leq n, i \neq j$), it follows from König's theorem that each of these graphs has a 1-factorization into γ_{n-1} 1-factors.

First suppose that n is even. Then the edges of G_n which are edges of one of the graphs $G_n(S^i_{n-1})$ ($1 \leq i \leq n$) can be partitioned into Δ_{n-1} 1-factors. Since n is even, the complete graph K_n has a 1-factorization, and it now follows that the edges of G_n which are edges of one of the graphs $G_n(S^i_{n-1}, S^j_{n-1})$ ($1 \leq i, j \leq n, i \neq j$) can be partitioned into $(n-1)\gamma_{n-1}$ 1-factors. Hence in this case G_n has a 1-factorization.

Now suppose that n is odd. Then the edges of K_n can be partitioned into n sets each being a matching with $(n-1)/2$ edges. Hence it follows that the edges of G_n which are edges of one of the graphs $G_n(S^i_{n-1}, S^j_{n-1})$ ($1 \leq i, j \leq n, i \neq j$) can be partitioned into $n\gamma_{n-1}$ matchings each with $(n-1)!\,(n-1)/2$ edges. Each of these matchings coupled with a 1-factor of one of the graphs $G_{n-1}(S^i_{n-1})$ gives a 1-factor of G_n. By Lemma 2.6, $\Delta_{n-1} - \gamma_{n-1} \geq 0$. Hence these $n\gamma_{n-1}$ matchings of G_n and γ_{n-1} 1-factors of a 1-factorization of each $G_n(S^i_{n-1})$ can be paired to give $n\gamma_{n-1}$ 1-factors of G_n. The remaining $\Delta_{n-1} - \gamma_{n-1}$ 1-factors of the 1-factorization of $G_n(S^i_{n-1})$ ($1 \leq i \leq n$) give $\Delta_{n-1} - \gamma_{n-1}$ 1-factors of G_n completing a 1-factorization of G_n. Hence in this case G_n also has a 1-factorization, and the theorem follows.

References

[1] M.L. Balinski and A. Russakoff, On the assignment polytope, SIAM Rev. 16 (1974) 516–525.
[2] C. Berge, Graphs and Hypergraphs (North-Holland, Amsterdam, 1973).
[3] R.A. Brualdi and P.M. Gibson, Convex polyhedra of doubly stochastic matrices II. Graph of Ω_n, J. Combinatorial Theory B 22 (1977) 175–198.
[4] W. Imrick, The connectivity of the graph of the assignment polytope, J. Combinatorial Theory B, to be published.

Annals of Discrete Mathematics 3 (1978) 55–74.

LOOPY GAMES

J.H. CONWAY

Department of Pure Mathematics and Mathematical Statistics, University of Cambridge, Cambridge CB2 1SB, England

0. Introduction

Two players, Left and Right, play the sum

$$A+B+C+\cdots$$

of a finite number of games A, B, C, ... as follows. They move alternately and the player whose turn it is to move selects at will just one of the component games A, B, C, ... and makes a move legal for him in that component. If our games are of the sort discussed in [1] then every sum of this kind necessarily ends after finitely many moves, and the winner is therefore completely determined by the *normal play convention* that a player who does not move when it is his turn to do so *loses*. But in this paper, we consider also games that might continue indefinitely, perhaps because they contain repetitive cycles of moves, or *loops*.

To specify a game G formally, we require to know its set of *positions*, one of which, usually given the same name as G, is called the *initial position*, and we shall also need to know the rules which determine what changes of position correspond to legal moves for Left or Right. When the number of positions is small, this information can conveniently be conveyed on a graph like that of Fig. 1, in which the nodes represent positions (the heavy node representing the initial position), and an arrow $P \to_L Q$ denotes a legal move from P to Q for Left, while $P \to_R Q$ would denote one for Right.

A *play* of G is then a finite or infinite sequence $G \to_X H \to_Y K \to_Z \cdots$ (each of $X, Y, Z, \ldots = \text{L}$ or R) of legal moves between positions of G, starting from the initial position. It is called an *alternating* play if the sequence $X, Y, Z, \ldots$ is alternating, either L, R, L, R, ... or R, L, R, L, ...

To complete the definition of G, we may add rules which say who wins for various infinite plays. To some of the infinite plays we shall attach the sign +, meaning that Left wins, to others the sign − (Right wins), and to all the rest the ambiguous sign ±, meaning that the game is declared drawn. G is called *fixed* if none of the infinite plays are drawn, *free* if all of them are.

If G is any game, we shall write G^+ for the modified game in which all infinite plays that are draws in G are redefined to be wins for Left, and G^- for that in which these plays are called wins for Right. $G(on)$ denotes the game obtained

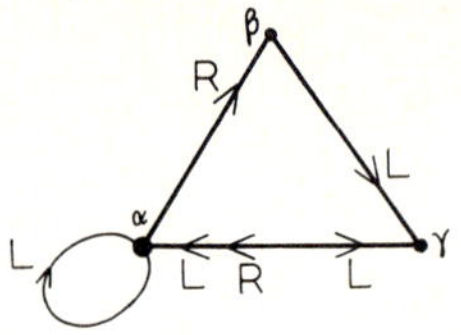

Fig. 1

from G by making *all* infinite plays wins for Left, and $G(\mathit{off})$ that for which all are wins for Right. For free games

$$G^{+}=G(on) \quad \text{and} \quad G^{-}=G(\mathit{off}),$$

while for fixed ones

$$G^{+}=G=G^{-}.$$

As a matter of notation, we shall write

$$G=\{A, B, C, \ldots \mid D, E, F, \ldots\}$$

to indicate that from the position G the legal moves for Left are to $A, B, C, \ldots$ only, while those for Right are to $D, E, F, \ldots$ only. Thus in Fig. 1,

$$\alpha=\{\alpha, \gamma \mid \beta\}, \qquad \beta=\{\gamma \mid \ \}, \qquad \gamma=\{\alpha \mid \alpha\}.$$

In these circumstances $A, B, C, \ldots$ are called the *Left options* of G, and G^{L} denotes a typical one of them, while $D, E, F, \ldots$, and typically G^{R}, are called the *Right options* of G. So in Fig. 1, α^{L} denotes either α or γ, α^{R} denotes only β, β^{L} denotes γ, while β^{R} has no meaning, and either γ^{L} or γ^{R} denotes α.

In this notation, the Left options of $G+H$ are all the games of the form

$$G^{\mathrm{L}}+H \quad \text{or} \quad G+H^{\mathrm{L}},$$

while its Right options are

$$G^{\mathrm{R}}+H \quad \text{or} \quad G+H^{\mathrm{R}},$$

so that we can write

$$G+H=\{G^{\mathrm{L}}+H, G+H^{\mathrm{L}} \mid G^{\mathrm{R}}+H, G+H^{\mathrm{R}}\}.$$

An *end* position, usually given the name 0, is one that has neither Left nor Right options

$$\{\,|\,\}=0.$$

It follows from the normal play convention that the player whose turn it is to move from an end position *loses*, and so of course a player who *reaches* an end position *wins*.

1. Enders and stoppers

A play in the sum of a number of games determines plays in the individual components in an obvious manner, for instance the play

$$G+H \to_L G+H^L \to_R G+H^{LR} \to_L G^L+H^{LR} \to_R G^L+H^{LRR} \to_L G^{LL} + H^{LRR} \to_R \cdots$$

determines the component plays

$$G \to_L G^L \to_L G^{LL} \to \cdots \quad \text{and} \quad H \to_L H^L \to_R H^{LR} \to_R H^{LRR} \to \cdots.$$

Notice that the play in the sum might be alternating even when (as here) the component plays are not — it is for this reason that we cannot afford to restrict ourselves to alternating plays only.

The game of Fig. 2 satisfies the *ending condition*: it has no infinite play, alternating or not. In such a game, called an *ender*, if the players keep on making moves, even in non-alternate fashion, they will necessarily arrive at an end position after a finite number of turns. A sum of games which individually satisfy the ending condition also satisfies that condition.

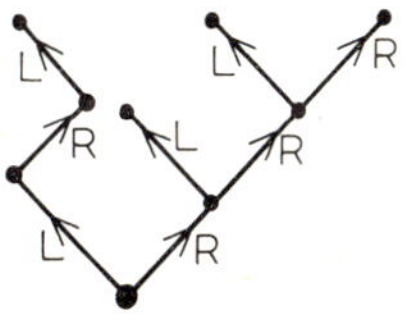

Fig. 2

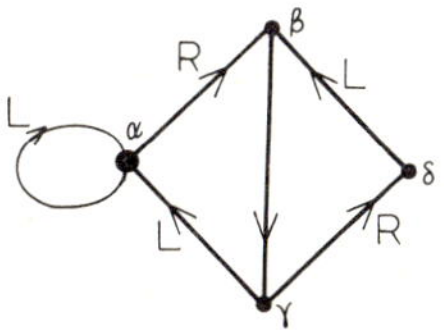

Fig. 3

The game of Fig. 3 is not an ender, but satisfies the weaker *stopping condition*, that there is no infinite *alternating* play from any position. Such games we call *stoppers* — if Left and Right play alternately in a stopper the game is sure to come to a *stop*, with the player whose turn it is unable to move, but it need not have *ended*, because there might be moves available to the player whose turn it *isn't*. A sum of components that are stoppers need not itself be a stopper — for example the games *on* and *off* of Fig. 4 are individually stoppers in which no alternating play has more than one move, but their sum is the third game of the figure, called *dud*, in which both Left and Right always have pass moves.

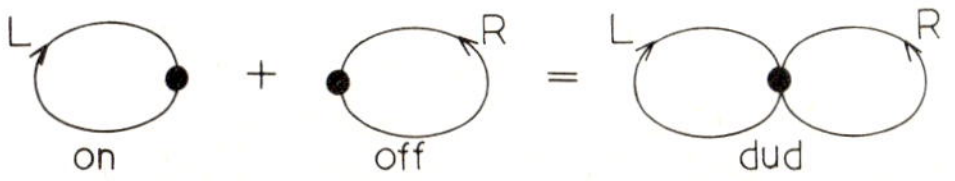

Fig. 4

2. Outcomes and order-relations for sums

To complete the definition of a sum, we shall need to say which infinite plays are wins for Left or Right, and which are drawn.

Any infinite play in the sum $G+H+K+\cdots$ determines plays in the components, at least one of which is infinite. If all the component plays that are infinite give wins for the same player, we shall say that that player wins the sum, and otherwise will agree that the sum is drawn. A play is drawn when some component is drawn, or when some two of the components are won by different players.

We shall define the *negative*, $-G$, of a game G by reversing the roles of Left and Right throughout—moves that were legal for one player in G become legal for the other in $-G$, and a player wins $-G$ just when his opponent would win G. We shall write $G-H$ for $G+(-H)$, even though in this theory there is no useful sense in which $G-G$ is equivalent to 0.

If A and B are *fixed* games (no draws) we shall define $A \geqslant B$ to mean that, supposing that Right starts, Left has a strategy which guarantees him either a draw or a win in the difference game $A-B$. We shall abbreviate this whole phrase, including the understanding that Right starts, to "Left can survive in $A-B$".

Theorem 2.1.

$$A \geqslant B \textit{ implies } (A+C)^+ \geqslant (B+C)^+,$$

$$A \geqslant B \textit{ implies } (A+C)^- \geqslant (B+C)^-.$$

Proof. Suppose $\aleph$ is the given survival strategy for $A-B$. We shall construct from it a strategy $\beth$ which enables Left to survive in both

$$(A+C)^+ - (B+C)^+ = (A+C)^+ + (-B-C)^- = \Delta_1$$

and

$$(A+C)^- - (B+C)^- = (A+C)^- - (-B-C)^+ = \Delta_2,$$

as follows.

In the compound game $A+C-B-C$ Left responds to any Right move in A or $-B$ with the response provided by $\aleph$, and to a Right move in C or $-C$ with the mirror image move in $-C$ or C. To see that this avoids loss (supposing Right starts), consider in Table 1 the signs of the resulting plays in all components (writing 0 for finite play).

If the play in C and $-C$ is finite (line 1 in Table 1), then

$$\operatorname{sign}(A+C)^+ = \operatorname{sign}(A+C)^- = \operatorname{sign}(A)$$

and

$$\operatorname{sign}(-B-C)^- = \operatorname{sign}(-B-C)^+ = \operatorname{sign}(-B),$$

Table 1

Line	A	$+C$	$-B$	$-C$	$(A+C)^+$	$(-B-C)^-$	$(A+C)^-$	$(-B-C)^+$
1	X	0	Y	0	X	Y	X	Y
2	+	+	?	−			+	?[a]
3	0	+	?	−	} + (lines 2–4)	− { (lines 2–4)	+	?[a]
4	−	+	+	−			−	+
5	?	−	+	+	?[a]	+		
6	?	−	0	+	?[a]	+	} − (lines 5–7)	+
7	+	−	−	+	+	−		

[a] Here the ? signs in a given line need not be identical.

so since Left avoids loss in $A-B$, he does so in Δ_1 and Δ_2. Otherwise *either*

$$\text{sign}\,(C)=+, \qquad \text{sign}\,(-C)=- \quad \text{(lines 2, 3, 4 in Table 1)}$$

or

$$\text{sign}\,(C)=-, \qquad \text{sign}\,(-C)=+ \quad \text{(lines 5, 6, 7 in Table 1).}$$

In the former case sign $(A+C)^+=+$ so that Left has avoided loss in Δ_1. In the latter case, if sign $(-B)=+$ or 0, then sign $(-B-C)^-=+$ (lines 5 and 6 in Table 1), while otherwise sign $(-B)=-$, so sign $(A)=+$ (since ℵ ensures that Left survives in $A-B$), and we can deduce that sign $(A+C)^+=+$ showing that Left has avoided loss in Δ_1 once again. The argument for Δ_2 is similar (see the right-hand portion of the table).

So far we have only defined inequalities between fixed games. If we define for more general games $A \geqslant B$ to mean that both $A^+ \geqslant B^+$ and $A^- \geqslant B^-$, then we can generalise the statement of Theorem 2.1 to read:

Theorem 2.2. *For all games, $A \geqslant B$ implies $A+C \geqslant B+C$.*

Proof. This is deduced immediately from the above definition and the formulae

$$(X+Y)^+=(X^+ + Y^+)^+, \qquad (X+Y)^- = X^- + Y^-,$$

which are obviously valid for all X, Y.

Theorem 2.3. *$A \geqslant B \geqslant C$ implies $A \geqslant C$.*

Proof. The theorem need only be proved for fixed games A, B, C. The proof is rather subtle, and Left must employ one of the servants, Mr. read, to help in constructing the desired strategy for $A-C$ from the given ones ℵ for $A-B$, ℶ for $B-C$.

[Footnote on footmen: their names are spelt with lower case letters (eg., r), to distinguish them from the real players Left and Right, for whom we use L and R.]

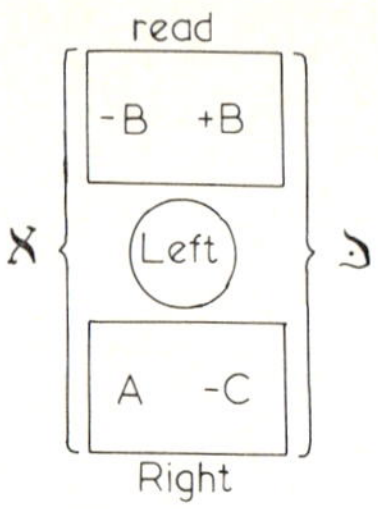

Fig. 5

Left should set up games $-B$ and B on the upper table (Fig. 5) across which he faces Mr. read, and A and $-C$ on the lower one, where his opponent is Right, and should sit on a swivel-chair between these two tables. He should then respond to a move made by either read or Right with the response given by the appropriate strategy ℵ or ב, and instruct Mr. read always to reply to a move in either $-B$ or B with the mirror-image move in B or $-B$.

We assert that the moves played by Left across the lower table will then constitute a loss-avoiding strategy for him in $A-C$, Right starting.

If the total play is finite, then since Left makes the last move in each of the pairs $A-B$ and $B-C$, he does so in the pair $A-C$, which he therefore wins (line 1 of Table 2). If not, but the play in $-B$ and B is finite, then one of the signs x and y of A and $-C$ must be +, so that Left survives in $A-C$ as desired (line 2 of Table 2).

Table 2

Line	A	$-B$	$+B$	$-C$	A	$-C$	
1	0	0	0	0	0	0	(and Left wins)
2	x	0	0	y	x	y	
3	+	−	+	?	+	?	
4	?	+	−	+	?	+	

In the cases when play in $-B$ and B is infinite (lines 3 and 4 of Table 2), we must ensure that Left makes infinitely many moves against his real opponent, Right, rather than entering an infinite huddle with the servant.

But if

$$\text{sign}(B)=+, \qquad \text{sign}(-B)=- \quad \text{(line 3 of Table 2)},$$

then ℵ ensures that sign $(A)=+$, while conversely if

$$\text{sign}(B)=-, \qquad \text{sign}(-B)=+ \quad \text{(line 4 of Table 2)},$$

then ב forces sign $(-C)=+$, so that in each case infinitely many moves are made in $A-C$, and Left has avoided loss therein.

3. Some results from the theory of ending games

In the next few sections, we outline the theory of ending games, as developed in [1]. For such games the relation $G \geqslant H$ means that, supposing Right starts, Left has a strategy which *wins* $G-H$ for him, since now no drawn games are possible. In terms of this, we can define four atomic relations

$G \doteq H$ means that $G \geqslant H$ and $H \geqslant G$ (G and H have the *same* value),

$G > H$ means that $G \geqslant H$ but $H \not\geqslant G$ (G's value *exceeds* H's),

$G < H$ means that $G \geqslant H$ but $G \not\geqslant H$ (G's value is *less* than H's),

$G \parallel G$ means that $G \not\geqslant H$ and $H \not\geqslant G$ (the values are *incomparable*).

The assertion that $G \doteq H$ means that Left can win $G-H$ if Right starts, while Right can win if Left starts, so that $G-H$ is a *second-player-win.* The reason why we say that G and H have the same value in this case is that it turns out that then G can be replaced by H in any sum of games without affecting the outcome. Since we often drop the distinction between games and their values, we shall often drop the dot from above the equality sign $\doteq$.

More generally, if $G \geqslant H$, then a term H in any sum may be replaced by G without destroying the existence of a strategy for *Left*, so that G is at least as valuable to Left as H is.

The basic idea of the theory is that from inequalities on the components of a sum we can derive inequalities for the total, and if we can deduce enough to decide its order-relation with 0, then we know who wins:

$G \doteq 0$ if and only if G is a *second-player-win*,
$G > 0$ if and only if G is a *win for Left* (no matter who starts),
$G < 0$ if and only if G is a *win for Right* (no matter who starts),
$G \parallel 0$ if and only if G is a *first-player-win*.

4. Simplifying ending games

There are many alterations we can make to the structure of a game which do not affect its value, or change it only in a restricted way:

The value of G is unaltered or increased when we

(α) increase the value of any option,

(β) insert a new option for Left,

(γ) delete one of the options for Right.

We examine (β) more closely — if the proposed new option H for Left satisfies $H \geqslant G$, then the value of G is *strictly increased*, but if not, i.e., $H \lhd| G$ (this means $H < G$ or $H \parallel G$), then the value of G is *unchanged.* This latter fact is called the *gift-horse principle*, and the new option H, when $H \lhd| G$, is called a *gift-horse* for

Left. Similarly, if $H \leqslant G$, then G's value will be strictly decreased by adding H as a new Right option, and otherwise ($H | \rhd G$), H will be a gift-horse for Right and leave the value unaltered.

If G^{L_1} and G^{L_0} are two different Left options of G that satisfy $G^{L_1} \leqslant G^{L_0}$, then we say that G^{L_1} is *dominated by* G^{L_0}, and in this case G's value will be unaffected when we delete G^{L_1} provided that we retain G^{L_0}. In general, we may omit many dominated options simultaneously, provided that we retain enough options to dominate them. For Right options, we say that G^{R_0} dominates G^{R_1} provided we have the reversed inequality $G^{R_1} \geqslant G^{R_0}$, and once again we can omit dominated options provided we retain enough to dominate them.

If G is an ending game with only finitely many positions, we can repeatedly apply these ideas until every position is free from dominated options. The finiteness restriction is necessary, because in the game

$$\omega = \{0, 1, 2, 3, \dots |\}$$

every option is dominated, and they plainly cannot all be omitted, although of course infinitely many can — for instance

$$\omega = \{0, 1, 4, 9, 16, 25, 36, \dots |\} = \{1, 2, 4, 8, 16, 32, 64, \dots |\}.$$

Despite its infinitude of positions, ω is an ending game — its tree is sketched in Fig. 6.

The following kind of simplification is both more subtle, and more generally applicable. If a particular Left option G^{L_0} of G has itself a Right option $G^{L_0R_0}$ which satisfies $G^{L_0R_0} \leqslant G$, then we say that the move to G^{L_0} is *reversible* (*through* $G^{L_0R_0}$). In these circumstances we can, without affecting the value of G, replace G^{L_0} as a Left option of G by all the Left options $G^{L_0R_0L}$ of $G^{L_0R_0}$. This is called *bypassing* G^{L_0}, and is illustrated in Fig. 7. Similarly, the Right option G^{R_1} is reversible (through $G^{R_1L_1}$) if $G^{R_1L_1} \geqslant G$, and is bypassed by replacing it by all the $G^{R_1L_1R}$.

It is legitimate to bypass a number of reversible moves at once, and in this way we can arrange that any ending game, whether finite or infinite, can be completely freed of positions with reversible moves. If G has only finitely many positions, we

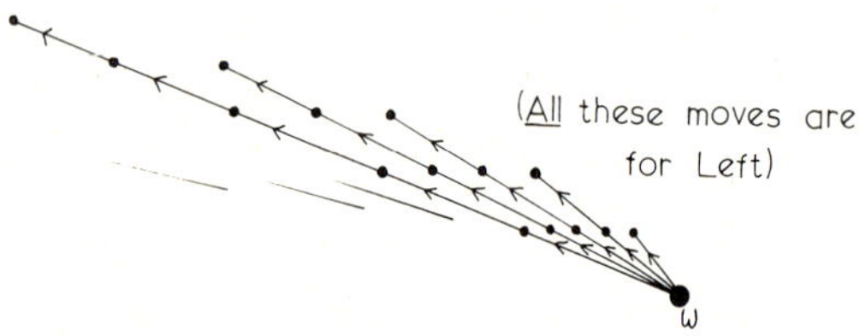

Fig. 6

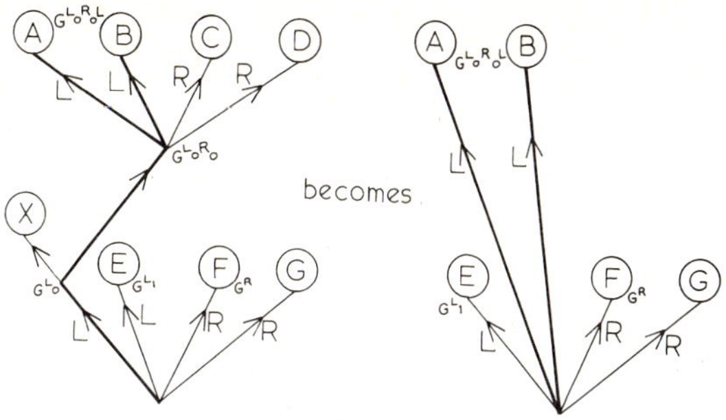

Fig. 7

can also free these from any dominated options, and so arrive at the *simplest form* of G. The point of this concept is illustrated by the following theorem.

Theorem 4.1. *If G and H are ending games in which no positions have either dominated or reversible options, and G and H have the same value, then G and H are isomorphic.*

This theorem provides us with an efficent way to work with sums of games with only finitely many positions. By eliminating dominated options and bypassing reversible moves we first find the simplest form, which is an exact invariant for the value. The value, by our earlier theorems, contains exactly that information about a game which is relevant in computing the outcome of sums involving it.

5. On numbers and games

Certain ending games can be identified with numbers. (For games with only finitely many positions only the *dyadic rational* numbers $m/2^n$ arise.) The numbers naturally form a tree, as shown in Fig. 8, and we say that x *is simpler than* y if the path from 0 to y includes x. (0 is the simplest number of all.) Each number y

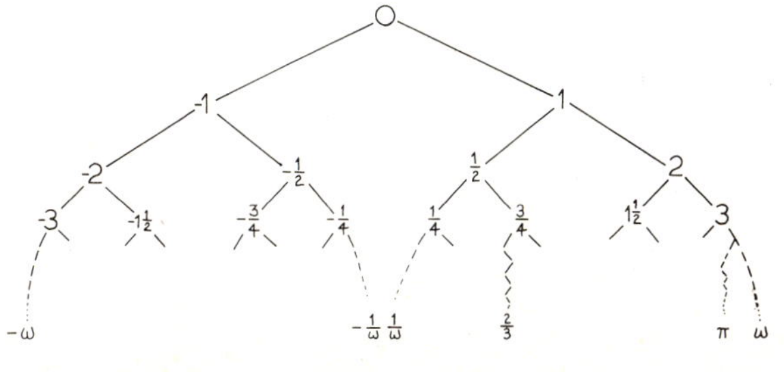

Fig. 8

has a *canonical form* in which the options are *all* the numbers that precede y along this path, for example:

$$0=\{\,|\,\},\quad 1=\{0\,|\,\},\quad 2=\{0,1\,|\,\},\quad 3=\{0,1,2\,|\,\},\ldots,\omega=\{0,1,2,\ldots\,|\,\},$$

$$-0=0,\quad -1=\{\,|\,0\},\quad -2=\{\,|\,0,-1\},\quad -3=\{\,|\,0,-1,-2\},\ldots,-\omega=\{\,|\,0,-1,\ldots\},$$

$$\tfrac{1}{2}=\{0\,|\,1\},\quad \tfrac{1}{4}=\{0\,|\,1,\tfrac{1}{2}\},\quad \tfrac{1}{8}=\{0\,|\,1,\tfrac{1}{2},\tfrac{1}{4}\},\ldots,\frac{1}{\omega}=\{0\,|\,1,\tfrac{1}{2},\tfrac{1}{4},\ldots\},$$

$$\tfrac{3}{4}=\{0,\tfrac{1}{2}\,|\,1\},\quad \tfrac{5}{8}=\{0,\tfrac{1}{2}\,|\,1,\tfrac{3}{4}\},\quad \tfrac{11}{16}=\{0,\tfrac{1}{2},\tfrac{5}{8}\,|\,1,\tfrac{3}{4}\},\ldots,\tfrac{2}{3}=\{0,\tfrac{1}{2},\tfrac{5}{8},\ldots\,|\,1,\tfrac{3}{4},\ldots\}$$

and it may also have a simplest form obtained by omitting the dominated options:

$$0=\{\,|\,\},\qquad 1=\{0\,|\,\},\quad 2=\{1\,|\,\},\quad 3=\{2\,|\,\},\ldots,\omega+1=\{\omega\,|\,\},$$
$$-1=\{\,|\,0\},\quad -2=\{\,|\,-1\},\quad -3=\{\,|\,-2\},\ldots,-\omega-1=\{\,|\,-\omega\},$$

$$\tfrac{1}{2}=\{0\,|\,1\},\quad \tfrac{1}{4}=\{0\,|\,\tfrac{1}{2}\},\quad \tfrac{1}{8}=\{0\,|\,\tfrac{1}{4}\},\ldots,\frac{1}{2\omega}=\left\{0\,\middle|\,\frac{1}{\omega}\right\},$$

$$\tfrac{3}{4}=\{\tfrac{1}{2}\,|\,1\},\quad \tfrac{5}{8}=\{\tfrac{1}{2}\,|\,\tfrac{3}{4}\},\quad \tfrac{11}{16}=\{\tfrac{5}{8}\,|\,\tfrac{3}{4}\},\quad 1\tfrac{1}{2}=\{1\,|\,2\}.$$

It can be shown that ω has no simplest form.

When the value of every position in G is a number, these numbers can be computed using the rule:

If every option of G is a number, and every Left option is strictly less than every Right option, then G is itself a number, namely the SIMPLEST number greater than every Left option and less that every Right one.

$$\{2,3\tfrac{1}{2}\,|\,\}=\{\pi\,|\,17\}=4,\quad \{1\,|\,e\}=\{1\tfrac{1}{2}\,|\,\}=2,\quad \{-1\,|\,\}=0,\quad \{\tfrac{1}{4}\,|\,\tfrac{7}{8}\}=\tfrac{1}{2}.$$

This rule generalises to give the *simplicity principle*, which sometimes assigns a numerical value to a position which has non-numerical options:

If there is some number, z say, that satisfies

$$\textit{every } G^{L} \lhd| z \lhd| \textit{ every } G^{R},$$

then the value of G is a number, namely the simplest such z.

To apply this, of course, we need techniques enabling us to compare games with numbers.

For every ending game G there are two Dedekind sections $\mathrm{L}(G)$ and $\mathrm{R}(G)$ of the number-line with the following properties:

$X>\mathrm{L}(G)$ if and only if $x \geq G$,

$x<\mathrm{R}(G)$ if and only if $x \leq G$,

$\mathrm{L}(G)>x>\mathrm{R}(G)$ if and only if $x \parallel G$.

When G is a number, z say, then $\mathrm{L}(G)$ is the section between z and all smaller numbers, $\mathrm{R}(G)$ that between z and all larger ones, as in Fig. 9. But usually we have $\mathrm{R}(G) \leq \mathrm{L}(G)$, as in Fig. 10.

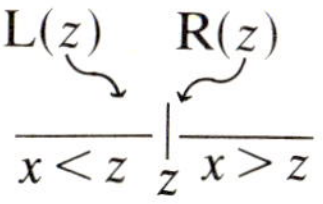

Fig. 9

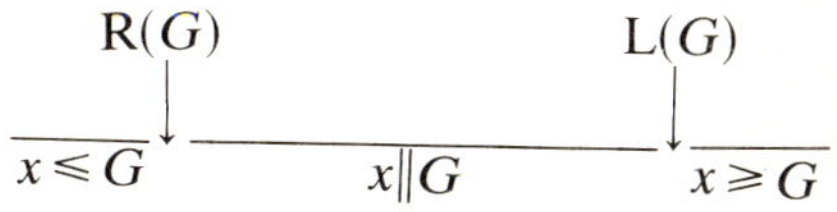

Fig. 10

To compute these sections, let Left and Right play G "intelligently", stopping play when the value first equals a number, z say, with player Y ($Y = \mathrm{L}$ or R) about to move. Then if player X started, we have

$$X(G) = Y(z).$$

Thus

$$\mathrm{L}(\{5 \mid 4\}) = \mathrm{R}(5), \qquad \mathrm{R}(\{5 \mid 4\}) = \mathrm{L}(4),$$
$$\mathrm{L}(\{7 \mid \{5 \mid 4\}\}) = \mathrm{R}(7), \qquad \mathrm{R}(\{7 \mid \{5 \mid 4\}\}) = \mathrm{R}(5).$$

It is important to realise that the only case when G is a number is when $\mathrm{L}(G) < \mathrm{R}(G)$ as sections, and that then its value is to be found by the simplicity rule. For instance if $G = \{4 \mid \{5 \mid 4\}\}$ the above rules suggest $\mathrm{L}(G) = \mathrm{R}(4)$ and $\mathrm{R}(G) = \mathrm{R}(5)$ and so $\mathrm{L}(G) < \mathrm{R}(G)$. In fact, G is a number, namely the simplest number z satisfying

$$\mathrm{R}(4) < z < \mathrm{R}(5),$$

namely 5 itself, and so $\mathrm{L}(G) = \mathrm{L}(5)$, $\mathrm{R}(G) = \mathrm{R}(5)$, and we should have stopped play at G itself.

6. Other ending games

Ending games can have many other values besides numbers, and we can afford here only to give a few simple examples:

(α) *Switches.* These are games $\{x \mid y\}$ for which x and y are numbers with $x \geqslant y$. When playing a sum of switches and numbers, move in that switch with the largest value of $x - y$.

(β) *Nim-numbers* or *nimbers.* These are defined inductively

$$*0 = 0 = \{\,\mid\,\},$$
$$*1 = * = \{0 \mid 0\},$$
$$*2 = \{0, * \mid 0, *\},$$
$$*3 = \{0, *, *2 \mid 0, *, *2\},$$
$$\vdots$$
$$*n = \{*0, *1, \ldots, *(n-1) \mid *0, *1, \ldots, *(n-1)\},$$
$$*\omega = \{*0, *1, *2, \ldots \mid *0, *1, *2, \ldots\},$$

and in fact for all ordinals

$$*\alpha = \{*\beta(\beta < \alpha) \mid *\beta(\beta < \alpha)\}.$$

The sum of two nimbers is another (use binary notation without carry — e.g. $*3 + *5 = *6$), and we can say that G will certainly have a nimber value $*n$ provided that

(i) every $G^L \lhd| *n \lhd|$ every G^R, *and*
every $*m(m < n)$ appears as both a Left and a Right option of G.

These observations contain the Sprague-Grundy theory of impartial games.

(γ) *Small games.* Nimbers are particular examples of games that are greater than every negative number $\left(\text{even } -\dfrac{1}{\omega}\right)$, and less than every positive one. There even exist positive games of this kind, notably $\uparrow = \{0 \mid *\}$ (pronounced "up").

7. Sums of free loopy games

We return to the topic of loopy games, more particularly the free games, in which infinite play is always declared a draw. What information do we need about the individual components $G, H, K, \ldots$ of a sum $G + H + K + \cdots$ of such games if we are to compute the outcome of the total?

We answer this as follows. If we want to see whether Left can at least draw, we might as well redefine all draws as wins for Left, so obtaining the game

$$(G + H + K + \cdots)^+ = (G^+ + H^+ + K^+ + \cdots)^+,$$

and so we need only know the values of G^+, H^+, $K^+, \ldots$. If, on the other hand, we want to know whether Left can win, we might as well redefine all draws as losses for Left, when we are considering the game

$$(G + H + K + \cdots)^- = (G^- + H^- + K^- + \cdots)^-,$$

and so need only know the values of G^-, H^-, $K^-, \ldots$.

To indicate that a game G has G^+ of value A and G^- of value B, we shall write merely $G = A \,\&\, B$.

For free games G, we have of course

$$G^+ = G(on), \quad G^- = G(\mathit{off}),$$

and so A and B in this case are often called the *onside* and *offside* of G.

The answer to our question is therefore this — we can work out the outcomes of sums of open games in terms only of the onsides and offsides of the individual components. Since it often happens that the two sides of G have values that are equal to ending games, and maybe to numbers, this is a very real simplification.

Consider for instance the following familiar little game from our infancy (Fig. 11). Left may say "'tis" and then Right may respond with "'tisn", and Left

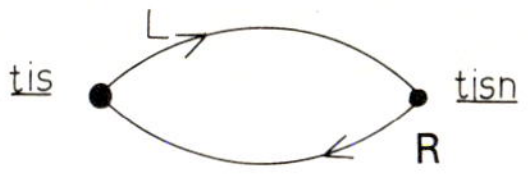

Fig. 11

follow with "'tis" again, and so on alternately. Formally, the game has just two positions, *tis* and *tisn*, with a Left move from *tis* to *tisn* and a Right one back from *tisn* to *tis*.

We assert that

$$tis = 1 \,\&\, 0, \qquad tisn = 0 \,\&\, {-1},$$

and so we can predict the outcomes of all sums of these games together with other open games whose onsides and offsides are numbers. The reader is recommended to check the equality $tis^+ = 1$ by the only method we have given him so far — showing that Left has survival strategies in both the differences $tis^+ - 1$ and $1 - tis^+$, Right starting.

8. The iteration method

We can of course always check any asserted inequality or equality about values of games by investigating suitably defined difference games. But we need a method for finding such equations as $G = A \,\&\, B$, as well as for checking them. Here is a method that often works, and is informative even if it does not.

Take the graph that defines the game, and against each node mark an upper bound for the game obtained by taking that node as the initial position. We shall use $[G]$ for the game marked at the node G. Now at any node $H = \{H^L \mid H^R\}$ we can if we like replace our present upper bound by the possibly better one given by the formula $\{[H^L] \mid [H^R]\}$.

It might well happen that by repeated use of such improvements, or possibly some more powerful techniques, that we eventually arrive at a system of upper bounds $[H]$ that satisfy the equations

$$[H] = \{[H^L] \mid [H^R]\}$$

for all positions H of G. We assert that then we have found the onsides of these positions $[H] = H(on)$.

What shall we use for the initial upper bounds? Here we have a simple choice:

on^+ is an upper bound for every game G,

and so if we like we can start with $[H] = on^+$ for every position H.

We might remark at this point that it is this property that prompted our choice of name for *on*. *On* has now become the standard set-theoretical name for the class of all *o*rdinal *n*umbers (the initial letters provide the reason), which in von Neumann's sense is also the *largest* ordinal number. We might call it *The*

Burali-Forti Number since the Burali-Forti paradox is customarily resolved by declaring that *on* is a Proper Class, rather than a proper set! In our language the Burali-Forti paradox corresponds to the assertion that *on* is an option of itself.

To prove our assertion about on^+, observe that Left certainly avoids defeat in $on^+ - G$ by the simple strategy of *always playing in the component* on^+.

Before we prove that the answer given by the iteration process is correct, we shall discuss a few examples to show how it works. In these examples, we abbreviate on^+ to *on*.

The games *tis* and *tisn*. Here the equations to be solved are

$$tis = \{tisn \mid \}, \qquad tisn = \{ \mid tis\}.$$

We start with the initial approximations

$$[tis] = [tisn] = on$$

and find as our next approximations

$$[tis] = \{on \mid \}, \qquad [tisn] = \{ \mid on\},$$

which we evaluate as

$$[tis] = on \text{ (again)}, \qquad tisn = 0,$$

using obvious generalisations of the simplest number rule. From these approximations in turn we derive further ones:

$$[tis] = \{0 \mid \} = 1, \qquad tisn = \{ \mid on\} = 0 \text{ (again)},$$

but now the process finishes, since the next approximations are

$$[tis] = \{0 \mid \} = 1 \text{ (again)}, \qquad [tisn] = \{ \mid 1\} = 0 \text{ (again)}.$$

Our theorem now asserts that the final approximations really are the respective onsides:

$$tis(on) = 1, \qquad tisn(on) = 0.$$

(It also explains why we use the notations $G(on)$ and $G(off)$.) We leave it to the reader to compute in a similar way the offsides

$$tis(off) = 0, \qquad tisn(off) = -1,$$

which are derived by starting at the universal lower bound $off^- = off$.

For our next example we take the games α, β, γ, δ shown in Fig. 12 so that we

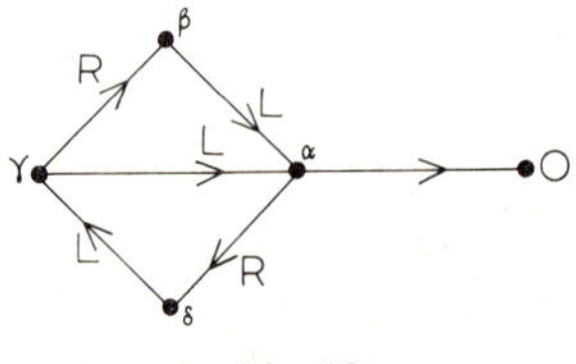

Fig. 12

Table 3

$\{0 \mid \delta\} = \alpha$	$\{\alpha \mid \} = \beta$	$\{\alpha \mid \beta\} = \gamma$	$\{\gamma \mid \} = \delta$
on	*on*	*on*	*on*
$\{0 \mid on\} = 1$			
	$\{1 \mid \} = 2$		
		$\{1 \mid 2\} = 1\frac{1}{2}$	
			$\{1\frac{1}{2} \mid \} = 2$
$\{0 \mid 2\} = 1$, so the process has converged.			
off	*off*	*off*	*off*
	$\{off \mid \} = 0$		
		$\{off \mid 0\} = -1$	
			$\{-1 \mid \} = 0$
$\{0 \mid 0\} = *$			
	$\{* \mid \} = 0$		
		$\{* \mid 0\} = \downarrow$	
			$\{\downarrow \mid \} = 0$

have to solve the equations:

$$\alpha = \{0 \mid \delta\}, \qquad \beta = \{\alpha \mid \}, \qquad \gamma = \{\alpha \mid \beta\}, \qquad \delta = \{\gamma \mid \}.$$

In Table 3, a single entry on a line indicates a new approximation and this process has converged as well.

We conclude that

$$\alpha = 1 \,\&\, *, \qquad \beta = 2 \,\&\, 0, \qquad \gamma = 1\tfrac{1}{2} \,\&\, \downarrow, \qquad \delta = 2 \,\&\, 0.$$

It might help the reader if we remark that $*$ is incomparable with 0, and that $\downarrow$ (pronounced "down"), being the negative of $\uparrow$, is a negative infinitesimal. So the simplicity principle, in its generalised form, tells us that $\{* \mid \} = \{\downarrow \mid \} = 0$. Since $tis = 1 \,\&\, 0$, we have actually shown that $\beta = \delta = tis + tis$.

For our last example, we take the game

$$dud = \{dud \mid dud\}$$

of Fig. 4. Here from the initial approximation *on* we derive $\{on \mid on\} = on$, and from *off* we similarly find $\{off \mid off\} = off$. So in the "&" notation:

$$dud = on \,\&\, off.$$

9. The approximation theorem

When the iteration process converges, it produces games $[H]$ that satisfy

$$[H] = \{[H^{L}] \mid [H^{R}]\}$$

for all positions H of G. We shall call these equations *the equations defining G.*

We are about to prove the *Approximation Theorem*, which asserts that any solution of them lies between the on and offsides:

$$H(off) \leqslant [H] \leqslant H(on)$$

for every position H of G. There is actually a slight generalisation:

Theorem 9.1. *If the games $[H]$ form a subsolution of the equations defining G, in the sense that*

$$[H] \leqslant \{[H^L] \mid [H^R]\}$$

for all positions H of G, then they also satisfy

$$[H] \leqslant H(on)$$

for all such positions.

Proof. It will not harm us to suppose that the games $[H]$ are fixed, for we can replace them by the $[H]^+$ if not, and we can also suppose that $H = H(on)$. Then the hypotheses tell us that Left can survive in each of the games

$$\{[H^L] \mid [H^R]\} - [H],$$

while the conclusion requires him to survive in $H - [H]$. We need only produce a strategy for $G - [G]$.

Since this strategy is quite hard to find, we shall suppose that Right kindly places a potential infinity of his more mathematically inclined servants, Messrs. rado (r_0), radon (r_1), . . . , rademacher (r_m), . . . at the disposal of Left, and allows him to use the Great Hall, and various furnishings, of *The Wright House*, which is a rather grand establishment.

On the far table in Fig. 13 is set up the real game $G - [G]$, which Left is to play against his real opponent, Right. But even before play starts, Left instructs r_0 to bring in an additional table on which is set up the difference game

$$\{[G^L] \mid [G^R]\} - \{[G^L] \mid [G^R]\} = X_0 - X_0,$$

and a chair labelled $\aleph_0$, to be placed near the games $\{[G^L] \mid [G^R]\}$ and $-[G]$.

Left has, by the hypotheses of the theorem, a survival strategy, which we also call $\aleph_0$, for the sum of these two games. The chair marked $\beth$, which was already in the hall, is placed near the games $G = \{G^L \mid G^R\}$ and $-\{[G^L] \mid [G^R]\}$.

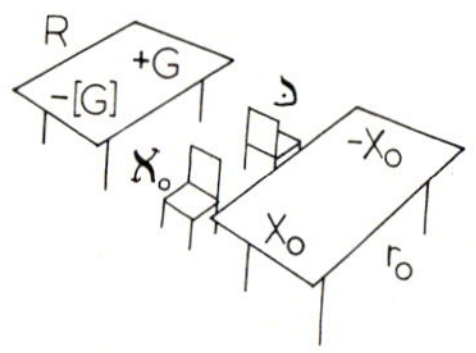

Fig. 13

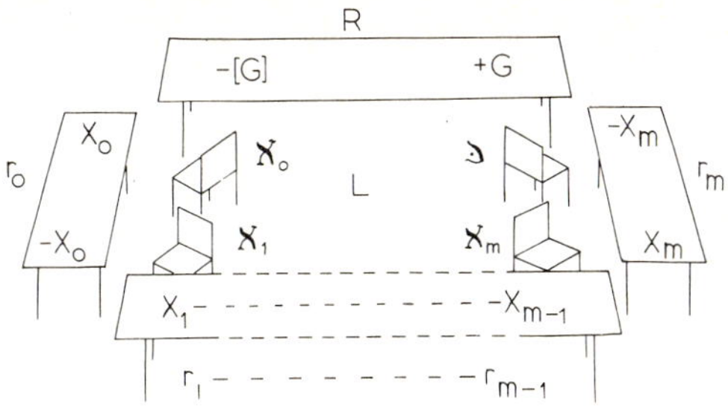

Fig. 14

As the game proceeds, Left occasionally instructs a new footman (r_i) to bring in a new chair ($\aleph_i$), and a new table on which is set up a position of the form $X_i - X_i$. Footman r_i is detailed from then on to respond to a Left move in either X_i or $-X_i$ with the mirror image move in the other. In Fig. 14 we show a number of these tables, all marked with the positions in which they were originally set up.

The chairs $\aleph_0, \aleph_1, \ldots, \aleph_m$, and $\beth$ are placed *between* adjacent tables, and each corresponds to a strategy, of sorts, for playing the two games nearest to it. The strategies $\aleph_i$ are easiest to describe. When chair $\aleph_i$ was first brought in, the games it was put next to were in a position of the form

$$\{[H^{\mathrm{L}}] \mid [H^{\mathrm{R}}]\} - [H]$$

for some position H of G—strategy $\aleph_i$ is Left's survival strategy for this game given by our hypotheses.

The two games nearest to chair $\beth$ will usually have the form

$$H = \{H^{\mathrm{L}} \mid H^{\mathrm{R}}\} \quad \text{and} \quad -\{[H^{\mathrm{L}}] \mid [H^{\mathrm{R}}]\}$$

for some position H of G. The "strategy" $\beth$ is then the following sequence of actions. If Right makes a move in either of these games, Left is to make the corresponding move in the other, making the compound position have the form $K - [K]$ for some position $K = H^{\mathrm{L}}$ or H^{R} of G. He then instructs a new footman, r_{m+1}, to bring in a new table on which is set up the difference game

$$\{[K^{\mathrm{L}}] \mid [K^{\mathrm{R}}]\} - \{[K^{\mathrm{L}}] \mid [K^{\mathrm{R}}]\},$$

and a new chair, $\aleph_{m+1}$, to be placed near to the games $\{[K^{\mathrm{L}}] \mid [K^{\mathrm{R}}]\}$ and $-[K]$ for whose sum he has a survival strategy we shall also call $\aleph_{m+1}$.

The chair $\beth$ is then repositioned next to

$$K = \{K^{\mathrm{L}} \mid K^{\mathrm{R}}\} \quad \text{and} \quad -\{[K^{\mathrm{L}}] \mid [K^{\mathrm{R}}]\}.$$

Left's total strategy is therefore this. To any move, whether played by his real opponent Right, or one of the footmen $r_0, r_1, \ldots$ he replies with the response given by the strategy corresponding to the nearest chair. The strategies $\aleph_i$ are those for various differences $X - Y$ with $X \geqslant Y$ that are given us by the hypotheses, while the strategy $\beth$ requires just one "imitation" move, and a call for a new table and chair to be inserted. It is plain that this compound strategy always gives Left a reply in a game somewhere in the Hall, but not entirely clear that he will eventually respond to any move in the real game with another move in that game, and perhaps even less clear that he avoids loss in that game if it continues indefinitely. We now proceed to establish these facts.

If infinitely many tables are brought in in the course of play, both results are easy. Each new table was only brought in after a move in G had been made by either Left or Right, so infinitely many moves have been made in G, and since $G = G(on)$, all infinite plays in G count as wins for Left, who therefore wins or draws the compound game $G-[G]$. So we shall suppose that only the finitely many tables shown in Fig. 14 were brought in.

In this case the total play in G and $-X_m$ has been finite (sign 0) as must have been that in X_m, if footman r_m has correctly obeyed his orders. Let i be the greatest number, if any, for which the play in X_i was infinite. Then strategy $\aleph_{i+1}$ avoids loss in $X_{i+1}-X_i$, so that the sign of the play in $-X_i$ must be +, whence sign $(X_i) = -$. To have avoided the threatened loss for Left in $X_i - X_{i-1}$, strategy $\aleph_i$ must have given the sign + for $-X_{i-1}$, whence sign $(X_{i-1}) = -$, and so on. Eventually, we see that strategy $\aleph_0$ forces the sign of the play in $-[G]$ to be +, showing both that Left made infinitely many moves in the real game $G-[G]$ and that he avoided loss in that game.

In the final case, the *total play* was finite, and Left's strategies $\aleph_0 \ldots \aleph_m$ ensure that he made the last move in it. This cannot have been against any of the footmen $r_0, \ldots, r_m$ since they have always the mirror-image reply to make, and so it must have been played in the real game against Right, who has not replied and therefore *loses*, by the normal play convention. The approximation theorem is therefore established.

In any convergent case of the iteration process, the Approximation Theorem tells us that the upper bounds $[H]$ it gives for the games $H(on)$ are also lower bounds, and therefore are the correct answers.

10. Some results about stoppers

Stoppers have some special properties which make them particularly easy to handle.

Theorem 10.1. *If G is a stopper, then*

(α) $G(on) \geq H(off)$ *if and only if* $G(off) \geq H(off)$,

(β) $G(off) \leq H(on)$ *if and only if* $G(on) \leq H(on)$.

Proof. We need only prove (α), by symmetry. In the difference game $G-H$ the condition that G is a stopper ensures that if there is infinite play in G, there must be infinite play in $-H$. Since $-H(off) = (-H)(on)$ this will count as a win in *either*

of $G(on)-H(off)$ and $G(off)-H(off)$. If there has only been finitely much play in G, then of course the signs attached to infinite plays in it are irrelevant.

Corollary 10.2. *If G and H are both stoppers, then the inequalities*

$$G(off) \geqslant H(off), \qquad G(on) \geqslant H(off), \qquad G(on) \geqslant H(on)$$

are all equivalent, and any one of them suffices to prove $G \geqslant H$.

It can also be shown that

Theorem 10.3. *Any stopper can be put into a form free of reversible moves, and, if it has only finitely many positions, can also be freed of dominated options, and so has a simplest form.*

We do not give the proof here, since it follows closely the corresponding result for enders, which is given in [1]. The condition that G is a stopper prevents the possibility of a non-terminating sequence of bypasses of reversible options. It also justifies the omission of dominated options, which is not quite so obvious as it is for enders.

Theorem 10.4. *If G and H are fixed stoppers in simplest form, and $G=H$, then G and H are isomorphic games.*

Proof. (Patterned on the corresponding proof for enders, in [1].) Left has a survival strategy in $G-H$. If Right moves in this, say to $G^{R}-H$, what can Left's reply be? Plainly not $G^{RL}-H$, for this requires $G^{RL} \geqslant H=G$, showing that G had a reversible option. So Left's reply is to some game $G^{R}-H^{R}$, showing that for every G^{R} there must be some $H^{R} \leqslant G^{R}$. But for similar reasons there must be some $G^{R'} \leqslant H^{R}$, and so $G^{R'} \leqslant G^{R}$. Since G has no dominated options, this entails that $G^{R'}=H^{R}=G^{R}$, and we have shown that for every Right option of either game there is an equal Right option of the other.

Since a similar statement holds of Left options, we have established a 1-1 correspondence between the options of G and those of H. We can now proceed to establish similar correspondences between options of these options, and so on, identifying the entire set of positions of one game with those of the other. The argument also shows that the only way to survive in the difference game $G-H$ is to play the "mirror-image" strategy.

We must now show that the signs attached to infinite plays of G and H also correspond. But if, say, a certain play of H were given the sign $+$ while the corresponding play of G received $-$, then Right, starting in $G-H$, could play so that both components received the sign $-$, so forcing Left to lose, and showing that $G \neq H$. This concludes the proof.

The value of these results follows from the fact that very many free games can be written in the form A & B, where A and B are stoppers, for which Corollary

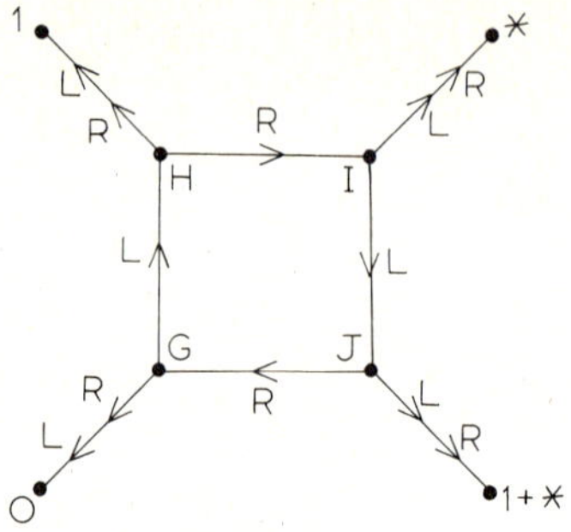

Fig. 15

10.2 shows us we need not distinguish between various notions of inequality. (This was why we allowed ourselves to abbreviate on^+ to *on* in an earlier section.) If this happens for a game G with only finitely many positions, then we can put A and B into simplest form, by Theorem 10.3 and by Theorem 10.4, the resulting pair of graphs is a complete invariant for G.

Unfortunately, not everything in the garden is quite so lovely. Bach has produced the game G shown in Fig. 15 (the *carousel*) which has a number of disturbing features:

(i) Its onside and offside are *not* equivalent to stoppers. This can be proved by a method like that of Theorem 10.4.)

(ii) The Left option O of G is dominated by H, but cannot be omitted. (The reason, roughly, is that Left always arrives at a better position by taking another trip round the carousel than he does by stepping off to 0 now. But there are circumstances in which he can win by stepping off sometime, but will not win by going round and round forever.)

(iii) If in $G-G$, the first player always moves round the carousel, then the second player cannot afford to do anything but make the corresponding move in the other component. (This is proved by case-by-case analysis.)

Question. Is there a wider notion than that of stopper which will allow us to find a simplest form theorem for all free games with finitely many positions?

Acknowledgement

To Simon Norton, who dramatically simplified the theory by the systematic introduction of fixed games, to Clive Bach, for the carousel and a very large number of arguments, and to Bob Li for the paper that started it all [2]. To Richard Guy, for his never failing interest in the subject of this paper.

References

[1] J.H. Conway, On Numbers and Games (Academic Press, New York, 1976).
[2] R. Li, Sums of zugswang games, J. Combinatorial Theory.

Annals of Discrete Mathematics 3 (1978) 75–92.

HAMILTON CIRCUITS AND LONG CIRCUITS

G.A. DIRAC

Matematisk Institut, Ny Munkegade, 8000 Aarhus C, Denmark

1. Introduction and terminology

In this paper the term *graph* will denote an undirected graph without loops or multiple edges which may be finite or infinite. The set of vertices of a graph Γ will be denoted by $\mathcal{V}(\Gamma)$, the set of its edges by $\mathcal{E}(\Gamma)$, and $|\mathcal{V}(\Gamma)|$ will be denoted by n_Γ. The *length* of a path or a circuit Δ is the number of edges it contains and is denoted by $l(\Delta)$. If $\mathbf{Y}$ is a path and $X, Y \in \mathcal{V}(\mathbf{Y})$, then the unique (X, Y)-path contained in $\mathbf{Y}$, will be denoted by $\mathbf{Y}(X, Y)$. If A and B are two non-empty disjoint subgraphs of a graph (for example, sets of vertices), then a path of the graph which has a vertex of A and a vertex of B as its end-vertices and has nothing else in common with $A \cup B$ will be called an *(A, B)-path*. The number of (A, B)-paths of length 1 will be denoted by $e(A, B; \Gamma)$. A path of a graph will be called *terminated* in the graph if no path of the graph contains it as a proper subgraph. If Γ is a graph and $V \in \mathcal{V}(\Gamma)$, then the set of those vertices of Γ which are joined by an edge to V in Γ will be denoted by $\mathcal{N}_\Gamma(V)$, and $|\mathcal{N}_\Gamma(V)|$ is the *valency* of V in Γ, denoted by $v_\Gamma(V)$. A graph will be called *k-connected*, where $k > 0$, if corresponding to each pair of distinct vertices V and W the graph contains a set of k or more (V, W)-paths such that each pair of them have nothing except V and W in common. If Δ is a subgraph of the graph Γ then $\Gamma(\Delta)$ denotes $\Gamma - (\Gamma - \Delta)$.

The main purpose of this paper is to establish weak necessary conditions for the existence of a Hamilton circuit in a graph, partly in terms of the valencies of the vertices, and for any given integer $f > 3$ to establish weak necessary conditions for the existence of a circuit of length $\geq f$ in a 2-connected graph which may be finite or infinite, also partly in terms of the valencies of the vertices.

2. A theorem of Menger type

It is convenient to prefix a Menger type result which will be used later. It holds also for graphs with multiple edges.

Theorem 2.1. *Suppose that Γ is a graph, $\mathcal{A}$ and $\mathcal{B}$ are disjoint sets of vertices of Γ, $A \in \mathcal{A}$ and $B \in \mathcal{B}$ and Γ contains an $(\mathcal{A}, \mathcal{B})$-path whose end-vertices are A and B,* *and Γ contains a set $\mathcal{P}$ of two or more pairwise disjoint $(\mathcal{A}, \mathcal{B})$-paths. Then there exists a subset $\mathcal{P}_1$ of $\mathcal{P}$ and a set $\mathcal{P}_2$ of $(\mathcal{A}, \mathcal{B})$-paths of Γ such that $|\mathcal{P}_1| = |\mathcal{P}_2| \leq 2$, $\mathcal{P} \cap \mathcal{P}_2 = \emptyset$, and $(\mathcal{P} - \mathcal{P}_1) \cup \mathcal{P}_2$ is a set of pairwise disjoint paths of Γ whose union includes A and B.*

Proof. Let $\mathbf{Y}$ denote an $(\mathcal{A}, \mathcal{B})$-path of Γ whose end-vertices are A and B and Π the union of all paths of $\mathcal{P}$. Suppose $\{A, B\} \not\subseteq \Pi$ and $\Pi \cap \mathbf{Y} \neq \emptyset$ (otherwise there is nothing to prove).

As $\mathbf{Y}$ is followed from A to B let A' denote the first vertex in Π and B' the last, and let $\mathbf{Y}_{A'}$, $\mathbf{Y}_{B'}$, respectively, denote the paths of $\mathcal{P}$ to which A', B' belong.

If $Y_{A'} = Y_{B'}$ then Theorem 2.1 holds with $\mathcal{P}_1 = \{\mathbf{Y}_{A'}\}$ and

$$\mathcal{P}_2 = \{\mathbf{Y}(A, A') \cup \mathbf{Y}_{A'}(A', B') \cup \mathbf{Y}(B', B)\}.$$

If $\mathbf{Y}_{A'} \neq \mathbf{Y}_{B'}$, then put $\mathcal{A} \cap \mathbf{Y}_{B'} = X$ and $\mathcal{B} \cap \mathbf{Y}_{A'} = Y$. Now Theorem 2.1 holds with $\mathcal{P}_1 = \{\mathbf{Y}_{A'}, \mathbf{Y}_{B'}\}$ and

$$\mathcal{P}_2 = \{\mathbf{Y}(A, A') \cup \mathbf{Y}_{A'}(A', Y), \mathbf{Y}_{B'}(X, B') \cup \mathbf{Y}(B', B)\}.$$

3. Hamilton circuits in graphs

Notations. The edge joining two distinct vertices V and W will be denoted (V, W). If Γ is a graph and $V \in \mathcal{V}(\Gamma)$, then the set of all vertices $U \neq V$ of Γ with the property that Γ contains a circuit Θ such that $U, V \in \Theta$ and at least one of the two (U, V)-paths whose union is Θ has length 2, will be denoted by $\mathscr{C}(V)$, and the set of all vertices U of $\mathscr{C}(V)$ such that $v_\Gamma(U) \leq v_\Gamma(V)$ by $\mathscr{C}'(V)$. If W is a vertex of Γ other than V, then the set of all vertices of $\mathscr{C}(V)$ other than W which are not joined to W by an edge will be denoted by $\mathscr{C}(V, W)$. $|\mathscr{C}(V)|$, $|\mathscr{C}'(V)|$, $|\mathscr{C}(V, W)|$ and $|\mathscr{C}'(V) \cap \mathscr{C}(V, W)|$ will be denoted by $c(V)$, $c'(V)$, $c(V, W)$ and $c'(V, W)$, respectively. Let $\mathbf{Y}$ denote any path of length ≥ 1 in Γ and V an end-vertex of $\mathbf{Y}$. Then if $V_1, \ldots, V_n$ denote the vertices of $\mathbf{Y}$ in order from end to end with $V_1 = V$, the set of all vertices V_i of $\mathbf{Y}$ such that $2 \leq i \leq n-1$ and $(V, V_{i+1}) \in \Gamma$ will be denoted by $\mathcal{T}(V, \mathbf{Y})$, and $|\mathcal{T}(V, \mathbf{Y})|$ by $t(V, \mathbf{Y})$. If $n = 2$ then $\mathcal{T}(V, \mathbf{Y}) = \emptyset$.

The following statement follows from the definitions of $\mathscr{C}(V)$ and $\mathcal{T}(V, \mathbf{Y})$:

(1)(A) *Let Γ denote a graph, $\mathbf{Y}$ any path of length ≥ 2 in Γ, and V and W the end-vertices of $\mathbf{Y}$. Then $\mathcal{T}(V, \mathbf{Y}) \subseteq \mathscr{C}(V)$ and $\mathcal{T}(W, \mathbf{Y}) \subseteq \mathscr{C}(W)$.*

(B) *Let U denote the vertex of $\mathbf{Y}$ such that $l(\mathbf{Y}(U, V)) = 2$. Suppose that X is any vertex of $\mathbf{Y} - V$ such that $(V, X) \in \Gamma$. Then $\forall Y \in \mathcal{V}(\mathbf{Y}(V, X)) - V$: $Y \in \mathscr{C}(V)$ if and only if Γ contains a (V, Y)-path of length 2 (which is the case if e.g. $(X, Y) \in \Gamma$, in particular if $U \in \mathscr{C}(V)$.*

(C) *If, in addition,* $(\mathbf{Y}(V, X) - X) \cap \mathcal{N}_\Gamma(W) \neq \emptyset$, *then* $\forall Y \in \mathcal{V}(\mathbf{Y}) - V$: $Y \in \mathcal{C}(V)$ *if and only if* Γ *contains a* (V, Y)*-path of length* 2 *(which is the case if e.g.* $(X, Y) \in \Gamma$*).*

The following easily proved statement follows from a result of the writer [2]:

(2) *A circuit of a connected graph is a Hamilton circuit of the graph if and only if it includes all the vertices of a longest path of the graph.*

Next, an observation due to Ore [4].

(3) *If* Γ *is a graph and* $\mathbf{Y}$ *is a path of length* ≥ 2 *in* Γ *and* V *and* W *are the end-vertices of* $\mathbf{Y}$, *then each vertex* X *of* $\mathcal{T}(V, Y)$ *is the end-vertex of an* (X, W)*-path of* Γ *which includes every edge of* $\mathbf{Y}$ *except one, and whose vertex-set is* $\mathcal{V}(\mathbf{Y})$. *Furthermore,*

$$\mathcal{N}_\Gamma(W) \cap \mathcal{T}(V, Y) \neq \emptyset \Leftrightarrow \mathcal{N}_\Gamma(V) \cap \mathcal{T}(W, \mathbf{Y}) \neq \emptyset,$$

and if this is the case then Γ *contains a circuit which includes every edge of* $\mathbf{Y}$ *except possibly one intermediate edge, and whose vertex-set is* $\mathcal{V}(\mathbf{Y})$.

The following will also be used:

(4) *If* Γ *is a graph and* $\mathbf{Y}$ *is any terminated path of* Γ *with two distinct end-vertices* V *and* W, *then* $t(V, \mathbf{Y}) = v_\Gamma(V) - 1$ *and* $t(W, \mathbf{Y}) = v_\Gamma(W) - 1$.

Proof of (4). $\mathcal{N}_\Gamma(V) \cup \mathcal{N}_\Gamma(W) \subseteq \mathcal{V}(\mathbf{Y})$ since $\mathbf{Y}$ is terminated in Γ. One of the vertices of $\mathcal{N}_\Gamma(V)$ is the vertex first after V on $\mathbf{Y}$ (this vertex exists because $V \neq W$), the other vertices of $\mathcal{N}_\Gamma(V)$, if any, are each preceded (as $\mathbf{Y}$ is followed from V to W) by a vertex of $\mathcal{T}(V, \mathbf{Y})$. Thus $t(V, \mathbf{Y}) = v_\Gamma(V) - 1$. Similarly $t(W, \mathbf{Y}) = v_\Gamma(W) - 1$.

We now come to a simple necessary and sufficient condition for a graph to contain a Hamilton circuit.

Theorem 3.1. *Any graph* Γ *with* ≥ 3 *vertices contains a Hamilton circuit if and only if it is connected, and there exists a longest path* $\mathbf{Y}$ *of* Γ *such that if* V *and* W *denote the end-vertices of* $\mathbf{Y}$, *then* $\mathcal{N}_\Gamma(W) \cap \mathcal{T}(V, \mathbf{Y}) \neq \emptyset$ *(which is equivalent to* $\mathcal{N}_\Gamma(V) \cap \mathcal{T}(W, \mathbf{Y}) \neq \emptyset$*).*

Proof. If Γ contains a Hamilton circuit, then Γ is connected, and any path obtained from a Hamilton circuit of Γ by deleting one of its edges clearly has the property required.

To prove the converse suppose that Γ is connected and $n_\Gamma \geq 3$ and Y is a longest path of Γ as described in the theorem. $n_{\mathbf{Y}} \geq 3$ because $n_\Gamma \geq 3$ and Γ is connected. Therefore by (3), Γ contains a circuit which includes every vertex of $\mathbf{Y}$. By (2) this circuit is a Hamilton circuit of Γ.

The symmetry mentioned exists by (3). Now Theorem 3.1 is proved.

Theorem 3.2. *Any graph Γ with ≥ 3 vertices contains a Hamilton circuit if and only if it is connected and there exists a longest path $\mathbf{Y}$ of Γ such that if V and W denote the end-vertices of $\mathbf{Y}$, then either $(V, W) \in \Gamma$ or $v_\Gamma(V) + v_\Gamma(W) \geq n_{\mathbf{Y}}$ or $c(V, W) \leq v_\Gamma(V) - 2$ or $c(W, V) \leq v_\Gamma(W) - 2$.*

Proof. If Γ contains a Hamilton circuit then the condition stated in the theorem clearly holds.

To prove the converse, suppose that Γ is connected and $n_\Gamma \geq 3$ and $\mathbf{Y}$ is a longest path of Γ with end-vertices V and W and with the property stated in Theorem 3.2. $n_{\mathbf{Y}} \geq 3$ because Γ is connected and $n_\Gamma \geq 3$. Therefore if $(V, W) \in \Gamma$ then $\mathbf{Y} \cup \{(V, W)\}$ is a Hamilton circuit of Γ by (2).

In what follows assume that $(V, W) \notin \Gamma$.

Suppose now that $v_\Gamma(V) + v_\Gamma(W) \geq n_{\mathbf{Y}}$. Because $\mathbf{Y}$ is a longest path of Γ, $\mathcal{N}_\Gamma(V) \cup \mathcal{N}_\Gamma(W) \subseteq \mathcal{V}(\mathbf{Y})$. Therefore the number of vertices of $\mathbf{Y} - W$ not joined by an edge to W in Γ is $n_{\mathbf{Y}} - 1 - v_\Gamma(W)$. Since $v_\Gamma(V) + v_\Gamma(W) \geq n_{\mathbf{Y}}$ this number is $\leq v_\Gamma(V) - 1$. Therefore the number of vertices of $\mathbf{Y} - V - W$ not joined by an edge to W is $\leq v_\Gamma(V) - 2$ since $(V, W) \notin \Gamma$. Now $\mathcal{T}(V, \mathbf{Y}) \subseteq \mathcal{V}(\mathbf{Y}) - V - W$ by definition and $t(V, Y) = v_\Gamma(V) - 1$ by (4). From the last two statements it follows that $\mathcal{N}_\Gamma(W) \cap \mathcal{T}(V, Y) \neq \emptyset$. Therefore by Theorem 3.1 Γ contains a Hamilton circuit.

Suppose next that for example $c(V, W) \leq v_\Gamma(V) - 2$. Because $\mathbf{Y}$ is a longest path of Γ, $t(V, \mathbf{Y}) = v_\Gamma(V) - 1$ by (4). From this $\mathcal{N}_\Gamma(W) \cap \mathcal{T}(V, \mathbf{Y}) \neq \emptyset$, for otherwise by (1A) $\mathcal{T}(V, \mathbf{Y}) \subseteq \mathcal{C}(V, W)$, from which and $t(V, Y) = v_\Gamma(V) - 1$ it follows that $c(V, W) \geq v_\Gamma(V) - 1$, which is contrary to hypothesis. Since $\mathcal{N}_\Gamma(W) \cap \mathcal{T}(V, \mathbf{Y}) \neq \emptyset$, by Theorem 3.1 Γ contains a Hamilton circuit. Now Theorem 3.2 is proved.

From Theorem 3.2 a result can be deduced in which connectedness is not assumed.

Theorem 3.3. *If a graph Γ is finite and contains ≥ 3 vertices, and for each pair of distinct vertices V and W such that each of them is the end-vertex of some terminated path in Γ (but not necessarily of the same one) either $(V, W) \in \Gamma$ or $v_\Gamma(V) + v_\Gamma(W) \geq n_\Gamma$ or $c(V, W) \leq v_\Gamma(V) - 2$ or $c(W, V) \leq v_\Gamma(W) - 2$, then Γ contains a Hamilton circuit.*

Proof. Show that Γ is connected, and then apply Theorem 3.2.

The following Theorem can be used to deduce that a given graph contains a Hamilton circuit.

Theorem 3.4. *If a graph Γ is finite and connected and contains $\geqslant 3$ vertices, and for each pair of distinct vertices V_1 and V_2 of Γ such that $(V_1, V_2) \notin \Gamma$ either $v_\Gamma(V_1)+v_\Gamma(V_2) \geqslant n_\Gamma$ or $c'(V_1 V_2) \leqslant v_\Gamma(V_1)-2$ or $c'(V_2, V_1) \leqslant v_\Gamma(V_2)-2$ or for $i=1$ or 2*

$$\exists \{X_1, X_2\} \subseteq \mathcal{N}_\Gamma(V_i):$$

$$\forall X \in (\mathcal{N}_\Gamma(X_1) \cup \mathcal{N}_\Gamma(X_2)) \cap \mathcal{C}'(V_i, V_{i\pm 1}): v_\Gamma(X) > v_\Gamma(V_i),$$

then Γ contains a Hamilton circuit.

Proof. The proof will be based on a method first used by Pósa [6]. Suppose that Γ does not contain a Hamilton circuit (reductio ad absurdum). Among the longest paths of Γ let $\mathbf{Y}$ be one such that the sum of the valencies of its end-vertices in Γ is maximum, and let V_1 and V_2 denote the end-vertices of $\mathbf{Y}$. $n_{\mathbf{Y}} \geqslant 3$ because Γ is connected and $n_\Gamma \geqslant 3$.

Since the sum of the valencies of the end-vertices of $\mathbf{Y}$ is maximum, it follows from (3) that $\mathcal{T}(V_i, \mathbf{Y}) \subseteq \mathcal{C}'(V_i)$ for $i=1,2$.

From Theorem 3.1 $(V_1, V_2) \notin \Gamma$ and

$$\mathcal{T}(V_1, \mathbf{Y}) \cap \mathcal{N}_\Gamma(V_2) = \mathcal{T}(V_2, \mathbf{Y}) \cap \mathcal{N}_\Gamma(V_1) = \emptyset.$$

From this and the above we have that $\mathcal{T}(V_1, \mathbf{Y}) \subseteq \mathcal{C}'(V_1, V_2)$ and $\mathcal{T}(V_2, \mathbf{Y}) \subseteq \mathcal{C}'(V_2, V_1)$. Hence, by (4), $c'(V_1, V_2) \geqslant v_\Gamma(V_1)-1$ and $c'(V_2, V_1) \geqslant v_\Gamma(V_2)-1$.

By Theorem 3.2, $v_\Gamma(V_1)+v_\Gamma(V_2) \leqslant n_\Gamma - 1$.

Therefore the last alternative of Theorem 3.4 holds. Let the notation be chosen so that $X_1, X_2 \in \mathcal{N}_\Gamma(V_1)$ and

$$\forall X \in (\mathcal{N}_\Gamma(X_1) \cup \mathcal{N}_\Gamma(X_2)) \cap \mathcal{C}'(V_1, V_2): v_\Gamma(X) > v_\Gamma(V_1).$$

$\mathcal{N}_\Gamma(V_1) \subseteq \mathcal{V}(Y)$ because $\mathbf{Y}$ is a longest path of Γ, so $X_1, X_2 \in \mathcal{V}(\mathbf{Y})$. Let the notation be chosen so that $l(\mathbf{Y}(V_1, X_2)) \geqslant 2$. A $\mathbf{Y}$ is followed from V_1 to V_2 let Z be the last vertex before X_2. Then $Z \in \mathcal{N}_\Gamma(X_2) \cap \mathcal{T}(V_1, \mathbf{Y})$, therefore $Z \in \mathcal{N}_\Gamma(X_2) \cap \mathcal{C}'(V_1, V_2)$, so by hypothesis $v_\Gamma(Z) > v_\Gamma(V_1)$. But since among the longest paths of Γ, $\mathbf{Y}$ is one, the sum of the valencies of whose end-vertices is maximum, $v_\Gamma(Z) \leqslant v_\Gamma(V_1)$. This contradiction proves Theorem 3.4.

Note. The proof shows that Γ contains a Hamilton circuit if a pair of vertices V_1, V_2 exist which are not joined by an edge and are joined by a longest path of Γ and have the property stated in the theorem. However, to apply this stronger form of the theorem would require checking which pairs of vertices are joined by longest paths.

It is easy to deduce Chvátal's theorem on Hamilton circuits [3]:

Suppose that a graph Γ is finite and contains $\geqslant 3$ vertices, and $s(\leqslant j)$ and $s(\geqslant j)$ denote the number of vertices with valency $\leqslant j$ and $\geqslant j$ in Γ, respectively. If for each integer i with $1 \leqslant i \leqslant \frac{1}{2}(n-1)$ either $s(\leqslant i) \leqslant i-1$ or $s(\geqslant n-i) \geqslant i+1$, then Γ contains a Hamilton circuit. For suppose that this is note true, then it may be assumed that adding any new edge to Γ creates a Hamilton circuit. But then clearly Γ is connected, and if V_1 and V_2 are any two independent vertices the sum of whose valencies in Γ is maximal, then this sum is $\leqslant n_\Gamma - 1$, and also $c'(V_1, V_2) \geqslant v_\Gamma(V_1) - 1$ and $c'(V_2, V_1) \geqslant v_\Gamma(V_2) - 1$ and by assuming w.l.g. that $v_\Gamma(V_1) \leqslant v_\Gamma(V_2)$ we arrive at a contradiction.

Unlike Chváral's theorem and Pósa's theorem [5], Theorems 3.1–3.4 set no upper bound for any $i > 0$ on the number of vertices having valency $\leqslant i$ in the graph.

4. Long circuits in 2-connected graphs

The only 2-connected graphs which contain no circuit of length >3 are the 3-circuits, as may be easily verified. Also, in any graph each circuit is contained in a 2-connected spanned subgraph. In this section some sufficient conditions will be established for any integer $f \geqslant 4$ that a finite or infinite 2-connected graph with $>f$ vertices should contain a circuit of length $\geqslant f$. In addition the graphs which satisfy such conditions and contain no circuit of length $>f$ will be classified. For this purpose some special results will be proved. We require some more.

Definition 4.1. A graph which contains x vertices, and each pair of distinct vertices are joined by one edge, where x may be finite or infinite, will be denoted by $\langle x \rangle$. A graph which consists of an $\langle x \rangle$, and of y further vertices each of them joined by an edge only to every vertex of the $\langle x \rangle$, where x and y may be finite or infinite, will be denoted by $\langle\langle x \rangle, y\rangle$, and the $\langle x \rangle$ will be called the *core*. If $y > 1$ and a new edge is added to a $\langle\langle x \rangle, y\rangle$, the resulting graph will be denoted by $\langle\langle x \rangle, y+\rangle$ and the $\langle x \rangle$ will be called the *core*. (The new edge is not incident with any vertex of the core.) If Θ is a circuit and $\mathbf{Y}$ is any path of length $\geqslant 1$ such that $\mathbf{Y} \cap \Theta$ consists of the two end-vertices of $\mathbf{Y}$, then $\mathbf{Y}$ will be called a *chord* of Θ.

It is easy to verify the following two results (5) and (6).

(5) *Suppose that Γ is any graph and Θ is any longest circuit of Γ and $\mathbf{Y}$ is any chord of Θ with end-vertices M and N. Let $\mathbf{Y}_1$ and $\mathbf{Y}_2$ denote the two (M, N)-paths whose union is Θ. Then:*

(A) *$(M, N) \notin \Theta$ and $l(\mathbf{Y}) \leqslant l(\mathbf{Y}_1), l(\mathbf{Y}_2)$.*

(B) *If R_1 is an interior vertex of $\mathbf{Y}_1$ and R_2 is an interior vertex of $\mathbf{Y}_2$, and*

$$l(\mathbf{Y}_1(M, R_1)) + l(\mathbf{Y}_2(N, R_2)) < l(\mathbf{Y}) + g,$$

where g is an integer $\geqslant 1$, *then* Γ *does not contain any* (R_1, R_2)-*chord of* Θ *which has length* $\geqslant g$ *and is disjoint with* $\mathbf{Y}$. *In particular, if*

$$l(Y_1(M, R_1)) + l(\mathbf{Y}_2(N, R_2)) \leqslant l(\mathbf{Y}),$$

then Γ *does not contain any* (R_1, R_2)-*chord of* Θ *disjoint with* $\mathbf{Y}$, *so* $(R_1, R_2) \notin \Gamma$.

(6) *Suppose that* Γ *is a connected graph and a longest circuit of* Γ *has the same length as a longest path of* Γ. *Then if* Θ *is any longest circuit of* Γ, $\Gamma - \Theta$ *consists of one or more isolated vertices, and if two vertices are joined by an edge of* Θ, *then at least one of them is joined to no vertex of* $\Gamma - \Theta$ *in* Γ.

Theorem 4.2. *Suppose that* Γ *is a graph and* Θ *is a longest circuit of* Γ *and* $n_\Theta = f$, *where* f *is even. Then*

$$\forall V \in \mathcal{V}(\Gamma) - \mathcal{V}(\Theta)\colon\ e(V, \Theta; \Gamma) \leqslant \tfrac{1}{2}f.$$

Suppose that $V \in \mathcal{V}(\Gamma) - \mathcal{V}(\Theta)$ *and* $e(V, \Theta; \Gamma) = \frac{1}{2}f$. *Then the notation can be chosen so that the vertices of* Θ *are in cyclic order* $V_1, W_1, \ldots, V_{f/2}, W_{f/2}$ *and* V *is joined to* $W_1, \ldots, W_{f/2}$. *If in addition* Γ *is 2-connected, then* $\Gamma \subseteq \langle\langle \frac{1}{2}f\rangle, n_\Gamma - \frac{1}{2}f\rangle$, *the vertices of the core being* $W_1, \ldots, W_{f/2}$.

Proof. It follows at once from (5A) that $e(V, \Theta; \Gamma) \leqslant \frac{1}{2}f$ and that if $e(V, \Theta; \Gamma) = \frac{1}{2}f$ then V is joined to every other vertex of Θ and the notation can be chosen as described. Suppose now that this is the case, and that Γ is 2-connected. Then:

(i) The only (V, Θ)-paths in Γ are in the $\frac{1}{2}f$ paths of length 1 which join V to $W_1, \ldots, W_{f/2}$.

For by (5A) no (V, Θ)-path ends in $V_1, \ldots, V_{f/2}$ and no (V, Θ)-path of length >1 ends in $W_1, \ldots, W_{f/2}$.

(ii) No chord of Θ joins two of $V_1, \ldots, V_{f/2}$.

For suppose that a chord of Θ joins V_1 and V_i. By (i) this chord does not contain V. But then we have a contradiction to (5B) with $\mathcal{V}(\mathbf{Y}) = \{W_1, V, W_i\}$. Now (ii) is proved.

(iii) No chord of Θ of length $\geqslant 2$ joins one of $V_1, \ldots, V_{f/2}$ to one of $W_1, \ldots, W_{f/2}$.

For suppose that such a chord joins V_1 and W_i. Then $i \neq 1, \frac{1}{2}f$ from the maximality of Θ, and V does not belong to the chord by (i). But then we have a contradiction to (5B) with $\mathcal{V}(\mathbf{Y}) = \{W_1, V, W_i\}$. Now (iii) is proved.

(iv) No chord of Θ of length $\geqslant 3$ joins two of $W_1, \ldots, W_{f/2}$.

For suppose that such a chord joins W_1 and W_i. Then $3 \leqslant i \leqslant \frac{1}{2}f - 1$ from the maximality of Θ, and V does not belong to the chord by (i). But then we have a contradiction to (5B) with $\mathcal{V}(\mathbf{Y}) = \{W_2, V, W_{i+1}\}$. Now (iv) is proved.

(v) No two vertices of $\mathcal{V}(\Gamma) - W_1 - \cdots - W_{f/2}$ are joined by an edge.

For suppose on the contrary that X and Y are to such vertices and $(X, Y) \in \Gamma$.

$X \neq V$ and $X \neq V$. For if e.g. $X = V$ then $Y \notin \Theta$ by (i). Therefore, since Γ is

2-connected, $\Gamma - V$ contains a (Y, Θ)-path of length ≥ 1. Consequently Γ contains a (V, Θ)-path of length ≥ 2. But this is contrary to (i), therefore $X \neq V$ and $Y \neq V$.

$X \notin \Theta$ and $Y \notin \Theta$. For if e.g. $X \in \Theta$ then by (ii) and the above, $Y \notin \mathcal{V}(\Theta) \cup \{V\}$. It may be assumed that $X = V_1$. $\Gamma - X$ contains a (Y, Θ)-path of length ≥ 1 since Γ is 2-connected. Consequently Γ contains a chord of Θ of length ≥ 2 with V_1 as an end-vertex. But this is not the case by (ii) and (iii), so $X \notin \Theta$ and $Y \notin \Theta$.

Since $\{X, Y\} \cap (\Theta \cup \{V\}) = \emptyset$ and $(X, Y) \in \Gamma$ and Γ is 2-connected, there exists a chord of Θ containing (X, Y), and this has length ≥ 3. But by (ii), (iii) and (iv) no chord of Θ has length ≥ 3. This contradiction proves (v).

By (v), $\Gamma \subseteq \langle\langle \frac{1}{2}f \rangle, n_\Gamma - \frac{1}{2}f \rangle$. Now Theorem 4.2 is proved.

If the longest circuit of the graph has odd length, then there are more alternatives:

Theorem 4.3. *Suppose that Γ is a graph and Θ is a longest circuit of Γ and $n_\Theta = f$, where f is odd. Then*

$$\forall V \in \mathcal{V}(\Gamma) - \mathcal{V}(\Theta)\colon e(V, \Theta; \Gamma) \leq \tfrac{1}{2}(f-1).$$

Suppose that $V \in \mathcal{V}(\Gamma) - \mathcal{V}(\Theta)$ and $e(V, \Theta; \Gamma) = \frac{1}{2}(f-1)$. Then the notation may be chosen so that the vertices of Θ in cyclic order are $V_1, W_1, \ldots, V_{(f-1)/2}, W_{(f-1)/2}, V_{(f+1)/2}$ and V is joined to $W_1, \ldots, W_{(f-1)/2}$. If in addition Γ is 2-connected, then

either $\Gamma \subseteq \langle\langle \frac{1}{2}(f-1) \rangle, n_\Gamma - \frac{1}{2}(f-1) + \rangle$,

or $f \geq 5$ and Γ is a subgraph of a graph obtained from a $\langle\langle \frac{1}{2}(f-1) \rangle, m \rangle$ with $\frac{1}{2}(f-1) \leq m \leq n_\Gamma - \frac{1}{2}(f+3)$ by selecting one vertex in the core and one vertex not in the core, and taking two or more new vertices and joining each of them to just the two selected vertices,

or $f \geq 7$ and Γ is a subgraph of a graph obtained from a $\langle\langle \frac{1}{2}(f-1) \rangle, m \rangle$ with $\frac{1}{2}(f-3) \leq m \leq n_\Gamma - \frac{1}{2}(f-1) - 4$ by selecting two distinct vertices W_1 and $W_{(f-1)/2}$ in its core, and taking two or more trees, each consisting of a vertex joined to ≥ 1 others (pairwise disjoint and disjoint with the $\langle\langle \frac{1}{2}(f-1) \rangle, m \rangle$), and if such a tree has just 2 vertices then joining both of them to W_1 and to $W_{(f-1)/2}$, while if a tree has ≥ 3 vertices then joining the vertex having valency > 1 in the tree to both W_1 and $W_{(f-1)/2}$ and joining all the vertices having valency 1 in the tree to just one of W_1 and $W_{(f-1)/2}$.

If in addition Γ is 3-connected or contains at most one vertex of valency ≤ 3, then

$$\Gamma \subseteq \langle\langle \tfrac{1}{2}(f-1) \rangle, n_\Gamma - \tfrac{1}{2}(f-1) + \rangle.$$

Proof. From (5A) it follows that $e(V, \Theta; \Gamma) \leq \frac{1}{2}(f-1)$ and that if $e(V, \Theta; \Gamma) = \frac{1}{2}(f-1)$, then the notation may be chosen as described. Suppose now that this is the case. Then

(i) The only (V, Θ)-paths in Γ are the $\frac{1}{2}(f-1)$ paths of length 1 which join V to $W_1, \ldots, W_{(f-1)/2}$.

The reason is the same as in the proof of Theorem 4.2.

(ii) No chord of Θ joins two of $V_1, \ldots, V_{(f+1)/2}$.

For suppose that such a chord joins V_i and V_j. By (i) it does not contain V. If $i=1$ then $2 \leqslant j \leqslant \frac{1}{2}(f-1)$, therefore (5B) with $\mathscr{V}(\mathbf{Y})=\{W_1, V, W_j\}$ is contradicted, therefore $i \neq 1$, and similarly $j \neq 1$. But then (5B) with $\mathscr{V}(\mathbf{Y})=\{W_i, V, W_j\}$ is contradicted. Now (ii) is proved.

(iii) No chord of Θ of length $\geqslant 2$ joins one of $V_1, \ldots, V_{(f+1)/2}$ to one of $W_1, \ldots, W_{(f-1)/2}$ except possibly when $f \geqslant 5$, $(V, W_{(f-1)/2})$-chords of length 2 or $(V_{(f+1)/2}, W_1)$-chords of length 2, and if Γ contains a $(V_1, W_{(f-1)/2})$-chord of Θ of length 2 then no chord of Θ has $V_{(f+1)/2}$ as an end-vertex, and if Γ contains a $(V_{(f+1)/2}, W_1)$-chord of Θ of length 2 then no chord of Θ has V_1 as an end-vertex.

For suppose that a chord of Θ of length $\geqslant 2$ joins V_i and W_j. Then $f \geqslant 5$, $(V_i, W_j) \notin \Theta$ by (5A). By (i) the chord does not contain V.

$i=1$ or $i=\frac{1}{2}(f+1)$. For suppose that $2 \leqslant i \leqslant \frac{1}{2}(f-1)$. Then if $i>j$ we have a contradiction to (5B) with $\mathscr{V}(\mathbf{Y})=\{W_{j+1}, V, W_i\}$ and if $i<j$ we have a contradiction to (5B) with $\mathscr{V}(\mathbf{Y})=\{W_{i-1}, V, W_{j-1}\}$. Hence $i=1$ or $i=\frac{1}{2}(f+1)$.

If $i=1$, then $j=\frac{1}{2}(f-1)$. For if $2 \leqslant j \leqslant \frac{1}{2}(f-3)$, then (5B) with $\mathscr{V}(\mathbf{Y})=\{W_1, V, W_j\}$ is contradicted. Similarly, if $i=\frac{1}{2}(f+1)$, then $j=1$.

All $(V_1, W_{(f-1)/2})$-chords of Θ and all $(V_{(f+1)/2}, W_1)$-chords of Θ (if any) have length $\leqslant 2$ by (5A).

It has now been proved that no chord of Θ of length $\geqslant 2$ joins one of $V_1, \ldots, V_{(f+1)/2}$ to one of $W_1, \ldots, W_{(f-1)/2}$ except possibly when $f \geqslant 5$ $(V_1, W_{(f-1)/2})$-chords of length 2 or $(V_{(f+1)/2}, W_1)$-chords of length 2.

Suppose that Γ contains a $(V_1, W_{(f-1)/2})$-chord of Θ of length 2, and let U denote the intermediate vertex of any such chord. Then no chord of Θ has $V_{(f+1)/2}$ as an end-vertex. For suppose on the contrary that Y' is such a chord of Θ. By (5A) $U \notin Y'$ and the other end-vertex of Y' is neither V_1 nor $W_{(f-1)/2}$. Therefore by (ii) the other end-vertex of Y' is one of $W_1, \ldots, W_{(f-3)/2}$, say W_i. But then the union of the path whose vertices in order are $V_1, U, W_{(f-1)/2}, V_{(f-1)/2}, \ldots, W_{i+1}, V, W_1, V_2, \ldots, W_i$, and of Y' and $(V_1, V_{(f+1)/2})$ is a circuit in Γ of length $>f$ by (i) and since $U \notin Y'$. But this is contrary to the maximality of Θ. Thus if Γ contains a $(V, W_{(f-1)/2})$-chord of length 2 of Θ then no chord of Θ has $V_{(f+1)/2}$ as an end-vertex. Similarly, if Γ contains a $(V_{(f+1)/2}, W_1)$-chord of Θ of length 2 then no chord of Θ has V_1 as an end-vertex. Now (iii) is proved.

(iv) No chord of Θ of length $\geqslant 3$ joins two of $W_1, \ldots, W_{(f-1)/2}$ except possibly when $f \geqslant 7$, $(W_1, W_{(f-1)/2})$-chords of length 3, and if Γ contains such a chord then no chord of Θ has V_1 as an end-vertex except possibly $(V_1, W_{(f-1)/2})$-chords of length $\leqslant 2$, and no chord of Θ has $V_{(f+1)/2}$ as an end-vertex except possibly $(V_{(f+1)/2}, W_1)$-chords of length $\leqslant 2$.

For suppose that a chord of Θ of length $\geqslant 3$ joins W_i and W_j, where $i<j$. By (i) it does not contain V. Then by (5A), $f \geqslant 7$ and $j-i \geqslant 2$. $i=1$, because if $i \neq 1$ then $2 \leqslant i \leqslant j-2$, and therefore (5B) with $\mathscr{V}(\mathbf{Y})=\{W_{i-1}, V, W_{j-1}\}$ is contradicted. Similarly $j=\frac{1}{2}(f-1)$. Then the length of the chord is 3 by (5A).

Suppose that Γ contains a $(W_1, W_{(f-1)/2})$-chord of Θ having length 3, and let S and T denote the two intermediate vertices of any such chord. No chord of Θ has V_1 as an end-vertex except possibly $(V_1, W_{(f-1)/2})$-chords of length ≤ 2. For if this is false, then by (5A) and (ii) there is a (V_1, W_i)-chord of Θ with $2 \leq i \leq \frac{1}{2}(f-3)$, say Y'. By (5A), $V, S, T \notin \mathbf{Y}'$. Therefore the union of $\mathbf{Y}'$ and of the path whose vertices are $W_i, V_{i+1}, \ldots, W_{(f-3)/2}, V, W_{i-1}, V_{i-1}, \ldots, W_1, S, T, W_{(f-1)/2}, V_{(f+1)/2}$ is a circuit in Γ with length $>f$. But this is contrary to the maximality of Θ. Thus no chord of Θ has V_1 as an end-vertex except possibly $(V_1, W_{(f+1)/2})$-chords of length ≤ 2. Similarly no chord of Θ has $V_{(f+1)/2}$ as an end-vertex except possibly $(V_{(f+1)/2}, W_1)$-chords of length ≤ 2. Now (iv) is proved.

Suppose now that Γ is 2-connected.

If Γ contains neither a $(V_1, W_{(f-1)/2})$-chord of Θ of length 2, nor a $(V_{(f+1)/2}, W_1)$-chord of Θ of length 2, nor a $(W_1, W_{(f-1)/2})$-chord of Θ of length 3, then by a reasoning similar to that used to establish (v) in the proof of Theorem 4.2 it is seen that

$$\mathscr{E}(\Gamma - W_1 - \cdots - W_{(f-1)/2}) = \{(V_1, V_{(f+1)/2})\}.$$

From this it follows that in this case

$$\Gamma \subseteq \langle\langle \tfrac{1}{2}(f-1)\rangle, n_\Gamma - \tfrac{1}{2}(f-1) + \rangle,$$

the vertices of the core being $W_1, \ldots, W_{(f-1)/2}$.

Suppose that Γ contains one or more $(V_1, W_{(f-1)/2})$-chords of Θ of length 2, but no $(W_1, W_{(f-1)/2})$-chord of length 3. Then by (i), (ii), (iii) and (5B) $V_{(f+1)/2}$ is joined only to V_1 and to $W_{(f-1)/2}$ in Γ, and the same is true for the intermediate vertex of each $(V_1, W_{(f-1)/2})$-chord of Θ of length 2. Therefore in this case the second alternative of Theorem 4.3 holds. Similarly, if Γ contains one or more $(V_{(f+1)/2}, W_1)$-chords of Θ of length 2 but no $(W_1, W_{(f-1)/2})$-chord of length 3, then the second alternative of Theorem 4.3 holds. The vertices of the core are $W_1, \ldots, W_{(f-1)/2}$.

Suppose that Γ contains one or more $(W_1, W_{(f-1)/2})$-chords of Θ of length 3. Let $\mathbf{Y}_0$ be any such chord, and the vertices of $\mathbf{Y}_0$ in order $W_1, V', V'', W_{(f-1)/2}$. By considering the circuit $(\Gamma - V_1 - V_{(f+1)/2}) \cup \mathbf{Y}_0$ in place of Θ it is seen that (i)–(iv) with V' in place of V_1 and V'' in place of $V_{(f+1)/2}$ apply to it. It follows that *either* the connected component of $\Gamma - W_1 - W_{(f+1)/2}$ containing V' and V'' is a $\langle 2\rangle$ and V', V'' are joined to no vertex of Γ other than W_1 and $W_{(f-1)/2}$, *or* this connected component is a tree with ≥ 3 vertices in which V' is joined to all the other vertices, and in Γ V' is joined only to W_1 and perhaps to $W_{(f-1)/2}$ as well, while the other vertices of the tree are joined only to $W_{(f-1)/2}$ *or* we have the situation just described but with V' replaced by V'' and $W_{(f-1)/2}$ by W_1. The same applies to the connected component of $\Gamma - W_1 - W_{(f-1)/2}$ containing V_1 and $V_{(f+1)/2}$ of course. Thus in this case the third alternative of Theorem 4.3 holds, the vertices of the core being again $W_1, \ldots, W_{(f-1)/2}$.

It is easy to see that if the second or the third alternative of Theorem 4.3 holds, then Γ is not 3-connected, and contains two or more vertices having valency ≤ 3. Now Theorem 4.3 is proved.

It is worth mentioning that:

(7) *If Γ is a graph and Θ is a longest circuit of Γ and V and W are two vertices of $\Gamma-\Theta$ such that $\Gamma-\Theta$ contains a (V, W)-path of length p and $e(V, \Theta; \Gamma)>0$ and $e(W, \Theta; \Gamma)>0$, then*

$$e(V, \Theta; \Gamma)+e(W, \Theta; \Gamma)\leq \tfrac{1}{2}n_{\Theta}-p.$$

The next theorem is concerned with longest paths and circuits, with particular attention to the extreme cases.

Theorem 4.4. *Let f be any integer >4. Suppose that Γ is a (finite or infinite) graph, and $n_\Gamma>f$, and $\mathbf{Y}$ is a longest path of Γ such that if A and B denote the end-vertices of $\mathbf{Y}$ and a and b denote $v_\Gamma(A)$ and $v_\Gamma(B)$, respectively, then $a+b=f$. As $\mathbf{Y}$ is followed from A to B let A' denote the last vertex which belongs to $\mathcal{N}_\Gamma(A)$ and B' the first vertex which belongs to $\mathcal{N}_\Gamma(B)$.*

(A) *If $a=f-1$ or $b=f-1$, then Γ contains a circuit of length $\geq f$.*

(B) *If $A'=B'$ and Γ is 2-connected, then Γ contains a circuit of length $\geq f+1$.*

(C) *If $Y(A, A')\cap Y(B, B')=\emptyset$ and Γ is 2-connected, then Γ contains a circuit of length $\geq f+2$.*

(D) *If $\mathbf{Y}(A, A')$ and $\mathbf{Y}(B, B')$ have more than one vertex in common, then $\Gamma(\mathbf{Y})$ contains a circuit of length $\geq f$. Moreover if Γ contains no circuit of length $>f$ and Γ is 2-connected, then $\Gamma(Y)$ contains a circuit Θ of length f such that $\Theta\cap Y= Y(A, P)\cup Y(B, Q)$ with $P\neq A, B$ and $Q\neq A, B$ and $l(Y(P, Q)\geq 2$, and either*

(1) *$f=l(\mathbf{Y})$ and (6) holds for Γ, or*

(2) *$8\leq f$ and $3\leq l(\mathbf{Y}(P, Q))\leq l(\mathbf{Y}(A, B'))+1$, $l(\mathbf{Y}(B, A'))+1$, $\mathbf{Y}(P, Q)$ is a longest (P, Q) path of $\Gamma-(\Theta-P-Q)$ and $\Gamma-\Theta$ contains no circuit of length $\geq 2l(Y(P, Q))-3$, and $\Gamma(\Theta)$ contains circuits of all lengths from 3 to f, and either*

(2.1) *$P=B'$ and $Q=A'$ and $\Gamma=\bigcup_{i\in\mathcal{I}}\Gamma_i$, where $|\mathcal{I}|\geq 3$, $\forall i\in\mathcal{I}:\Gamma_i$ is a subgraph of Γ such that $P, Q\in\Gamma_i$ and Γ_i-P-Q is connected and non-empty, and*

$$\forall i_1\neq i_2\in\mathcal{I}: \mathcal{V}(\Gamma_{i_1}\cap\Gamma_{i_2})=\{P, Q\}$$

and if $\mathbf{Y}(A, P)$, $\mathbf{Y}(B, Q)$, $\mathbf{Y}(P, Q)$ belong to Γ_{i_1}, Γ_{i_2}, Γ_{i_3}, respectively, then $i_1\neq i_2\neq i_3\neq i_1$ and $\Gamma(\Theta)=\Gamma_{i_1}\cup\Gamma_{i_2}$, or

(2.2) *$\{A', B'\}\neq\{P, Q\}$ and $\Gamma=\Gamma_1\cup\Gamma_2$, where Γ_1 and Γ_2 are subgraphs of Γ, $V(\Gamma_1\cap\Gamma_2)=\{P, Q\}$, Γ_1-P-Q and Γ_2-P-Q are connected, and $\Gamma_1=\Gamma(\Theta)$ and $Y(P, Q)\subseteq\Gamma_2$.*

If in addition Γ contains no circuit of length $>f$ and Γ is 3-connected, then (1) is the case. Then if Γ contains a vertex with valency $[\frac{1}{2}f]$ which does not belong to every longest circuit of Γ, then

$$\Gamma\subseteq\langle\langle\tfrac{1}{2}f\rangle, n_\Gamma-\tfrac{1}{2}f\rangle$$

if f is even, and

$$\Gamma \subseteq \langle\langle \tfrac{1}{2}(f-1)\rangle\rangle,\ n_\Gamma - \tfrac{1}{2}(f-1)+\rangle$$

if f is odd.

In all four cases (A), (B), (C), (D), Γ *contains circuits of* $\geq f-2$ *different lengths.*

Proof. $l(\mathbf{Y}(A, A')) \geq a$ and $l(\mathbf{Y}(B, B')) \geq b$, since $\mathbf{Y}$ is a longest path and so $\mathcal{N}_\Gamma(A) \cup \mathcal{N}_\Gamma(B) \subseteq \mathbf{Y}$.

(A) This is obvious.

(B) Suppose that $A' = B'$ and Γ is 2-connected. Then $a, b \geq 2$. $\Gamma - A'$ is connected because Γ is 2-connected. Hence $\Gamma - A'$ contains a $(\mathbf{Y}(A, A') - A',\ \mathbf{Y}(B, B') - B')$-path. Let $\mathbf{Y}_1$ be any such path in $\Gamma - A'$ and let A'' denote the end-vertex of $\mathbf{Y}_1$ on $\mathbf{Y}(A, A')$ and B'' the other end-vertex of $\mathbf{Y}_1$, and as we follow $\mathbf{Y}$ from A to A' let A_1 denote the first vertex after A'' which belongs to $\mathcal{N}_\Gamma(A)$, and similarly define the vertex B_1 on $\mathbf{Y}(B, B')$. Clearly A_1 and B_1 exist. Then

$$\mathbf{Y}(A'', A) \cup \{(A, A_1)\} \cup \mathbf{Y}(A_1, A') \cup \mathbf{Y}(A', B_1) \cup \{(B, B_1)\} \cup \mathbf{Y}(B, B'') \cup \mathbf{Y}_1$$

is a circuit of Γ of length $\geq f+1$. Now (B) is proved.

(C) Suppose that $\mathbf{Y}(A, A') \cap \mathbf{Y}(B, B') = \emptyset$ and Γ is 2-connected. Then $a, b \geq 2$. So, since Γ is 2-connected, Γ contains two disjoint $(\mathbf{Y}(A, A'), \mathbf{Y}(B, B'))$-paths. In addition Γ contains $\mathbf{Y}(A', B')$. Therefore by Theorem 2.1, Γ contains two disjoint $(\mathbf{Y}(A, A'), \mathbf{Y}(B, B'))$-paths whose union includes A' and B'. Let $\mathbf{Y}_1$ and $\mathbf{Y}_2$ be any two such paths, and put $(\mathbf{Y}_1 \cup \mathbf{Y}_2) \cap \mathbf{Y}(A, A') = \{A', A''\}$ and $(\mathbf{Y}_1 \cup \mathbf{Y}_2) \cap \mathbf{Y}(B, B') = \{B', B''\}$. As we follow Y from A to A' let A_1 denote the first vertex after A'' which belongs to $\mathcal{N}_\Gamma(A)$, and as we follow $\mathbf{Y}$ from B to B' let B_1 denote the first vertex after B'' which belongs to $\mathcal{N}_\Gamma(B)$. Clearly A_1 and B_1 exist. Then

$$\mathbf{Y}(A'', A) \cup \{(A, A_1)\} \cup \mathbf{Y}(A_1, A') \cup \mathbf{Y}(B'', B) \cup \{(B, B_1)\} \cup \mathbf{Y}(B_1, B') \cup \mathbf{Y}_1 \cup \mathbf{Y}_2$$

is a circuit of Γ with length $\geq f+2$. Now (C) is proved.

(D) Suppose that $\mathbf{Y}(A, A') \cap \mathbf{Y}(B, B')$ contains more than one vertex. Then clearly

(i) $a \geq 2$ and $b \geq 2$.

If $\mathcal{N}_\Gamma(A) \cap \mathcal{T}(B, \mathbf{Y}) \neq \emptyset$, i.e. if $\mathcal{N}_\Gamma(B) \cap \mathcal{T}(A, \mathbf{Y}) \neq \emptyset$ (which is the case if $(A, B) \in \Gamma$), then $\Gamma(\mathbf{Y})$ contains a Hamilton circuit of Γ by Theorem 3.1, and such a circuit has length $n_\Gamma > f$. Therefore from now on suppose that:

(ii) $(A, B) \notin \Gamma$ and $\mathcal{N}_\Gamma(A) \cap \mathcal{T}(B, \mathbf{Y}) = \mathcal{N}_\Gamma(B) \cap \mathcal{T}(A, \mathbf{Y}) = \emptyset$.

Among all ordered pairs $[X, Y] \subseteq \mathcal{V}(\mathbf{Y})$ such that $X \in \mathbf{Y}(A, Y)$ and (A, Y), $(B, X) \in \Gamma$ and

$$(\mathbf{Y}(X, Y) - X - Y) \cap (\mathcal{N}_\Gamma(A) \cup \mathcal{N}_\Gamma(B)) = \emptyset$$

let $[P, Q]$ be one such that $l(\mathbf{Y}(P, Q))$ is a minimum. $[P, Q]$ exists because $B' \in \mathbf{Y}(A, A') - A'$. If $\mathbf{Y}$ is followed from A to B let P' denote the first vertex after P and Q' the last vertex before Q. From the definition of P and Q and (ii) there follows:

(iii) A, P, P', Q, B are five distinct vertices of $\mathbf{Y}$ and occur on $\mathbf{Y}$ in this order, similarly for A, P, Q', Q, B, furthermore $l(\mathbf{Y}(P, Q)) \geqslant 2$ and

$$(\mathcal{N}_\Gamma(A) \cup \mathcal{N}_\Gamma(B)) \cap \mathbf{Y}(P', Q') = \emptyset.$$

Now put

$$\mathbf{Y}(A, P) \cup \mathbf{Y}(B, Q) \cup \{(A, Q), (B, P)\} = \Theta,$$

then Θ is a circuit of Γ. Since

$$(\mathcal{N}_\Gamma(A) \cup \mathcal{N}_\Gamma)) \cap \mathbf{Y}(P', Q') = \emptyset,$$

we have

(iv) $\{A\} \cup \mathcal{N}_\Gamma(A) \cup \{B\} \cup (\mathcal{T}(B, \mathbf{Y}) - P') \subseteq \mathcal{V}(\Theta),$

$\{B\} \cup \mathcal{N}_\Gamma(B) \cup \{A\} \cup (\mathcal{T}(A, \mathbf{Y}) - Q') \subseteq \mathcal{V}(\Theta).$

From (ii) and (iv) we have that $l(\Theta) \geqslant a + b = f$.

It has now been proved that $\Gamma(\mathbf{Y})$ contains a Hamilton circuit of Γ or the circuit Θ of length $\geqslant f$.

Suppose that Γ contains no circuit of length $> f$. Then (ii), (iii) and (iv) hold, and Θ is a longest circuit of Γ and $l(\Theta) = f$. Therefore

(v) $\{A\} \cup \mathcal{N}_\Gamma(A) \cup \{B\} \cup (\mathcal{T}(B, Y) - P') = \mathcal{V}(\Theta),$

$\{B\} \cup \mathcal{N}_\Gamma(B) \cup \{A\} \cup (\mathcal{T}(A, Y) - Q') = \mathcal{V}(\Theta).$

For convenience put $l(Y(P, Q)) = r$. $r \leqslant l(Y(A, B')) + 1$ and $r \leqslant l(Y(B, A')) + 1$ because Θ is a longest circuit of Γ.

The alternatives (1) $r = 2$, (2) $r \geqslant 3$ will be considered in turn.

(1) Suppose that $r = 2$. Then $l(\Theta) = l(\mathbf{Y})$ and (6) applies to Γ.

(2) Suppose that $r \geqslant 3$. Since $\mathbf{Y}$ is a longest path of Γ there is no (P, Q)-path in $\Gamma - (\Theta - P - Q)$ longer than $\mathbf{Y}(P, Q)$.

Next it will be shown that

(vi) $B' \in \mathbf{Y}(A, P) - A, \quad (A, B') \in \Gamma;$

$\forall V \in \mathcal{V}(\mathbf{Y}(A, B')) - A - B' : (A, V) \in \Gamma \quad \text{and} \quad (B, V) \notin \Gamma;$

if $P \neq B'$ then

$\forall V \in \mathcal{V}(\mathbf{Y}(B', P)) - B' : \quad (A, V) \notin \Gamma \quad \text{and} \quad (B, V) \in \Gamma;$

$A' \in \mathbf{Y}(Q, B) - B, \quad (B, A') \in \Gamma;$

$\forall V \in \mathcal{V}(Y(B, A')) - B - A' : (B, V) \in \Gamma \quad \text{and} \quad (A, V) \notin \Gamma;$

if $Q \neq A'$ then

$\forall V \in \mathcal{V}(Y(A', Q)) - A' : \quad (B, V) \notin \Gamma \quad \text{and} \quad (A, V) \in \Gamma.$

For $B' \in \mathbf{Y}(A, P) - A$ from the definition of B' and P and since $(A, B) \notin \Gamma$. From (v),

$$\forall V \in \mathscr{V}(\Theta) - A - B: \qquad (A, V) \in \Gamma \quad \text{or} \quad V \in \mathscr{T}(B, \mathbf{Y}).$$

If $V \in \mathscr{V}(\mathbf{Y}(A, B')) - A$, then $V \notin \mathscr{T}(B, \mathbf{Y})$ from the definition of B', hence $(A, V) \in \Gamma$. From this and (ii),

$$\forall V \in \mathscr{V}(\mathbf{Y}(A, B')) - B': \qquad (B, V) \notin \Gamma.$$

The first two statements if (vi) are now proved.

Suppose that $P \neq B'$. If $\mathbf{Y}$ is followed from A to B let C denote the first vertex after B' and D the first vertex after C. $(A, C) \notin \Gamma$ by (ii). If $C = P$, then $(B, C) \in \Gamma$ of course. If $C \neq P$, then $D \in \Theta$, hence by (v), $(A, D) \in \Gamma$ or $D \in \mathscr{T}(B, Y)$. Therefore if $(B, C) \notin \Gamma$, then $(A, D) \in \Gamma$. But if $(A, C) \notin \Gamma$ and $(B, C) \notin \Gamma$ and $(A, D) \in \Gamma$, then, since $(B, B') \in \Gamma$, the definition of $[P, Q]$ is contradicted. This proves that $(B, C) \in \Gamma$. Thus $(A, C) \notin \Gamma$ and $(B, C) \in \Gamma$. From this and (ii) $(A, D) \notin \Gamma$. If $D \neq P$, then the above argument with B' replaced by C and C by D shows that $(B, D) \in \Gamma$. Repeating this step by step until P is reached proves the third statement of (vi).

The remaining statements of (vi) follow by symmetry.

It follows from (vi) that $\Gamma(\Theta)$ contains circuits of all lengths from 3 to f.

We will now consider the two cases (2.1) $P = B'$ and $Q = A'$, and (2.2) $P \neq B'$ or $Q \neq A'$.

(2.1) Suppose that $P = B'$ and $Q = A'$ and Γ is 2-connected.

We will first show that:

(vii) $\mathbf{Y}(A, P) - P$, $\mathbf{Y}(B, Q) - Q$ and $\mathbf{Y}(P, Q) - P - Q$ belong to three different connected components of $\Gamma - P - Q$, and each connected component of $\Gamma - P - Q$ is in Γ joined to P and to Q.

Proof of (vii). $\Gamma - P - Q$ contains no $(\Theta, \mathbf{Y}(P, Q))$-path. For suppose that $\mathbf{Y}'$ is such a path. It may be assumed that the end-vertices of $\mathbf{Y}'$ are $V \in \mathbf{Y}(A, P) - P$ and $W \in \mathbf{Y}(P, Q) - P - Q$. When $\mathbf{Y}$ is followed from A to B let V' denote the first vertex after V. By (vi), Γ contains the circuit

$$\mathbf{Y}(V, A) \cup \{(A, V')\} \cup \mathbf{Y}(V', P) \cup \{(B, P)\} \cup \mathbf{Y}(B, W) \cup \mathbf{Y}'$$

and this circuit contains $\mathscr{V}(\Theta)$ and W, therefore it is longer than Θ, which is contrary to hypothesis. Hence $\Gamma - P - Q$ contains no $(\Theta, \mathbf{Y}(P, Q))$-path.

$\Gamma - P - Q$ contains no $(\mathbf{Y}(A, P), \mathbf{Y}(B, Q))$-path. For suppose that $\mathbf{Y}''$ is such a path and let X and Y denote the end-vertices of $\mathbf{Y}''$, where $X \in \mathbf{Y}(A, P) - P$. As $\mathbf{Y}$ is followed from A to B let X' denote the first vertex after X and Y' the last vertex before Y. Then, since $\mathbf{Y}(P, Q) \cap \mathbf{Y}'' = \emptyset$ from what has just been proved, by (vi),

$$\mathbf{Y}(X, A) \cup \{(A, X')\} \cup \mathbf{Y}(X', Y') \cup \{(Y', B)\} \cup \mathbf{Y}(B, Y) \cup \mathbf{Y}''$$

is a circuit of Γ which contains $\mathscr{V}(\mathbf{Y})$ and is therefore longer than Θ. This is contrary to hypothesis, therefore $\Gamma - P - Q$ contains no $(\mathbf{Y}(A, P), \mathbf{Y}(B, Q))$-path.

This shows that $\mathbf{Y}(A, P)-P$, $\mathbf{Y}(B, Q)-Q$ and $\mathbf{Y}(P, Q)-P-Q$ belong to three different connected components of $\Gamma-P-Q$. Since $\Gamma-P-Q$ is disconnected and Γ is 2-connected, each connected component of $\Gamma-P-Q$ is joined by an edge to P and to Q. Now (vii) is proved.

Let the set of connected components of $\Gamma-P-Q$ be $\{\Gamma'_i : i\in\mathscr{I}\}$ and suppose that $\mathbf{Y}(A, P)-P\subseteq\Gamma'_{i_1}$, $\mathbf{Y}(B, Q)-Q\subseteq\Gamma'_{i_2}$ and $\mathbf{Y}(P, Q)-P-Q\subseteq\Gamma'_{i_3}$. Then $i_1\neq i_2\neq i_3\neq i_1$ from (vii). Also, $\mathbf{Y}(A, P)-P=\Gamma'_{i_1}$, for otherwise Γ'_{i_1} contains a vertex Z such that $Z\in\mathbf{Y}$ and Z is joined to a vertex Z' of $\mathbf{Y}(A, P)-P$ by an edge, but by (vi) either $Z'=A$ or $Z'\notin\mathscr{T}(A, \mathbf{Y})$, from which it follows by (3) that $\mathbf{Y}$ is not a longest path of Γ, which is contrary to hypothesis. Similarly $\mathbf{Y}(B, Q)-Q=\Gamma'_{i_2}$. Now put $\Gamma(\Gamma'_i\cup\{P, Q\})=\Gamma_i$ $\forall i\in\mathscr{I}$, and it is seen that (2.1) of (D) holds.

(2.2) Suppose that $P\neq B'$ or $Q\neq A'$, and Γ is 2-connected.

We will first show that:

(viii) $\Gamma-P-Q$ has exactly two connected components, one of them is $\Theta-P-Q$, and the other contains $\mathbf{Y}(P', Q')$, where P' is the first vertex after P and Q' is the last vertex before Q as $\mathbf{Y}$ is followed from A to B.

Proof of (viii). $\Theta-P-Q$ is connected, because if $P\neq B'$, then $B'\in\mathbf{Y}(A, P)-P$, and therefore $\Theta-P-Q$ contains

$$(\mathbf{Y}(A, P)-P)\cup(\mathbf{Y}(B, Q)-Q)\cup\{(B, B')\},$$

which is a connected graph containing $V(\Theta)-P-Q$, and similarly if $Q\neq A'$.

$\Theta-P-Q$ and $\mathbf{Y}(P', Q')$ belong to two different connected components of $\Gamma-P-Q$. For if not, then $\Gamma-P-Q$ contains a $(\Theta, \mathbf{Y}(P', Q'))$-path $\mathbf{Y}'$. The notation can be chosen so that the end-vertices of $\mathbf{Y}'$ are X and Y, where $X\in\mathbf{Y}(P', Q')$ and $Y\in\mathbf{Y}(B, Q)-Q$. As $\mathbf{Y}$ is followed from A to B let Y' be the last vertex before Y. If $Y\in\mathbf{Y}(Q, A')-Q$ then by (vi),

$$\mathbf{Y}'\cup\mathbf{Y}(Y, B)\cup\{B, P)\}\cup\mathbf{Y}(A, P)\cup\{(A, Y')\}\cup\mathbf{Y}(X, Y')$$

is a circuit in Γ which contains $\mathscr{V}(\Theta)$ and X, which contradicts the extremal property of Θ. If on the other hand $Y\in\mathbf{Y}(B, A')-A'$ then by (vi),

$$\mathbf{Y}'\cup Y(Y, B)\cup\{(B, Y')\}\cup\mathbf{Y}(Y', Q)\cup\{(Q, A)\}\cup\mathbf{Y}(A, X)$$

is a circuit of Γ longer than Θ, contrary to hypothesis. These contradictions prove that $\Theta-P-Q$ and $\mathbf{Y}(P', Q')$ belong to two different connected components of $\Gamma-P-Q$.

At most one of P, Q is joined to a vertex $\notin\mathbf{Y}$ in Γ. For if $P\neq B'$, then by (vi), $P\in\mathscr{T}(B, \mathbf{Y})$, and therefore P is not joined to a vertex $\notin\mathbf{Y}$ by (3), since $\mathbf{Y}$ is a longest path of Γ. Similarly if $Q\neq A'$, then Q is not joined to a vertex $\notin\mathbf{Y}$. But by hypothesis $P\neq B'$ or $Q\neq A'$. This proves the statement.

It follows that $\Gamma-P-Q$ has exactly two connected components, the one which contains $\Theta-P-Q$, say Γ'_1, and the one which contains $\mathbf{Y}(P', Q')$, say Γ'_2. For any other connected component of $\Gamma-P-Q$ would contain no vertex of $\mathbf{Y}$, and it would be joined to P and to Q since Γ is 2-connected.

$\Gamma_1' = \Theta - P - Q$. For otherwise there exists a vertex $Z \notin \mathbf{Y}$ in Γ_1' joined by an edge to a vertex Z' of $\Theta - P - Q$. In this case if $Z' = A'$, then $A' \in \mathbf{Y}(B, Q) - Q$, so if A'' is the last vertex before A' when $\mathbf{Y}$ is followed from A to B, then by (vi),

$$\{Z, Z', (Z, Z')\} \cup \mathbf{Y}(A', B) \cup \{(B, P)\} \cup \mathbf{Y}(P, A'') \cup \{(A, A'')\} \cup \mathbf{Y}(A, P) - P$$

is a path of Γ which contains $\mathscr{V}(\mathbf{Y})$ and Z. But this is contrary to the definition of $\mathbf{Y}$, therefore $Z' \neq A'$. Similarly $Z' \neq B'$.

Since $Z' \neq A', B'$, by (vi), $Z' = A$ or $Z' = B$ or $Z' \in \mathscr{T}(A, \mathbf{Y})$ or $Z' \in \mathscr{T}(B, \mathbf{Y})$, and therefore by (3) again Y is not a longest path in Γ. This contradiction proves that $\Gamma_1' = \Theta - P - Q$. Now (viii) is proved.

From (viii) it follows that (2.2) of (D) holds with $\Gamma_1 = \Gamma - \Gamma_2'$ and $\Gamma_2 = \Gamma - \Gamma_1'$.

In both cases, (2.1) and (2.2), if Φ is any circuit of $\Gamma - \Theta$, then Φ is in a connected component of $\Gamma - \Theta$ which is joined by an edge to P and to Q and to no other vertex of Γ. It follows from this, since Γ is 2-connected by hypothesis, that Γ contains a (Φ, P)-path $\mathbf{Y}_P$ and a (Φ, Q)-path $\mathbf{Y}_Q$ such that $\mathbf{Y}_P \cap \mathbf{Y}_Q = \emptyset$. Clearly $\mathbf{Y}_P \cup \Phi \cup \mathbf{Y}_Q$ contains a (P, Q)-path of length $\geqslant \frac{1}{2}l(\Phi) + 2$. Therefore $l(\Phi) \leqslant 2r - 2$.

In case (2), $\Gamma - P - Q$ is disconnected. Therefore if Γ is 3-connected then (1) is the case. The last but one statement of Theorem 4.4 follows directly from Theorems 4.2 and 4.3. The last statement is easily verified.

There follows a theorem giving necessary conditions, involving also the valencies of the vertices, for a finite or infinite 2-connected graph to contain a circuit of prescribed minimal length.

Theorem 4.5. *Let f denote an integer $\geqslant 4$ and Γ any 2-connected graph such that $n_\Gamma > f$, and for each pair of distinct vertices V_1 and V_2 of Γ such that $(V_1, V_2) \notin \Gamma$ either $v_\Gamma(V_1) + v_\Gamma(V_2) \geqslant f$, or $c'(V_1, V_2) \leqslant v_\Gamma(V_1) - 2$, or $c'(V_2, V_1) \leqslant v_\Gamma(V_2) - 2$, or for $i = 1$ or 2,*

$$\exists \{X_1, X_2\} \subseteq \mathscr{N}_\Gamma(V_i)\ \forall X \in (\mathscr{N}_\Gamma(X_1) \cup \mathscr{N}_\Gamma(X_2)) \cap \mathscr{C}'(V_i, V_{i\pm 1}): \qquad v_\Gamma(X) > v_\Gamma(V_i).$$

Then either Γ contains arbitrarily long circuits, or Γ contains a Hamilton circuit, or there exist longest paths in Γ and if $\mathbf{Y}$ is one such that the sum of the valencies of its end-vertices is maximum and $= d$, then $d \geqslant f$. In the latter case Γ contains circuits of $\geqslant f - 2$ different lengths, and Γ contains a circuit of length $> f$ except if $d = f$ and we have the extreme situation of Theorem 4.4(D). If in addition Γ is 3-connected and contains $> f$ vertices of valency $\geqslant [\frac{1}{2}f]$ or some vertex of valency $\geqslant [\frac{1}{2}f] \notin$ the intersection of all longest circuits, then Γ contains a circuit of length $> f$, or f is even and

$$\Gamma \subseteq \langle\langle \tfrac{1}{2}f \rangle, n_\Gamma - \tfrac{1}{2}f \rangle$$

or f is odd and

$$\Gamma \subseteq \langle\langle \tfrac{1}{2}(f-1)\rangle, n_\Gamma - \tfrac{1}{2}(f-1)+\rangle.$$

Proof. Either Γ contains arbitrarily long paths, or Γ contains a longest path. In the first case Γ contains arbitrarily long circuits because it is 2-connected [1, Theorem 1]. In what follows suppose that Γ contains neither arbitrarily long circuits nor a Hamilton circuit. Then among the longest paths of Γ let $\mathbf{Y}$ be one such that the sum of the valencies of its end-vertices is maximum; let this sum be denoted by d. Obviously $l(\mathbf{Y})>1$.

We prove that $d \geq f$. Suppose that $d<f$ (reductio ad absurdum). Let V_1 and V_2 denote the two end-vertices of $\mathbf{Y}$. $v_\Gamma(V_1)\geq 2$ and $v_\Gamma(V_2)\geq 2$ since Γ is 2-connected. $(V_1, V_2)\notin \Gamma$ by Theorem 3.1 because Γ is assumed not to contain a Hamilton circuit. $v_\Gamma(V_1)+v_\Gamma(V_2)=d<f$ by hypothesis. From the definition of $\mathbf{Y}$ and (3) we have $\mathcal{T}(V_i, Y)\subseteq \mathscr{C}'(V_i)$ for $i=1,2$. From Theorem 3.1,

$$\mathcal{T}(V_1, \mathbf{Y})\cap \mathcal{N}_\Gamma(V_2)=\emptyset, \qquad \mathcal{T}(V_2, \mathbf{Y})\cap \mathcal{N}_\Gamma(V_1)=\emptyset.$$

Thus $\mathcal{T}(V_1, \mathbf{Y})\subseteq \mathscr{C}'(V_1, V_2)$ and $\mathcal{T}(V_2, \mathbf{Y})\subseteq \mathscr{C}'(V_2, V_1)$. Hence by (4), $c'(V_1, V_2)\geq v_\Gamma(V_1)-1$ and $c'(V_2, V_1)\geq v_\Gamma(V_2)-1$. Therefore Γ has the last of the list of alternative properties assumed in the theorem. From here a contradiction is obtained in the same way as in the proof of Theorem 3.4. Consequently $d\geq f$.

Since $d\geq f$, by Theorem 4.4, contains a circuit of length $\geq f$, and in fact circuits of $\geq f-2$ different lengths.

If Γ contains no circuit of length $>f$ then $d=f$ and Theorem 4.4(D) applies to Γ.

If Γ is also 3-connected and contains no circuit of length $>f$, then by Theorem 4.4, $f=l(\mathbf{Y})$ and (6) applies to Γ. If in addition more than f vertices have valency $\geq[\frac{1}{2}f]$, so if there is a vertex of valency $\geq[\frac{1}{2}f]$ which is not contained in every longest circuit of Γ, then if f is even

$$\Gamma \subseteq \langle\langle \tfrac{1}{2}f\rangle, n_\Gamma - \tfrac{1}{2}f\rangle$$

by Theorem 4.2, and if f is odd

$$\Gamma \subseteq \langle\langle \tfrac{1}{2}(f-1)\rangle, n_\Gamma - \tfrac{1}{2}(f-1)+\rangle.$$

Now Theorem 4.5 is proved.

Theorem 4.5 implies some results of Pósa [6] and of the author [1] on the existence of circuits of given minimum length f in 2-connected graphs in terms of fixed bounds on the number of vertices with valencies $<\frac{1}{2}f$, however Theorem 4.5 makes no such restrictions.

References

[1] G.A. Dirac, Some theorems on abstract graphs, Proc. London Math. Soc. 2 (3) (1952) 69–81.

[2] G.A. Dirac, Paths and circuits in graphs: extreme cases, Acta Math. Acad. Sci. Hungar. 10 (1959) 357–362.

[3] V. Chvátal, On Hamilton's ideals, J. Combinatorial Theory 12(B), 163–168.
[4] O. Ore, Note on Hamilton circuits, Am. Math. Monthly 67 (1960) 55–57.
[5] L. Pósa, A theorem concerning Hamilton lines, Magyar Tud. Akad. Mat. Kutató Int. Közl. 7 (1962) 225–226.
[6] L. Pósa, On the circuits of finite graphs, Magyar Tud. Akad. Mat. Kutató Int. Közl. 8 (1964) 355–361.

Annals of Discrete Mathematics 3 (1978) 93–109.

SIMPLICIAL DECOMPOSITIONS OF INFINITE GRAPHS*

R. HALIN

Mathematisches Seminar, Universität Hamburg, 2000 Hamburg 13, W. Germany

1. Introduction

Simplicial decompositions were used the first time by Wagner [19, 20] in his beautiful characterizations of all (finite) maximal graphs not contractible onto K_5 and, respectively, onto $K_{3,3}$. Later Wagner and the writer [21, 6, 7, 10] determined the homomorphism-bases (i.e. the elementary brickstones in the decompositions of the maximal graphs in question) for several other graphs, or classes of graphs. But the determination of the homomorphism base of a graph seems to be only possible if the latter is relatively small (so far, no bases are known of graphs with more than 6 vertices). Nevertheless this form of decomposition is attractive by itself, and therefore the writer tried again and again to investigate its general properties and to apply it to other problems in graph theory. Surprisingly, especially in the case of infinite graphs, simplicial decompositions turn out to be a useful tool. It was shown in [11, 12] that under rather general assumptions an uncountable graph G has a simplicial decomposition whose members are all of "small" cardinality. This made it for instance possible to tackle a generalization of Hadwiger's conjecture to graphs with infinite chromatic number [12].

In the present paper we summarize and extend the previous methods and results. A general decomposition theorem is proved and applied to several problems. For example we get insight in the structure of the graphs which do not contain a subdivision of a complete graph of a given uncountable order. Further it will be shown that an n-connected graph of regular order $\mathfrak{a} > \aleph_0$ contains a subdivision of a $K_{\mathfrak{a},n}$ (n a positive integer); this generalizes a result of Dirac [1]. Also we obtain the following: If G has uncountable regular order $\mathfrak{a}$ and does not contain a subdivision of an infinite complete graph, then there is a finite subgraph F of G such that $G-F$ has $\mathfrak{a}$ components. Finally it is shown that every connected graph which does not contain a subdivision of an infinite complete graph has a normal rooted (spanning) tree in the sense of Jung.

2. Terminology, notation

In general we follow the standard notation as it is used, or at least understood, by all graph-theorists.

* This paper was written while the author was visiting the University of Aarhus, Denmark.

The *order* of a graph G is $|V(G)|$, denoted briefly by $|G|$. By $G[T]$ we denote the subgraph of G *induced* by some $T \subseteq G$ or $\subseteq V(G)$. By $\supset$ we always indicate proper inclusion. Most times we simply write $v \in G$ instead of $v \in V(G)$.

In this context we find it more convenient to say "*simplex*" instead of "complete graph"; $S(\mathfrak{a})$ denotes a simplex of order $\mathfrak{a}$. In general the letter S is reserved to denote simplices.

$G > H$ means that G contains H as a subgraph of a contraction (more precisely, that there is a bijection of $V(H)$ to a family of disjoint connected subgraphs Z_h of G $(h \in V(H))$ such that $[h, h'] \in E(H)$ implies the existence of an edge (in G) between Z_h and $Z_{h'}$); we then say G is *homomorphic* to H. If G contains a subdivision of H, we write $G >_u H$. Mind that for $\mathfrak{a} > \aleph_0$, $\mathfrak{a}$ regular

$$G > S(\mathfrak{a}) \Leftrightarrow G >_u S(\mathfrak{a})$$

(Jung [14]).

If P is a path, say connecting the vertices a, b, by $\mathring{P}$ we denote its "inner part", i.e. the graph $P - a - b$, and if u, $v \in V(P)$, by P_{uv} we denote the (unique) u, v-path contained in P.

We write $a \cdot T \cdot b(G)$ if T *separates* a, b in G (i.e. if the vertices a, b are not in T and every a, b-path in G meets T).

For $T \subset G$ and $v \in V(G - T)$ we define the *connection graph* $G(v \to T)$ from v to T (in G) as the subgraph of G induced by the union of all paths P in G starting in v and having $\mathring{P} \cap T = \emptyset$. (Thus $G(v \to T)$ is the subgraph induced by the connected component C of $G - T$ which contains v plus all the vertices of T which are adjacent to a vertex of C.) T is called an *inward subgraph* of G if there is a $v \in G - T$ such that $G(v \to T) \cap T = T$ (i.e. if every $t \in V(T)$ can be reached from v by a path P with $\mathring{P} \cap T = \emptyset$).

If x, y are distinct vertices of G, their Menger number $\mu_G(x, y)$ is the maximum number of internally disjoint x, y-paths in G. (Here, as throughout in this paper, "number" is used in the sense of "cardinal".) If $\mathfrak{a}$ is a cardinal, $\mathfrak{a}^+$ denotes its immediate successor. $\omega(\mathfrak{a})$ is the initial ordinal of cardinality $\mathfrak{a}$; for an ordinal σ, $W(\sigma)$ denotes the set of ordinals $< \sigma$. The axiom of choice is assumed throughout.

If M is a set with a relation $<$ which is irreflexive, asymmetric and transitive (i.e. it is the irreflexive kernel of a partial order) such that for every $x \in M$, the set of $y \in M$ with $y < x$ is a chain with respect to $<$ (that means: if y, $z < x$, then $y = z$, $y < z$ or $z < y$), then M with $<$ is called an *order-theoretical tree*.

3. Simplicial decompositions

Let G be a graph, and let G_λ $(\lambda < \sigma)$ be a family of subgraphs of G, where $\sigma > 0$ is an ordinal number (and λ runs through all ordinals smaller than σ). We say these G_λ form a *simplicial decomposition* of G if G is the union of the G_λ and for every τ, $0<\tau<\sigma$, holds:

$$\bigcup_{\lambda<\tau} G_\lambda \cap G_\tau = S_\tau$$

is a simplex properly contained in both $\bigcup_{\lambda<\tau} G_\lambda$ and G_τ.

Thus G is built up from the G_λ by a transfinite process (if σ is infinite), in each step of which G_λ is pasted to the part constructed before, along a simplex, like an infinite cactus is composed of its branches.

The subgraphs $G'_\tau := \bigcup_{\lambda<\tau} G_\lambda$ are called *partial sums* of the above decomposition. An S_λ may be the empty graph $\emptyset$, but no G_λ can be empty if $G \neq \emptyset$ (which will be assumed throughout).

Further, each G_λ and each G'_τ $(0<\tau<\sigma)$ must be an induced subgraph of G, and G is connected if and only if all G_λ are connected and no S_λ is empty. One has

$$a \cdot S_\tau \cdot b(G) \qquad (*)$$

for every $a \in G'_\tau - S_\tau$, $b \in G_\tau - S_\tau$ (see [8, (1.1)]).

To every simplicial decomposition there is associated a tree structure which reflects the manner G is composed of the G_λ: For $v \in G$, let λ_v denote the smallest λ such that v belongs to G_λ. Then for $\lambda, \kappa < \sigma$ we set $\lambda \lhd \kappa$ if $\lambda < \kappa$ and there is a vertex $v \in G_\kappa$ such that $\lambda_v = \lambda$. $\lhd$ is an asymmetric relation in $W(\sigma)$. Let $\ll$ denote its transitive closure, i.e. $\lambda \ll \kappa$ if and only if there is a finite chain

$$\lambda = \lambda_0 \lhd \lambda_1 \lhd \cdots \lhd \lambda_n = \kappa.$$

Proposition 3.1. *$W(\sigma)$ is an order-theoretical tree with respect to $\ll$.*

Proof. Since $\lambda \ll \kappa$ implies $\lambda < \kappa$, $\ll$ is asymmetric and irreflexive, hence, as it is transitive, a partial order (more precisely, the irreflexive kernel of such a relation). Assume that $\ll$ does not define an order theoretical tree. Then there is a smallest τ and λ, κ such that $\lambda \ll \tau$, $\kappa \ll \tau$ but λ, κ are not comparable with respect to $\ll$. There exist chains $\lambda = \lambda_0 \lhd \lambda_1 \lhd \cdots \lhd \lambda_n = \tau$, $\kappa = \kappa_0 \lhd \kappa_1 \lhd \cdots \lhd \kappa_m = \tau$ $(n, m > 0)$. By definition there are $a, b \in S_\tau$ with $\lambda_{n-1} = \lambda_a$, $\kappa_{m-1} = \lambda_b$; say $\lambda_a \leq \lambda_b$. Since $[a, b] \in E(G)$ and G'_{λ_b+1} is an induced subgraph containing a, b, it follows $[a, b] \in E(G_{\lambda_b})$, i.e. $a \in G_{\lambda_b}$; therefore $\lambda_a = \lambda_b$ or $\lambda_a \lhd \lambda_b$. Thus $\lambda, \kappa \ll \lambda_b = \kappa_{m-1} < \tau$ and λ, κ not comparable under $\ll$, which contradicts the minimal choice of τ.

If for τ, $0<\tau<\sigma$, S_τ is contained in some G_λ, $\lambda<\tau$, then we can replace the above tree structure by a graph theoretical tree (on $W(\sigma)$): Namely let then τ^- denote the smallest λ of that kind, and draw all the edges $[\lambda, \lambda^-]$.

Especially if all the S_τ are finite, each S_τ is contained in a G_λ, $\lambda<\tau$, and τ^- is the largest λ such that $\lambda \lhd \tau$. Then the simplicial decomposition gives good insight in the structure of the graph G; for instance the chromatic number $\chi(G)$ of G is the supremum of all the $\chi(G_\lambda)$. This is, however, not the case in general. For instance several of the S_λ (or parts of them) may sum up to form an $S(\aleph_1)$ though all the G_λ are at most countable. The tree structure associated with the simplicial decomposition then cannot be described by a graph; we then face extreme difficulties if we want to carry over the methods which apply in the case of finite S_λ.

4. Separation-invariant subgraphs

Every (induced) subgraph H of G which occurs as a member in some simplicial decomposition of G is called a *simplicial summand* of G. In our next proposition we characterize the simplicial summands of G by separation properties.

A subgraph H of G is called *separation invariant* in G if for any $x, y \in H$, $T \subset H$

$$x \cdot T \cdot y(H) \Rightarrow x \cdot T \cdot y(G).$$

Of course every simplex in G is separation invariant in G, though it need not be a simplicial summand. But we have (see [11, Section 2]):

Proposition 4.1. *Let H be an induced subgraph of G and not a simplex. Then the following statements are equivalent:*

(i) *H is a simplicial summand of G;*

(ii) *H is separation-invariant in G;*

(iii) *For every $a \in G - H$, $G(a \to H) \cap H$ is a simplex;*

(iv) *If P is a path in G connecting vertices a, b of H with $\mathring{P} \cap H = \emptyset$, then $[a, b] \in E(G)$.*

Proof. Is not difficult to show that each of the statements (ii) and (iii) is equivalent with (iv). If (iii) holds we get a simplicial decomposition with H and the graphs $G(a \to H)$ as members, thus (iii)$\Rightarrow$(i). — Assume (i). Then there is a simplicial decomposition $G_\lambda (\lambda < \sigma)$ of G such that H occurs as some G_τ. If (iv) does not hold there is a path P connecting non-adjacent a, b of H such that $\mathring{P} \cap H = \emptyset$. Let $\zeta(\tau < \zeta < \sigma)$ be the smallest cardinal such that $\bigcup_{\lambda \leq \zeta} G_\lambda$ contains such a path P. $\zeta > \tau$ by the definition of simplicial decomposition. P has a first vertex f and a last vertex l in common with G_ζ, $f \neq l$. It follows $f, l \in S_\zeta$ by the relation $(*)$ above. Now replacing P_{fl} by $[f, l]$ shows that there is a path of the kind in question in a sum $\bigcup_{\lambda \leq \zeta'} G_\lambda$ for some $\zeta' < \zeta$, which contradicts the choice of ζ. Hence (i) implies (iv).

Proposition 4.2. *Let G_i $(i \in I)$ be a family of separation-invariant subgraphs of G. Then also $H := \bigcap_{i \in I} G_i$ is separation-invariant in G.*

Proof. Assume H to be not separation-invariant in G. Let P be a path of minimal length such that

(i) P connects non-adjacent vertices a, b of H,

(ii) $\mathring{P} \cap H = \emptyset$.

(Of course $\mathring{P} \neq \emptyset$ by (i)).

Let $v \in V(\mathring{P})$. There exists an $i \in I$ such that $v \notin G_i$. Let x, y be the first vertex of G_i on P_{va}, P_{vb}, respectively. Then $x, y \in G(v \to G_i) \cap G_i$, hence $[x, y] \in E(G)$ since G_i is separation-invariant. By replacing P_{xy} in P by $[x, y]$ we get a shorter a, b-path P' with $\mathring{P}' \cap H = \emptyset$, which is a contradiction.

Since G is trivially separation-invariant in itself, we can form, for every $H \subseteq G$, the *separation-invariant closure* $C_G(H)$ of H in G, defined as the intersection of all separation-invariant subgraphs of G containing H.

Now we give an estimate of the order of $C_G(H)$ in terms of the Menger numbers of non-adjacent vertices in G, which is crucial for what follows.

Theorem 4.3. *Let $\mathfrak{a}$ be a regular cardinal $> \aleph_0$ such that for every pair x, y of non-adjacent vertices of G holds $\mu_G(x, y) < \mathfrak{a}$. Then for any $H \subseteq G$ with $|H| < \mathfrak{a}$ we have*

$$|C_G(H)| < \mathfrak{a}.$$

Proof. Let $H_0 := G[H]$. Assume then that, for some $n \in \mathbf{N}$, $H_{n-1} \subseteq G$ is already defined. For every pair x, y of distinct non-adjacent vertices in H_{n-1} choose a maximal (with respect to inclusion) system $\mathcal{P}_{xy}$ of $\mu_G(x, y)$ internally disjoint x, y-paths in G and let H_n be the subgraph of G induced by H_{n-1} and the union of all these $\mathcal{P}_{xy}$. If $|H_{n-1}| < \mathfrak{a}$ then there are less than $\mathfrak{a}$ systems $\mathcal{P}_{xy}$ each with less than $\mathfrak{a}$ elements, hence also $|H_n| < \mathfrak{a}$ by the regularity of $\mathfrak{a}$.

In this way we get a sequence

$$G[H] = H_0 \subseteq H_1 \subseteq H_2 \subseteq \cdots$$

of induced subgraphs of G with $|H_i| < \mathfrak{a}$. Let $H^* := \bigcup_{n=0}^{\infty} H_n$. Again by the regularity of $\mathfrak{a}$ and $\mathfrak{a} > \aleph_0$ we conclude $|H^*| < \mathfrak{a}$. H^* is separation-invariant (hence contains $C_G(H)$). Otherwise there were a path P in G connecting non-adjacent vertices a, b of H^* with $\mathring{P} \cap H^* = \emptyset$. There exists $n \in \mathbf{N}$ such that $a, b \in H_{n-1}$. But then the system $\mathcal{P}_{ab}$ added to H_{n-1} in the nth step would not have been maximal, with contradiction.

Let us denote by $\tilde{\mu}(G)$ the supremum of all the cardinals $\mu_G(x, y)$ where x, y run through all pairs of non-adjacent vertices of G. (If G is a simplex then we set $\tilde{\mu}(G) = 0$.)

Thus, if $\mu_G(x, y) > \tilde{\mu}(G)$, then x, y must be adjacent in G.

By the same argument as in the proof of Theorem 4.3 we obtain:

Theorem 4.4. *For $H \subseteq G$ the inequality holds*

$$|C_G(H)| \leq \max(|H|, \tilde{\mu}(G), \aleph_0).$$

Further we get

Corollary 4.5. *If G is uncountable and $\tilde{\mu}(G) < |G|$, then G has a separating simplex of order $\leq \max(\tilde{\mu}(G), \aleph_0)$, unless G itself is a simplex.*

Proof. If G is not a simplex, let H consist of two non-adjacent vertices and set $\mathfrak{a} = \max(\tilde{\mu}(G), \aleph_0)^+$. The result follows from Theorem 4.3 and Proposition 4.1.

5. The decomposition theorem

We can now prove

Theorem 5.1. *Let G be a graph and $\mathfrak{a}$ a regular cardinal with $|G| \geq \mathfrak{a} > \aleph_0$. Assume that $\mu_G(x, y) < \mathfrak{a}$ for every pair of non-adjacent vertices x, y in G and that G does not contain an $S(\mathfrak{a})$. Then G has a simplicial decomposition G_λ $(\lambda < \sigma)$, where σ is the initial ordinal of $|G|$, in which all G_λ have cardinality less than $\mathfrak{a}$.*

In addition this decomposition can be chosen in such a way that every S_λ is an inward simplex in G_λ, moreover such that $G_\lambda - S_\lambda$ is connected for every $\lambda < \sigma$.

Proof. Let $V(G)$ be well-ordered according to σ:

$$V(G) = \{x_\nu : \nu < \sigma\}.$$

G is not complete by hypothesis. Choose non-adjacent y', y'' in G and set

$$G_0 := C_G(x_0, y', y'').$$

Then by Theorem 4.3

$$|G_0| < \mathfrak{a}.$$

Now let $0 < \tau < \sigma$ and assume that separation-invariant subgraphs G_ζ of order less than $\mathfrak{a}$ for all $\zeta < \tau$ are already defined such that these G_ζ from a simplicial decomposition of their union, further the following conditions hold:

(1) Each $x_\nu (\nu < \zeta)$ is in $\bigcup_{\lambda<\zeta} V(G_\lambda)$,

(2) Each $G_\zeta - S_\zeta$ is connected,

(3) For at least one (then by (2) for all) $v \in V(G_\zeta - S_\zeta)$ there holds

$$S_\zeta = G\Big(v \to \bigcup_{\lambda<\zeta} G_\lambda\Big) \cap \Big(\bigcup_{\lambda<\zeta} G_\lambda\Big).$$

We shall now construct G_τ. First we state that $G'_\tau = \bigcup_{\lambda<\tau} G_\lambda$ is separation-invariant in G.

Otherwise there are non-adjacent a, b in G'_τ such that there is an a, b-path $P \subseteq G$ with $\mathring{P} \cap G'_\tau = \emptyset$. Let λ_a, λ_b be the smallest λ with $a \in G_\lambda$, respectively $b \in G_\lambda$; without loss of generality $\lambda_a \leq \lambda_b$. If $\lambda_a = \lambda_b$, then G_{λ_a} were not separation-invariant; hence $\lambda_a < \lambda_b$.

Since $a \in G(b \to \bigcup_{\lambda<\lambda_b} G_\lambda)$, $a \in \bigcup_{\lambda<\lambda_b} G_\lambda$ we have by (3)

$$a \in S_{\lambda_b} \subset G_{\lambda_b}$$

which would imply

$$P \cap G_{\lambda_b} = \{a, b\},$$

contradicting the separation-invariance of G_{λ_b}.

Now, to construct G_τ, let x be the x_ν with smallest ν such that $x_\nu \notin G'_\tau$ (it exists because of $|G'_\tau| < |G|$). By Proposition 4.1, $G(x \to G'_\tau) \cap G'_\tau =: S_\tau$ is a simplex; $|S_\tau| < \mathfrak{a}$ by hypothesis, and $S_\tau \neq G'_\tau$ since y', y'' are non-adjacent vertices of G'_τ.

For every $s \in V(S_\tau)$ choose an x, s-path P_s in G with $P_s \cap S_\tau = \{s\}$. Let now $G_\tau := C_G(S_\tau \cup \bigcup_{s \in S_\tau} P_s)$. Then G_τ has the required properties (i.e. it also fulfills the conditions stated for the G_ζ above).

In this way, by transfinite induction a sequence of $G_\lambda (\lambda < \sigma)$ is constructed which has the desired properties.

6. The $\mathfrak{a}$-saturation of a graph

We want to apply Theorem 5.1 to study the structure of graphs which do not contain a given graph (or any member of a given class of graphs) as a homomorphic image or as a subdivision. At this point, in actually establishing a simplicial decomposition, the condition on the Menger numbers may cause difficulties. These are taken care of by the considerations of this section.

We say that a graph G is $\mathfrak{a}$-*saturated* ($\mathfrak{a}$ a cardinal) if $\mu_G(a, b) \geq \mathfrak{a}$ implies $[a, b] \in E(G)$. The $\mathfrak{a}$-*saturation* $[G]_\mathfrak{a}$ of a graph G is the graph arising from G by adding all the edges $[a, b]$ where a, b are non-adjacent vertices of G with $\mu_G(a, b) \geq \mathfrak{a}$. (If $|G| < \mathfrak{a}$, then $[G]_\mathfrak{a} = G$.)

Proposition 6.1. *For any $T \subseteq G$ with $|T| < \mathfrak{a}$ and a, $b \in G$ we have*

$$a \cdot T \cdot b(G) \Leftrightarrow a \cdot T \cdot b([G]_\mathfrak{a}).$$

Especially, the non-adjacent pairs of vertices a, b with Menger number $<\mathfrak{a}$ are the same in G and $[G]_\mathfrak{a}$ and we have

$$\mu_{[G]_\mathfrak{a}}(a, b) < \mathfrak{a}$$

for all non-adjacent a, b in $[G]_\mathfrak{a}$.

Proof. Assume $a \cdot T \cdot b(G)$, $|T| < \mathfrak{a}$, and let C denote the component of $G-T$ which contains a. If there were an a, b-path in $[G]_\mathfrak{a}$ avoiding T, on this path there had to occur an edge $[x, y]$ with $x \in C$, $y \notin C$; hence $[x, y] \notin E(G)$, which implies $\mu_G(x, y) \geq \mathfrak{a}$. But $x \cdot T \cdot y(G)$ because x, y belong to different components of $G-T$, hence $\mu_G(x, y) \leq |T| < \mathfrak{a}$, with contradiction.

The other direction of the asserted equivalence is trivial.

Proposition 6.2. *Let H be a graph of order $< \mathfrak{a}$, $\mathfrak{a}$ a cardinal $> \aleph_0$. Then for any graph G*

$$[G]_\mathfrak{a} >_u H \Leftrightarrow G >_u H$$

$$[G]_\mathfrak{a} > H \Leftrightarrow G > H.$$

Proof. If $[G]_\mathfrak{a} >_u H$ or $> H$ then there is also a subgraph $H^* \subseteq [G]_\mathfrak{a}$ with $H^* >_u H$, or $H^* > H$ respectively, such that $|H^*| \leq \max(|H|, \aleph_0) < \mathfrak{a}$. If $[x, y]$ is an edge of H^* which is not in G, then $\mu_G(x, y) \geq \mathfrak{a}$; hence, by reasons of cardinality, there is an x, y-path P in G with $\mathring{P} \cap H^* = \emptyset$, and we can replace $[x, y]$ by P. By a routine application of Zorn's lemma we can carry through this replacing procedure "step by step" for all the edges of H^* not in G, and we find a subdivision of H^* in G.

If we choose H as a simplex, we can prove the last proposition also for $|H| = \mathfrak{a}$. In fact we have

Proposition 6.3. *Let $\mathfrak{a}$ be an infinite cardinal and G be an arbitrary graph. Then the following statements are equivalent:*

(i) $[G]_\mathfrak{a} \supseteq S(\mathfrak{a})$,

(ii) $[G]_\mathfrak{a} >_u S(\mathfrak{a})$,

(iii) $G >_u S(\mathfrak{a})$.

Proof. (i)$\Leftrightarrow$(ii) follows from Proposition 6.1, and (ii)$\Rightarrow$(iii) by a standard application of Zorn's lemma onto the set of all those subdivisions of simplices in G which have their branch vertices in $V(H^*)$, where H^* is the subdivision of $S(\mathfrak{a})$ contained in $[G]_\mathfrak{a}$ by hypothesis. The other assertions of Proposition 6.3 are clear.

Let H be a graph and $\mathfrak{a} > 0$ be a cardinal. Replace each edge $[x, y]$ of H by a system $\mathcal{P}_{xy}$ of $\mathfrak{a}$ internally disjoint x, y-paths such that no two paths of any two $\mathcal{P}_{xy}$ have internal points in comnon. We call the graph arising in this way a *subdivision of H of strength* $\mathfrak{a}$; we denote such a configuration by $U_\mathfrak{a}(H)$. By routine methods (see [12, (15)]) we find:

Proposition 6.4. *Let $A \subseteq V(G)$ with $|A| = \mathfrak{a}$ such that $\mu_G(x, y) \geq \mathfrak{b} > \mathfrak{a} \geq \aleph_0$ for any distinct x, $y \in A$ holds. Then there is a $U_\mathfrak{b}(S(\mathfrak{a}))$ which has A as its set of branch vertices.*

Proposition 6.5. *Assume* $[G]_{\mathfrak{a}^+} \supseteq S(\mathfrak{a})$, $\mathfrak{a} \geq \aleph_0$. *Then* $G \supseteq S(\aleph_0)$ *or* $G \supseteq U_{\mathfrak{a}^+}(S(\mathfrak{a}))$, *and if especially* $\mathfrak{a} = \mathfrak{b}^+ = 2^{\mathfrak{b}}$ *for some cardinal* $\mathfrak{b}$, *then* $G \supseteq S(\mathfrak{a})$ *or* $G \supseteq U_{\mathfrak{a}^+}(S(\mathfrak{a}))$.

Proof. $E(S(\mathfrak{a}))$ decomposes into two disjoint classes, namely into the class of edges which are also in G and the class of edges $[x, y]$ with non-adjacent x, y of G and $\mu_G(x, y) \geq \mathfrak{a}^+$. The assertion follows from Proposition 6.4 together with [5, Theorem 3(i) and Theorem 4(iii)].

7. Graphs without forbidden configurations

For some class $\mathfrak{C}$ of graphs, by $\hat{\mathfrak{C}}$ we denote the class of maximal elements of $\mathfrak{C}$, i.e. the class of those $G \in \mathfrak{C}$ for which $G \cup [x, y] \notin \mathfrak{C}$, for any pair of vertices x, y which are not adjacent in G. $\hat{\mathfrak{C}}$ may be empty (but only if $\mathfrak{C}$ does not contain a finite graph).

Let Γ be a non-empty class of finite graphs; if Γ consists of one element A only, which will be the most important special case, we identify Γ with A. By the *homomorphism-class* $\mathfrak{H}^*\Gamma$ we denote the class of all graphs G with $G \not> H$ for all $H \in \Gamma$. Similarly, by the *subdivision-class* $\mathfrak{H}^*_u\Gamma$ we denote the class of all graphs G with $G \not>_u H$ for all $H \in \Gamma$.

$\hat{\mathfrak{H}}^*\Gamma$ and $\hat{\mathfrak{H}}^*_u\Gamma$ then denote the class of maximal elements of $\mathfrak{H}^*\Gamma$, and $\mathfrak{H}^*_u\Gamma$, respectively. Thus $G \in \hat{\mathfrak{H}}^*\Gamma$ (or $\hat{\mathfrak{H}}^*_u\Gamma$) if and only if G is not $>$ (or $>_u$, respectively) to a member of Γ, but becomes $>$ (or $>_u$, respectively) to some $H \in \Gamma$ if any new edge to G is added. Mind that $S(t) \in \hat{\mathfrak{H}}^*\Gamma$ for every $t < \min_{H \in \Gamma} |H|$.

Proposition 7.1. *Every graph* $G \in \mathfrak{H}^*\Gamma$ *(or* $\in \mathfrak{H}^*_u\Gamma$*) can be extended, by adding edges, to an element* $\hat{G} \in \hat{\mathfrak{H}}^*\Gamma$ *(or* $\in \hat{\mathfrak{H}}^*_u\Gamma$, *respectively).*

Proof. Routine application of Zorn's lemma onto the set of all graphs $G \cup E'$ where E' is any set of new edges to G such that $G \cup E'$ is still in $\mathfrak{H}^*\Gamma$ (in $\mathfrak{H}^*_u\Gamma$, respectively).

Let us call a graph G *prime* if there is not a separating simplex in G, i.e. if there is no proper simplicial decomposition of G. A simplicial decomposition G_λ $(\lambda < \sigma)$ of G is called a *prime-graph decomposition* if all the G_λ are prime; it is called *reduced* if no G_λ is contained in a G_κ with $\lambda \neq \kappa$. By [8, Satz 3], a prime-graph decomposition of G is reduced if and only if all its members are maximal prime (induced) subgraphs of G. In [8] it was shown:

Theorem 7.2. *Every graph G without an infinite simplex has a reduced prime-graph decomposition* G_λ $(\lambda < \sigma)$. *The* G_λ *are uniquely determined; they are just all maximal prime (induced) subgraphs of* G.

In [8] also an example is given that a graph containing infinite simplices need not have a prime graph decomposition.

If Γ is a class of finite graphs, every element of $\hat{\mathfrak{H}}^*\Gamma$ has a prime-graph decomposition, by Theorem 7.2. The class of all prime graphs, occurring in the prime-graph decomposition of any $G \in \hat{\mathfrak{H}}^*\Gamma$, is called the *base* of $\hat{\mathfrak{H}}^*\Gamma$ and denoted by $\mathfrak{B}\hat{\mathfrak{H}}^*\Gamma$. One has $\mathfrak{B}\hat{\mathfrak{H}}^*\Gamma \subseteq \mathfrak{H}^*\Gamma$, but not necessarily $\subseteq \hat{\mathfrak{H}}^*\Gamma$. Analogously the base $\mathfrak{B}\hat{\mathfrak{H}}^*_u\Gamma$ (of $\hat{\mathfrak{H}}^*_u\Gamma$) is defined. It is clear that for the knowledge of the graphs in $\hat{\mathfrak{H}}^*\Gamma$ or, respectively, $\hat{\mathfrak{H}}^*_u\Gamma$ it is decisive to know the base-elements of these classes. (The way how these base-elements have to be composed in order to give maximal elements of $\mathfrak{H}^*\Gamma$ or $\mathfrak{H}^*_u\Gamma$, follows certain rules whose determination is of minor difficulty and which is omitted here). In [11] the following estimate for the order of the base elements was given:

Theorem 7.3. *If Γ is a class of finite graphs then every element of the base of $\hat{\mathfrak{H}}^*\Gamma$ or $\hat{\mathfrak{H}}^*_u\Gamma$ is finite or countable.*

Proof. Let G be an element of $\mathfrak{H}^*\Gamma$ or $\mathfrak{H}^*_u\Gamma$. Since $G \not\geq S(\aleph_0)$, G has, by Theorem 7.2, a prime-graph decomposition P_λ $(\lambda < \sigma)$. What we have to show is that each P_λ has order $\leq \aleph_0$. By Proposition 6.2, $[G]_{\aleph_0} \in \mathfrak{H}^*\Gamma$ (respectively $\in \mathfrak{H}^*_u\Gamma$), which means $G = [G]_{\aleph_0}$ by the maximality of G. Especially we conclude $\mu_G(x, y) < \aleph_0$ for all non-adjacent x, y in G, and this property carries over to each P_λ, since each P_λ is an induced subgraph of G. Hence by Theorem 5.1 each P_λ has a simplicial decomposition in which all members are finite or countable. But since each P_λ is prime, each such decomposition has only one member, namely P_λ, which therefore itself must be finite or countable.

By Theorem 7.3 it was possible to extend the known characterizations of the (maximal) graphs G not homomorphic to some given graph H, by means of determining $\mathfrak{B}\hat{\mathfrak{H}}^*(H)$, also to infinite graphs G. Wagner's theorem [20], for instance, extends in the following way:

Theorem 7.4. *The base $\mathfrak{B}\hat{\mathfrak{H}}^*S(5)$ consists of the non-planar graph W (see Fig. 1) and all prime maximal planar graphs (which may be finite or countable).*

The maximal countable planar graphs form an interesting class of graphs onto which apparently almost no research has been done so far. In [11] some basic properties of these graphs were proved. It was shown that such a graph must be 2-connected but is not necessarily 3-connected and that it is prime if and only if it is 4-connected. It does not necessarily contain triangles. Its connectivity is finite but may be arbitrarily great, which was observed by Mader [18]. Each such graph contains one-way infinite paths, which follows for instance from [13, Satz 6]. (An independent proof was orally communicated to the writer by Dirac). Further it can be deduced from [9, Satz 3]: If G is maximal planar and countable and if G

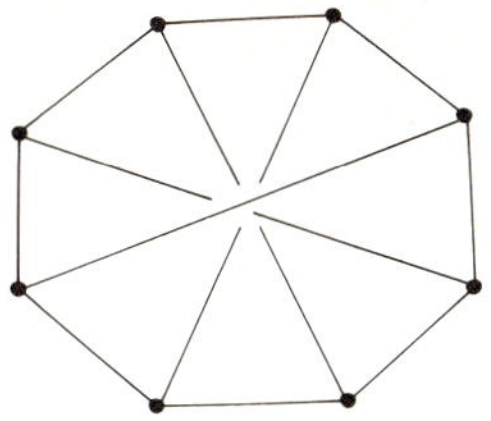

Fig. 1.

does not contain a two-way infinite path, then G has an edge which is contained in infinitely many separating triangles.

Also the characterization of all (finite) graphs not homomorphic to a 4-connected graph [10, Satz 6] easily carries over to infinite graphs. We have:

Theorem 7.5. *Let Γ_4 denote the class of (finite) 4-connected graphs. Then $\mathfrak{B}\hat{\mathfrak{H}}^*\Gamma_4$ consists of $S(\nu)$, $\nu \leq 4$, and the graphs W and P_ν of Figs.* 1 *and* 2.

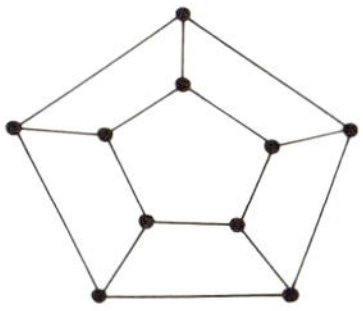

Fig. 2.

We saw that every graph G, which is not $>$ (or $>_u$) to some finite graph H, can be extended by adding edges to a graph $\hat{G}$ with the same property such that $\hat{G}$ has a simplicial decomposition with members of "small" (i.e. finite or countable) order. It is remarkable that this statement remains, in a modified version, valid if H is infinite, though in this case it makes no sense to consider maximal graphs with respect to the forbidden configuration.

Theorem 7.6. *Let G, H be graphs with $|G| > |H| \geq \aleph_0$ and assume $G \not>_u H$ (or $G \not> H$). Then G can be extended, by adding new edges, to a graph $\hat{G}$ which is also $\not>_u H$ (or $\not> H$, respectively) and which has a simplicial decomposition G_λ ($\lambda < \sigma = \omega(|G|)$) in which all members have order not greater than $|H|$. It may be chosen $\hat{G} = [G]_\mathfrak{a}$ with $\mathfrak{a} = |H|^+$.*

The proof follows from Theorem 5.1 in connection with Propositions 6.1 and 6.2.

A sharper result can be proved if especially H is an uncountable simplex.

Theorem 7.7. *Let $\mathfrak{a}$ be a regular, uncountable cardinal, and let G be a graph with $G \not>_u S(\mathfrak{a})$. Then $[G]_\mathfrak{a}$ is $\not>_u S(\mathfrak{a})$ and has a simplicial decomposition in which each member has order smaller than $\mathfrak{a}$.*

This follows from Theorem 5.1 in connection with Propositions 6.1 and 6.3.

8. Application to colouring problems

Erdös and Hajnal [4] define the *colouring number* $\zeta(G)$ of a graph G as the smallest ordinal ν such that there exists a well-ordering $<$ of $V(G)$ with respect to which each $v \in V(G)$ is adjacent to less than ν predecessors. The colouring number is related to the chromatic number of a graph by the inequality

$$\zeta(G) \geq \chi(G). \qquad (**)$$

It is natural to ask which configurations must be contained in G if $\zeta(G)$ is great. By Erdös and Hajnal [4, Theorem 7.1] it was shown that $\zeta(G) > \aleph_0$ implies the existence of an infinite path in G. In [12] the following stronger result is proved:

Theorem 8.1. *If* $\zeta(G) \geq \aleph_0$, *then for every* $\mathfrak{a} < \zeta(G)$ *there is a subdivision of* $S(\mathfrak{a})$ *contained in* G.

Proof. Assume first $\zeta(G) > \aleph_0$ and that $G \not>_u S(\mathfrak{a})$ for some $\mathfrak{a} < \zeta(G)$. Then, by Theorem 7.6, $[G]_{\mathfrak{a}^+} \not>_u S(\mathfrak{a})$ and $[G]_{\mathfrak{a}^+}$ has a simplicial decomposition G_λ ($\lambda < \omega(|G|)$) with $|G_\lambda| \leq \mathfrak{a}$ for all λ. Let $<_\lambda$ be a well-ordering of $G_\lambda - S_\lambda$ of order type $\omega(|G_\lambda - S_\lambda|)$. (Here $S_0 = \emptyset$ is to be understood). For $x \in V(G)$ let λ_x be the smallest λ such that $x \in G_\lambda$. Now for $x, y \in V(G)$ we set $x < y$ if either $\lambda_x < \lambda_y$, or if $\lambda_x = \lambda_y =: \lambda$ and $x <_\lambda y$. In this way a well-ordering of $V(G)$ is defined such that each vertex is adjacent to less than $\mathfrak{a}$ predecessors; this contradicts the hypothesis $\zeta(G) > \mathfrak{a}$.

If $\zeta(G) = \aleph_0$ the assertion is deduced from a result of Mader [17], using the fact that from $\zeta(G) \geq 2n - 2$ ($n \in \mathbf{N}$) it follows the existence of a finite subgraph F of G with $\zeta(F) \geq n$ (Erdös and Hajnal [4, Theorem 9.1]).

It cannot be shown that under the assumptions of Theorem 3.1 there must be a subdivision of $S(\zeta(G))$ in G, also if $\zeta(G)$ is not a limit cardinal. Namely, the complete bipartite graph $K_{\mathfrak{a},\mathfrak{a}^+}$ has colouring number $\mathfrak{a}^+$ but does not contain a subdivision of $S(\mathfrak{a}^+)$. But it can easily be shown that, under the aforesaid assumptions, G must contain $\zeta(G)$ disjoint subdivisions of $S(\mathfrak{a})$, since the deletion of less than $\zeta(G)$ such configurations results in a graph which, by cardinality reasons, has again colouring number $\zeta(G)$.

Hadwiger's famous conjecture asserts the implication

$$\chi(G) \geq \mathfrak{a} \Rightarrow G > S(\mathfrak{a}) \qquad (***)$$

for $\mathfrak{a} \in \mathbf{N}$. It makes sense to consider this statement also for infinite cardinals $\mathfrak{a}$. From the inequality $(**)$ and Theorem 3.1 we get immediately:

Corollary 8.2. *If* $\chi(G) \geq \aleph_0$, *then* $G >_u S(\mathfrak{a})$ *for every* $\mathfrak{a} < \chi(G)$.

The stronger implication $(***)$ is of course wrong if $\mathfrak{a}$ is a limit cardinal, since then the disjoint union of all $S(\mathfrak{b})$, $\mathfrak{b} < \mathfrak{a}$, has chromatic number $\mathfrak{a}$, without containing $S(\mathfrak{a})$ as a homomorphic image. It is, however, hopelessly difficult to decide $(***)$ if $\mathfrak{a}$ is of the form $\mathfrak{b}^+$ (see [12]).

Proposition 6.5 can be used to refine Theorem 8.1 and Corollary 8.2 (see [12, Satz 6]).

9. Connectivity and separability of uncountable graphs

Dirac [1, Theorem 10]† proved the interesting result that every 2-connected graph G of uncountable regular order contains a pair of vertices a, b such that $\mu_G(a, b) = |G|$ holds. The assertion is equivalent to $G >_u K_{2,|G|}$. In this form Dirac's theorem can be generalized as follows:

Theorem 9.1. *Let G be an n-connected graph ($n \in \mathbf{N}$) of uncountable regular order $\mathfrak{a}$. Then G contains a subdivision of the complete bipartite graph $K_{n,\mathfrak{a}}$.*

Proof. If $[G]_\mathfrak{a} \supseteq S(\mathfrak{a})$, then by Proposition 6.3 we have $G >_u S(\mathfrak{a}) \supseteq K_{n,\mathfrak{a}}$. Therefore let us assume $[G]_\mathfrak{a} \not\supseteq S(\mathfrak{a})$. By Proposition 6.1, $\mu_{[G]_\mathfrak{a}}(x, y) < \mathfrak{a}$ for all $x, y \in G$ with $[x, y] \notin E([G]_\mathfrak{a})$. Thus the hypotheses of Theorem 5.1 are fulfilled with respect to $[G]_\mathfrak{a}$. Therefore there exists a simplicial decomposition G_λ $(\lambda < \sigma)$ of $[G]_\mathfrak{a}$; $\sigma = \omega(\mathfrak{a})$. Since G is n-connected one easily shows that also every G_λ is n-connected and every S_λ has order at least n. For each λ, $0 < \lambda < \sigma$, choose an $x_\lambda \in G_\lambda - S_\lambda$ and n paths $P_{1\lambda}, \ldots, P_{n\lambda} \subseteq G_\lambda$ such that (i) each $P_{i\lambda}$ connects x_λ with a vertex of S_λ, (ii) $\mathring{P}_{i\lambda} \cap S_\lambda = \emptyset$, and (iii) for $i \neq j$, $P_{i\lambda}$ and $P_{j\lambda}$ have only x_λ in common. This choice is possible, by well-known properties of n-connected graphs. Let F_λ denote the union of $P_{1\lambda}, \ldots, P_{n\lambda}$ and set $F_\lambda \cap S_\lambda = T_\lambda$; $|T_\lambda| = n$ by the choice of the $P_{i\lambda}$.

For any λ, $0 < \lambda < \sigma$, let $f(\lambda)$ be the minimum of all ordinals κ such that $T_\lambda \subseteq G_\kappa$; $f(\lambda) < \lambda$ by the finiteness of T_λ.

By a theorem of Dushnik [3]‡ there is a λ_0 such that $|f^{-1}(\lambda_0)| = \mathfrak{a}$. For λ, $\kappa \in f^{-1}(\lambda_0)$, say $\lambda < \kappa$, we have

$$F_\lambda \cap F_\kappa \subseteq G_\lambda \cap G_\kappa \cap F_\kappa \subseteq S_\kappa \cap F_\kappa = T_\kappa \qquad (+)$$

Because the set of n-element subsets of $V(G_{\lambda_0})$ has cardinality at most $\max(\aleph_0, |V(G_{\lambda_0})|)$, there exist $\mathfrak{a}$ ordinals $\lambda \in f^{-1}(\lambda_0)$ such that all the corresponding T_λ coincide. Any two of the corresponding F_λ have only elements of T_λ in common, by (+). Hence $[G]_\mathfrak{a} >_u K_{n,\mathfrak{a}}$ from which the assertion follows by the same kind of argument as (ii)$\Rightarrow$(iii) in Proposition 6.3.

It is clear that Theorem 9.1 does not hold if $|G|$ is a singular cardinal. This is evident even in the case $n = 1$. All what we can say in this case is Theorem 9.4 below.

† See also [2] for a proof.

‡ We could also apply the well-known result that every uncountable connected graph of regular order $\mathfrak{a}$ contains a vertex of valency $\mathfrak{a}$, onto the graph with vertex set $W(\omega(\mathfrak{a}))$ and the edges $[\lambda, f(\lambda)]$.

Proposition 9.2. *If G is n-connected (n finite) and $H \subseteq G$, then there exists an n-connected subgraph H^* of G which contains H and has order at most* $\max(|H|, \aleph_0)$.

Proof. Set $H_0 = H$. Assume H_m to be already constructed. Then let H_{m+1} be the graph arising from H_m by adding n internally disjoint a, b-paths for every pair a, b of vertices of H_m with $\mu_{H_m}(a, b) < n$. Then it is easy to see that $\bigcup_{m=0}^{\infty} H_m = H^*$ is n-connected, and its order has the asserted bound.

From Proposition 9.2 we conclude,

Proposition 9.3. *If G is n-connected and infinite, then to every infinite cardinal $\mathfrak{a} \leq |G|$ there exists an n-connected subgraph H of G with $|H| = \mathfrak{a}$.*

From Proposition 9.3 and Theorem 9.1 we find,

Theorem 9.4. *If $|G| = \mathfrak{a} > \aleph_0$ is singular and G is n-connected, then there is a subdivision of $K_{n,\mathfrak{b}}$ in G for every $\mathfrak{b} < \mathfrak{a}$.*

Especially we have (for $n < \aleph_0$):

Corollary 9.5. *Every uncountable n-connected graph contains a subdivision of a finite n-connected graph.*

It is interesting that the analogous statement is not true for countable graphs and $n \geq 6$. For there are 6-connected countable planar graphs G; every finite graph H with $G >_u H$ or $G > H$ must have connectivity at most 5, by Euler's formula.

Next we investigate the alternative between the existence of a subdivision of an infinite simplex and of a finite set of vertices with high separation index. Again our result is obviously wrong for countable graphs.

Theorem 9.6. *Let G be a graph of uncountable regular order $\mathfrak{a}$ and $G \not>_u S(\aleph_0)$. Then there exists a finite $T \subset G$ such that $G - T$ has $\mathfrak{a}$ connected components.*

Proof. By Propositions 6.1 and 6.2 in connection with Theorem 5.1 we find a simplicial decomposition $G_\lambda (\lambda < \omega(\mathfrak{a}))$ of $[G]_{\aleph_1}$ in which all G_λ are finite or countable; especially all the S_λ must be finite. For $\lambda > 0$, let $f(\lambda)$ denote the smallest ordinal κ such that $S_\lambda \subseteq G_\kappa$. As in the proof of Theorem 9.1 there must be a λ_0 such that $f^{-1}(\lambda_0)$ has cardinality $\mathfrak{a}$, and there is a finite $T \subset V(G)$ such that $V(S_\lambda) = T$ for $\mathfrak{a}$ of the λ in $f^{-1}(\lambda_0)$. For these λ, all $G_\lambda - S_\lambda$ belong to different components of $[G]_{\aleph_1} - T$, by the separation-property $(*)$ stated in Section 3.

For the case of singular order we get by the same ideas as in the last proof:

Theorem 9.7. *Let G be a graph of uncountable singular order $\mathfrak{a}$ and $G \not\succ_u S(\aleph_0)$. Then for every $\mathfrak{b} < \mathfrak{a}$ there exists a finite $T \subset G$ such that $G - T$ has at least $\mathfrak{b}$ connected components.*

The existence of a T with the separation properties stated in Theorems 9.6 and 9.7 cannot be asserted if it is only stipulated $G \not\supseteq S(\aleph_0)$. This is shown by the minimal block that is obtained if every edge of an uncountable simplex is subdivided (at least) once.

By application of Proposition 6.5 we obtain the following sharpening of Theorem 9.6:

Theorem 9.8. *Let G be a graph of uncountable regular order $\mathfrak{a}$. Then (at least) one of the following statements holds:*

(i) $G \supseteq S(\aleph_0)$,

(ii) $G \supseteq U_{\aleph_1}(S(\aleph_0))$,

(iii) *There is a finite T such that $G - T$ has $\mathfrak{a}$ components.*

(If $|G| > \aleph_0$ is singular, (iii) has to be replaced by the statement: To every $\mathfrak{b} < |G|$ there is a finite T such $G - T$ has $\geq \mathfrak{b}$ components).

10. Normal rooted trees

Jung [15, 16] studied the following interesting concept: Let G be a connected graph. Then a rooted spanning tree T of G is called *normal* (with respect to G) if any two vertices which are adjacent in G are comparable with respect to the partial order of $V(G)$ determined by T and its root†.

A subset D of $V(G)$ is called *dispersed* (verstreut) if to every one-way infinite path U there exists a finite $F \subseteq G$ such that in $G - F$ there is no path connecting a $d \in D$ with a vertex $u \in U$. Jung [16, Satz 6′] proved that a necessary and sufficient condition for a connected graph G to have a normal rooted spanning tree (with given root) is that $V(G)$ is the union of a countable system of dispersed sets. Surprisingly, thus the choice of the root has no influence onto the existence or non-existence of such a spanning tree since the term "dispersed" is independent of the given root. It follows that, if G has a normal rooted spanning tree, then every connected subgraph of G has such a tree too. By Jung [16] every connected countable graph and every connected graph without an infinite system of two-way infinite paths has a normal rooted spanning tree. Using the preceding results we can sharpen the latter criterion.

† If r is the root, this partial order is defined by: $x \leq y$ if and only if the r, y-path in T contains x.

Theorem 10.1. *Every connected graph G which does not contain a subdivision of an infinite simplex has a normal rooted spanning tree.*

Proof. As in the proof of Theorem 9.6 we find a simplicial decomposition G_λ $(\lambda < \omega(|G|))$ of $[G]_{\aleph_1}$ where all $|G_\lambda| \leq \aleph_0$ and all S_λ are finite. In addition, by Theorem 5.1, each S_λ can be chosen in such a way that every $G_\lambda - S_\lambda$ is connected and S_λ is an inward simplex of G_λ.

By Jung's result, we can select a normal rooted spanning tree T_0 of G_0 (with respect to any given root $r_0 \in G_0$).

Now assume $0 < \tau < \omega(|G|)$ and suppose that for every $\lambda < \tau$ there are already determined T_λ, r_λ such that the following conditions are fulfilled:

(1) $r_\lambda \in S_\lambda$, if $\lambda > 0$,

(3) T_λ is a normal rooted spanning tree of $G_\lambda - (S_\lambda - r_\lambda)$, with respect to the root r_λ,

(3) $\bigcup_{\nu \leq \lambda} T_\nu =: T'_{\lambda+1}$ is a normal rooted spanning tree of $G'_{\lambda+1} = \bigcup_{\nu \leq \lambda} G_\nu$ (with root r_0).

Then $\bigcup_{\nu < \tau} T_\nu = \bigcup_{\lambda < \tau} T'_{\lambda+1} =: T$ is a normal rooted spanning tree of G'_τ. (For it is a tree, as the union of a chain of trees; it covers $V(G'_\tau)$; and finally, if there were adjacent x, y in G'_τ which are not comparable under the partial order $\leq$ of T with respect to r_0, the same situation would already occur in some $T'_{\lambda+1}$, $\lambda < \tau$, contradicting (3).)

Since S_τ is a simplex and $V(S_\tau) \subseteq V(T)$, any two vertices of S_τ are comparable under $\leq$, hence form a finite chain under $\leq$. Let r_τ be the maximum of this chain. By construction of G_τ, $G_\tau - (S_\tau - r_\tau)$ is connected and at most countable; hence by Jung's theorem we can choose a normal rooted spanning tree T_τ of $G_\tau - (S_\tau - r_\tau)$ with root r_τ.

We assert that $\bigcup_{\nu \leq \tau} T_\nu = T \cup T_\tau =: T'_{\tau+1}$ is a normal rooted spanning tree of $G'_{\tau+1}$ with root r_0.

First of course $T'_{\tau+1}$ is a tree spanning $G'_{\tau+1}$. Let now $[x, y] \in E(G'_{\tau+1})$. If x, y are both in G'_τ or both in $G_\tau - (S_\tau - r_\tau)$, it follows from the normality of T and from the choice of T_τ, respectively, that x, y are comparable under the partial order $\leq$ of $T'_{\tau+1}$ determined by the root r_0. If $x \in G_\tau - S_\tau$, $y \in G'_\tau$, then necessarily $y \in S_\tau$, hence $y \leq r_\tau \leq x$, and again x, y must be comparable under $\leq$.

In this way we get T_λ, r_λ for every $\lambda < \omega(|G|)$ such that conditions (1), (2), (3) are fulfilled, and we easily show (by the same argumentation as above concerning the union of the T_ν, $\nu < \tau$) that the union of all T_λ form a normal rooted spanning tree of G with respect to r_0.

By application of Proposition 6.5 we get the somewhat stronger result:

Theorem 10.2. *Every connected graph without an $S(\aleph_0)$ and without a $U_{\aleph_1}(S(\aleph_0))$ has a normal rooted spanning tree.*

It would be interesting to characterize the connected graphs which possess normal rooted spanning trees, in terms of forbidden configurations. We conclude with the following.

Conjecture 10.3. A connected graph G has a normal rooted spanning tree if and only if there is not an uncountable subset $X \subseteq V(G)$ such that $\mu_G(x, y) \geq \aleph_0$ for any $x \neq y$ of X holds.

References

[1] G.A. Dirac, Structural properties and circuits in graphs, Proc. 5th British Combinational Conference (1975) 135–140.

[2] G.A. Dirac, Cardinal-determining subgraphs of infinite graphs, Aarhus Universitet, Matematisk Institut, Preprint Series No. 27 (1976/77).

[3] B. Dushnik, A note on transfinite ordinals, Bull. Am. Math. Soc. 37 (1931) 860–862.

[4] P. Erdös and A. Hajnal, On chromatic number of graphs and set systems, Acta Math. Acad. Sci. Hungar. 17 (1966) 61–99.

[5] P. Erdös and R. Rado, A partition calculus in set theory, Bull. Am. Math. Soc. 62 (1956) 427–489.

[6] R. Halin, Über einen graphen-theoretischen Basisbegriff und seine Anwendung auf Färbungsprobleme, Dissertation Köln (1962).

[7] R. Halin, Über einen Satz von K. Wagner zum Vierfarben-problem, Math. Ann. 153 (1964) 47–62.

[8] R. Halin, Simpliziale Zerfällungen beliebiger (endlicher oder unendlicher) Graphen, Math. Ann. 156 (1964) 216–225.

[9] R. Halin, Charakterisierung der Graphen ohne unendliche Wege, Arch. Math. 16 (1965) 227–231.

[10] R. Halin, Zur Klassifikation der endlichen Graphen nach H. Hadwiger und K. Wagner, Math. Ann. 172 (1967) 46–78.

[11] R. Halin, Ein Zerlegungssatz für unendliche Graphen und seine Anwendung auf Homomorphiebasen, Math. Nachr. 33 (1967) 91–105.

[12] R. Halin, Unterteilungen vollständiger Graphen in Graphen mit unendlicher chromatischer Zahl, Abh. Math. Sem. Univ. Hamburg 31 (1967) 156–165.

[13] R. Halin und H.A. Jung, Über Minimalstrukturen von Graphen, insbesondere von n-fach zusammenhängenden Graphen, Math. Ann. 152 (1963) 75–94.

[14] H.A. Jung, Zusammenzüge und Unterteilungen von Graphen, Math. Nachr. 35 (1967) 241–268.

[15] H.A. Jung, Wurzelbäume und Kantenorientierungen in Graphen, Math. Nacht. 36 (1968) 351–360.

[16] H.A. Jung, Wurzelbäume und unendliche Wege in Graphen, Math. Nachr. 41 (1969) 1–22.

[17] W. Mader, Homomorphieeigenschaften und mittlere Kantendichte von Graphen, Math. Ann. 174 (1967) 265–268.

[18] W. Mader, Homomorphiesätze für Graphen, Math. Ann. 178 (1968) 154–168.

[19] K. Wagner, Über eine Erweiterung eines Satzes von Kuratowski, Deutsche Math. 2 (1937) 280–285.

[20] K. Wagner, Über eine Eigenschaft der ebenen Komplexe, Math. Ann. 114 (1937) 570–590.

[21] K. Wagner, Bemerkungen zu Hadwigers Vermutung, Math. Ann. 141 (1960) 433–451.

Annals of Discrete Mathematics 3 (1978) 111–123.

COMBINATORIAL COMPLETIONS*

Marshall HALL Jr.

Department of Mathematics, California Institute of Technology, Pasadena, CA, U.S.A.

1. Introduction

In combinatorial constructions, it is a matter of great interest to know when a partial system can be completed to a full one. Thus rows forming a Latin rectangle can always be completed to a full Latin square [4]. This and the related topic of completing nets to affine planes, studied by Bruck [1] are the subject of Section 2.

It has been shown by the writer and Ryser [8] that for rational matrices A satisfying either[1]

$$AA^{\mathrm{T}} = mI \tag{1}$$

or

$$AA^{\mathrm{T}} = (k-\lambda)I + \lambda J, \qquad AJ = kJ, \tag{2}$$

providing that some rational solution exists, any set of r initial rows satisfying the trivially necessary conditions will have a rational completion to a full matrix. These results are discussed in Section 3.

The more difficult question as to finding integral completions for (1) or (2) is discussed in Section 4. For (1) in results due to the writer [6] and to Verheiden [11] at least the last 7 rows may be added, but not in every case the last 8. For (2) it is shown in this paper for the first time that at least the last 4 rows may be added, but again an example shows that not in every case can the last 8 be added.

2. Completions of Latin squares and nets

It was shown by the writer [4] that a Latin rectangle can always be completed to a Latin square. An r by n Latin rectangle R on n letters $x_1, \ldots, x_n$ is a matrix of r rows and n columns such that each letter occurs exactly once in every row and no letter occurs twice in a column. The method of proof depends on considering the sets $S_1, \ldots, S_n$ which are subsets of $\{x_1, \ldots, x_n\}$ such that $x_j \in S_i$ if and only if x_j is not in the ith column of R. A further row which can be added to R to make an

* This research was supported in part by NSF Grant MPS-72-05035 A02.

[1] Equation (2) is satisfied by the incidence matrix of a symmetric block design. Equation (1) includes Hadamard matrices as special cases, and the case (2) with $\lambda = 0$ may also be thought of as a "null design".

$r+1$ by n Latin rectangle R' will consist of distinct representatives of $S_1, \ldots, S_n$ and conversely distinct representatives of $S_1, \ldots, S_n$ may be used to form such a row. The method of proof shows that such a row can be added in at least $(n-r)!$ ways. Then a further row can be added to R' in at least $(n-1-r)!$ ways. Hence if $n-r=s$, R can be completed to a Latin square in at least $s!(s-1)! \cdots 2!1!$ ways. In particular, even allowing for equivalence by permuting the last s rows arbitrarily, there are at least $(s-1)!(s-2)! \cdots 2!1!$ completions of R to a Latin square L.

The situation is quite different if we are given entries in a square not complete rows, such that no row or column contains a repeat. For example if the first row contains $x_1, \ldots, x_{n-1}$ in its first $n-1$ cells and the second row contains x_n in its last cell, then no completion exists. For the only possible entry in the nth cell of the first row is x_n and this yields a conflict in the last column. But a conjecture of Trevor Evans asserts that any start with no repeat in a row or column and at most $n-1$ entries can be completed. There is some information on this conjecture but it remains an open question.

Let L be a Latin square of order n. Let us take each of the n^2 cells of L as a point. From L we can define three families of parallel lines, each line containing n points. A line of the first family will be the points in a row of L. A line of the second family will be the points in a column of L. In the third family the ith line will consist of the cells of L containing x_i. Viewed in this way a Latin square is a 3-net N.

A k-net N is a system of points and lines such that

(i) N has at least one point;

(ii) the lines of N are partitioned into k disjoint, nonempty "parallel classes" such that

(a) each point of N is incident with exactly one line of each class;

(b) to two lines belonging to distinct classes there corresponds exactly one point of N which is incident with both lines.

Here we shall always suppose $k \geqslant 3$ and that the number of points in N is finite. To avoid trivial cases we also assume that every line has at least two points.

With these assumptions there is a positive integer n such that

(1) N has n^2 points;

(2) N has exactly kn distinct lines. These fall into k parallel classes of n lines each. Distinct lines of the same parallel class have no points in common. Two lines of different classes have exactly one common point.

(3) Every line of N contains exactly n points.

If we take two parallel classes and let the first correspond to the rows and the second to the columns of an n by n square, then let $K_1, \ldots, K_n$ be the lines of a third parallel class. If we place x_i in the cells corresponding to points of K_i, $i=1, \ldots, n$, we have a Latin square L. If there is a further parallel class corresponding to a Latin square L^*, then L and L^* are orthogonal which is to say that the pairs (a_{ij}, b_{ij}) with a_{ij} the entry in the cell of the ith row and jth column of L and b_{ij} the entry from L^*, then (a_{ij}, b_{ij}) $i=1, \ldots, n$; $j=1, \ldots, n$ gives all n^2

pairs (x_i, x_j) $i=1,\ldots,n$; $j=1,\ldots,n$. Conversely $k-2$ Latin squares any two of which are orthogonal yield a k-net.

A k-net on n^2 points can have k at most $n+1$ since there is at most one line joining two distinct points. An $(n+1)$-net on n^2 points is an affine plane of order n, designated $E(2, n)$, and conversely an affine plane of order n is an $(n+1)$-net on n^2 points. Thus a k-net N on n^2 points can be extended to an affine plane provided that $n+1-k=d$ further parallel classes can be adjoined to N to form a larger net N^*. Bruck [1] has considered the completion problem of extending a k-net to an affine plane. Here $d=n+1-k$ is called the deficiency of the net. His main results show that the completion is possible provided that d is small compared to n.

Theorem 2.1 (Bruck). *If $n>(d-1)^2$ and if N can be completed at all, then it can be completed uniquely.*

Theorem 2.2 (Bruck). *If $n>p(d-1)$, N can always be completed where $p(x)=\frac{1}{2}x^4+x^3+x^2+\frac{3}{2}x$.*

With a k-net N of order n (N has n^2 points) Bruck associates a graph G_1. The vertices of G_1 are the points of N and the (undirected) edges of G_1 are arcs joining P_i and P_j if and only if P_i and P_j are on a line of N. Let G_2 be the complementary graph of G_1. If G_2 can be shown to be the graph of a d-net N_2 then N and N_2 together form an affine plane. He can describe abstractly the properties of G_2 and define this as a *pseudo-net graph.* His proof consists in determining conditions so that G_2 will indeed be the graph of a d-net.

A transversal of a k-net N is a set of n points of N no two of which are on a line of N. He is able to prove that if $n>(d-1)^2$ two distinct transversals have at most one point in common. This proves his Theorem 2.1 above. The proof of Theorem 2.2 is more complicated.

3. Rational completions

The problems considered here can be described in terms of matrices. Let A be a rational matrix of order n such that for a positive integer m

$$AA^{\mathrm{T}}=mI. \tag{3}$$

Here A^{T} is the transpose of A. A necessary condition for the existence of a rational A satisfying (3) is given by

(1) If n is odd, m is a square.
(2) If $n\equiv 2 \pmod 4$, $=a^2+b^2$ for integers a, b.
(3) If $n\equiv 0 \pmod 4$, m is positive.

These necessary conditions for existence of a rational A are sufficient for the existence of an integral A satisfying (3).

For the second equation let $v > k > \lambda > 0$ be integers satisfying

$$k(k-1) = \lambda(v-1). \tag{4}$$

Let A be a matrix of order v satisfying

$$AA^{\mathrm{T}} = (k-\lambda)T + \lambda J = B, \tag{5}$$

where J is the matrix of order v with every entry a 1. This is the incidence equation satisfied by the incidence matrix A of a symmetric v, k, λ block design D. An incidence matrix A satisfies the further relations

$$AJ = kJ, \qquad JA = kJ. \tag{6}$$

$$A^{\mathrm{T}}A = (k-\lambda)I + \lambda J. \tag{7}$$

The following theorem due to Ryser may be found in the writer's book [5, p. 104].

Theorem 3.1. *A non-singular matrix A of order v satisfying either* (5) *or* (7) *and either of the relations in* (6) *will satisfy all four relations. Furthermore $k(k-1) = \lambda(v-1)$ will also be a consequence.*

As $AA^{\mathrm{T}} = A^{\mathrm{T}}A$ we shall refer to a matrix A satisfying all four equations as a normal matrix satisfying its incidence equations.

Necessary and sufficient conditions for the existence of a rational solution to the incidence equation (5) depend on the deep Hasse-Minkowski theory of quadratic forms. They are, however, easy to state and may be found in [5, p. 107].

Theorem 3.2 (Bruck, Ryser, Chowla [2, 3]). *Necessary and sufficient conditions for the existence of a rational matrix A satisfying* (5) *are*

(1) *If v is even, $k-\lambda$ is a square.*

(2) *If v is odd, $z^2 = (k-\lambda)x^2 + (-1)^{(v-1)/2}\lambda y^2$ has a solution in integers x, y, z not all zero.*

The completion problem of interest here is to be given an r by n (or r by v) matrix X to decide whether or not there is a square matrix A having X as its first r rows which satisfies (3) or (5). We shall treat the cases separately, though they have much in common.

For (3) X clearly must satisfy the condition

$$XX^{\mathrm{T}} = mI_r. \tag{8}$$

Two theorems are relevant here. The first comes directly from Hall and Ryser [8].

Theorem 3.3 (Hall and Ryser). *Suppose that A is a non-singular square matrix such that*

$$AA^{\mathrm{T}} = \left[\begin{array}{c|c} D_1 & 0 \\ \hline 0 & D_2 \end{array}\right] = D_1 \oplus D_2,$$

where D_1 is of order r, D_2 or order s and $r+s=n$. Let X be an arbitrary matrix of size r by n, such that $XX^{\mathrm{T}} = D_1$. Then there is an n by n matrix Z having X as its first r rows such that $ZZ^{\mathrm{T}} = D_1 \oplus D_2$. This result holds for all fields F of characteristic not 2.

Corollary 3.4. *An r by n matrix X such that $XX^{\mathrm{T}} = mI_r$ can be completed to an n by n matrix Z with X as its first r rows satisfying $ZZ^{\mathrm{T}} = mI$ providing that some matrix A exists with $AA^{\mathrm{T}} = mI$.*

Thus providing that the neccessary existence conditions hold for $AA^{\mathrm{T}} = mI$ then any initial r rows X satisfying $XX^{\mathrm{T}} = mI_r$ can be completed to a solution Z of $ZZ^{\mathrm{T}} = mI_n$.

The second theorem, a slight generalization of one in Hall and Ryser [8] may be found in [6].

Theorem 3.5. *Suppose that $AA^{\mathrm{T}} = D_1 \oplus D_2$ where A is of order n and nonsingular and D_1 and D_2 are of order r and $s = n - r$ and are nonsingular. Suppose further that X and Y are r by n matrices such that $XX^{\mathrm{T}} = YY^{\mathrm{T}} = D_1$. Then there exists an orthogonal matrix U of order n such that $XU = Y$. This result holds for all fields F of characteristic different from 2.*

Here an orthogonal matrix U is one satisfying $UU^{\mathrm{T}} = U^{\mathrm{T}}U = I$.

For the incidence equation (5) we shall assume that X and r by v matrix satisfies the two conditions

$$XX^{\mathrm{T}} = (k - \lambda)I_r + \lambda J_{rr}, \tag{9}$$

$$XJ = kJ_{rv}. \tag{10}$$

Here J_{rr} and J_{rv} are respectively matrices of sizes r by r and r by v in which every entry is 1.

Theorem 3.6 (Hall and Ryser). *Suppose that the conditions of Theorem 3.2 for the existence of a rational solution to* (3) *are satisfied. Then given an r by v matrix X the conditions of* (9) *and* (10) *are both necessary and sufficient for the existence of a rational v by v matrix Z with X as its first r rows which is a normal matrix satisfying the incidence equations* (5), (6) *and* (7).

Thus over the rational field providing (3) or (5) has any rational solution a matrix X of r rows satisfying the appropriate conditions (8) or (9) and (10) has a rational completion.

The key theorem is Theorem 3.3 whose proof depends on the Witt "subtraction theorem" for quadratic forms. This is given on p. 276 of (5).

4. Integral completions

Given an integral matrix X of size r by n (or r by v) satisfying the appropriate conditions (8) or (9) and (10), we ask here about the existence of an integral matrix Z satisfying (3) or (5), (6), (7), where X is the matrix of the first r rows of Z.

In these cases the completion can be shown to exist for small values of $s=n-r$ (or $s=v-r$). It may be true that, as in Bruck's results, the completion will exist if s is small compared to n (or v).

It was shown by the writer [6] that for (3) the solution exists providing $s=1$ or 2. This result has been improved by one of the writer's students, Verheiden [11].

Theorem 4.1 (Verheiden). *Let $r+s=n$. If X is an integral r by n matrix such that $XX^{\mathrm{T}}=mI_r$ and if some rational n by n matrix exists with $AA^{\mathrm{T}}=mI$, then providing $s\leq 7$ there exists a rational n by n matrix Z having X as its first r rows such that $ZZ^{\mathrm{T}}=mI$.*

This depends upon deep theorem on integral quadratic forms, in particular the result that an integral quadratic form which is positive definite and has determinant 1 will be integrally equivalent to a sum of squares if the dimension (number of variables) is at most 7.

For $s=8$ the conclusion of Theorem 4.1 is false. Suppose X is the 1 by 9 matrix consisting of 9 ones. Then there is no 9 by 9 matrix Z with its first row consisting of all 1's such that $ZZ^{\mathrm{T}}=9I$. If there were such a matrix the sum of the squares of the elements in the second row would be 9 and so the sum of these elements would be odd, conflicting with the fact that the inner product with the first row must be zero.

The results of Theorem 4.1 were obtained later independently by Hsia [9].

Using further deep properties of quadratic forms Verheiden proved the further interesting result.

Theorem 4.2 (Verheiden). *Under the hypotheses of Theorem* 4.1 *there exists, for any s, an n by n matrix Z having X as its first r rows such that $ZZ^{\mathrm{T}}=mI$ and such that there is some power of* 2, *say 2^e so that 2^eZ is integral.*

In the case of 9 by 9 matrices in which the first row consists of 9 1's there is a completion Z such that $2Z$ is integral.

$$2Z=\begin{bmatrix} 2 & 2 & 2 & 2 & 2 & 2 & 2 & 2 & 2 \\ 3 & 3 & -3 & -3 & 0 & 0 & 0 & 0 & 0 \\ 3 & -3 & 3 & -3 & 0 & 0 & 0 & 0 & 0 \\ 3 & -3 & -3 & 3 & 0 & 0 & 0 & 0 & 0 \\ 0 & 0 & 0 & 0 & 3 & 3 & -3 & -3 & 0 \\ 0 & 0 & 0 & 0 & 3 & -3 & 3 & -3 & 0 \\ 0 & 0 & 0 & 0 & 3 & -3 & -3 & 3 & 0 \\ 2 & 2 & 2 & 2 & -1 & -1 & -1 & -1 & -4 \\ 1 & 1 & 1 & 1 & -2 & -2 & -2 & -2 & 4 \end{bmatrix}. \tag{11}$$

This example also shows that 2 cannot be replaced by any other prime in Theorem 4.2 since the same difficulty remains when Z is multiplied by any odd number.

Now let us suppose that X is an integral matrix of size r by v satisfying the conditions of (9) and (10). Let $s=v-r$ and let Z be a rational normal completion of X

$$Z=\begin{bmatrix} X \\ Y \end{bmatrix}=\begin{bmatrix} x_{11} & \cdots & x_{1v} \\ x_{21} & \cdots & x_{2v} \\ \cdot\cdot & \cdots & \cdot \\ x_{r1} & \cdots & x_{rv} \\ y_{11} & \cdots & y_{1v} \\ y_{s1} & \cdots & y_{sv} \end{bmatrix}. \tag{12}$$

Here, being a rational normal solution, Z satisfies

$$ZZ^{\mathrm{T}}=Z^{\mathrm{T}}Z=(k-\lambda)I+\lambda J \tag{13}$$

and

$$ZJ=JZ=kJ. \tag{14}$$

Here (14) asserts that every row or column of Z has sum k while (13) asserts that the inner product of a row (or column) with itself is k and the inner product of two different rows (or columns) is λ.

Considering the ith column of Z we have

$$\begin{aligned} x_{1i}+x_{2i}+\cdots+x_{ri}+y_{1i}+\cdots+y_{si}&=k,\\ x_{1i}^2+x_{2i}^2+\cdots+x_{ri}^2+y_{1i}^2+\cdots+y_{si}^2&=k. \end{aligned} \tag{15}$$

Let us write

$$y_{1i}+\cdots+y_{si}=u_i, \qquad y_{1i}^2+\cdots+y_{si}^2=w_i. \tag{16}$$

Then subtracting the first equation of (15) from the second we have

$$\sum_{j=1}^{r}(x_{ji}^2-x_{ji})+w_i-u_i=0. \tag{17}$$

As the x_{ji} are rational integers we have $x_{ji}^2-x_{ji}\geqslant 0$ with equality only when $x_{ji}=0$ or 1. Thus

$$u_i\geqslant w_i \tag{18}$$

with equality only when every $x_{ji}=0$ or 1. On the other hand

$$\sum_{i=1}^{v}u_i=\sum_{i=1}^{v}\sum_{j=1}^{s}y_{ji}=\sum_{j=1}^{s}\sum_{i=1}^{v}y_{ji}=\sum_{j=1}^{s}k=sk. \tag{19}$$

Similarly

$$\sum_{i=1}^{v}w_i=\sum_{i=1}^{v}\sum_{j=1}^{s}y_{ji}^2=\sum_{j=1}^{s}\sum_{i=1}^{v}y_{ji}^2=\sum_{j=1}^{s}k=sk. \tag{20}$$

From (18), (19) and (20) it now follows that

$$u_i=w_i,\quad i=1,\ldots,v,\quad x_{ji}=0 \text{ or } 1,\quad j=1,\ldots,r,\quad i=1,\ldots,v. \tag{21}$$

Theorem 4.3. *Let us suppose that* (5) *has a rational solution and let X be an r by v integral matrix satisfying* (9) *and* (10) *and let*

$$Z=\begin{bmatrix}X\\Y\end{bmatrix}$$

be a normal rational completion of X satisfying (5), (6) *and* (7). *Then every $x_{ij}=0$ or 1 where $X=[x_{ij}]$ and in a column of Y (where $r+s=v$)*

$$y_{1j}+y_{2j}+\cdots+y_{sj}=y_{1j}^2+\cdots+y_{sj}^2=u_j$$

and $0\leqslant u_j\leqslant s$, $j=1,\ldots,v$ and $\sum_{j=1}^{v}u_j=sk$.

Proof. All parts of this theorem have already been proved except for the inequalities $0\leqslant u_j\leqslant s$. Trivially $0\leqslant u_j$ with equality only when every $y_{ij}=0$. From the Cauchy-Schwartz inequality $(y_1+\cdots+y_s)^2\leqslant s(y_1^2+\cdots+y_s^2)$ with equality only when $y_1=y_2=\cdots=y_s$. But in our case $y_1^2+\cdots+y_s^2=y_1+\cdots+y_s$ so that

$$(y_1+\cdots+y_s)^2\leqslant s(y_1+\cdots+y_s)$$

and $y_1+\cdots+y_s\leqslant s$ or $u_j\leqslant s$ with equality only when $y_1=y_2\cdots=y_s=1$.

We note that since every x is 0 or 1, that an integral completion, if it exists, will be a 0-1 matrix and so an incidence matrix.

We can now prove our main result on this subject.

Theorem 4.4. *Suppose that $v>k>\lambda>0$ are integers satisfying $k(k-1)=\lambda(v-1)$ and that there exists a rational matrix A of order v satisfying $AA^{\mathrm{T}}=(k-\lambda)I+\lambda J$. Suppose also that we are given an integral r by v matrix X such that*

$$XX^{\mathrm{T}}=(k-\lambda)I_r+\lambda J_{rr}, \tag{22}$$

$$XJ=kJ_{rv}. \tag{23}$$

Then with $s=v-r$, if $1\leqslant s\leqslant 4$ there exists an integral v by v matrix W with X as its first rows such that

$$WW^{\mathrm{T}}=W^{\mathrm{T}}W=(k-\lambda)I+\lambda J, \tag{24}$$

$$WJ=JW=kJ. \tag{25}$$

This conclusion is false if $s=8$.

If

$$Z=\begin{bmatrix} X \\ Y \end{bmatrix}$$

is a rational normal completion of X whose existence is assured by Theorem 3.6, then also

$$Z_1=\begin{bmatrix} X \\ UY \end{bmatrix}$$

is a rational normal solution providing that U is an s by s rational matrix satisfying

$$UU^{\mathrm{T}}=U^{\mathrm{T}}U=I_s, \qquad (1,\ldots,1)U=(1,\ldots,1)=e_s, \tag{26}$$

the vector e_s being a vector of s 1's. This last condition is necessary so that $JZ_1=kJ$. If $c_j=[y_{ij},\ldots,y_{sj}]^{\mathrm{T}}$ is a column of Y then Uc_j is the corresponding column of UY. Hence by Theorem 3.5 if $e_s^{\mathrm{T}}, c_{j_1}, c_{j_2},\ldots,c_{j_m}$ are linearly independent columns we can choose U so that $Ue_s^{\mathrm{T}}=e_s^{\mathrm{T}}, Uc_{j_1}, Uc_{j_2}, Uc_{j_m}$ are any rational columns with the same inner products as the original ones.

From (12), (13) and (14) we have the following relations on the columns of Z

$$\begin{aligned} &x_{1j}+x_{2j}+\cdots+x_{rj}+y_{1j}+\cdots+y_{sj}=k,\\ &x_{1j}^2+x_{2j}^2+\cdots+x_{rj}^2+y_{1j}^2+\cdots+y_{sj}^2=k,\\ &x_{1j}x_{1t}+x_{2j}x_{2t}+\cdots+x_{rj}x_{rt}+y_{1j}y_{1t}+\cdots+y_{sj}y_{st}=\lambda, \quad j\neq t. \end{aligned} \tag{27}$$

As the x's are integers it follows that the inner products of the columns of Y are all integers. We also recall from Theorem 4.3 that

$$y_{1j}+y_{2j}+\cdots+y_{sj}=y_{1j}^2+y_{2j}^2+\cdots+y_{sj}^2=u_j. \tag{28}$$

Proof of theorem 4.4.

Case $s=1$. Here from (28) $y_{1j}=u_j$ is an integer in every case and the theorem holds.

Case $s=2$. $0 \leq u_j \leq 2$, $j=1m\ldots, v$ from Theorem 4.3. If $u_j=0$ the jth column is $[0,0]^{\mathrm{T}}$, while if $u_j=2$ the jth column is $[1,1]^{\mathrm{T}}$. Hence if no $u_j=1$ the last two rows are identical, which is a conflict since they are linearly independent. Hence some $u_j=1$ and permuting the columns we may assume $u_1=1$. Choosing U appropriately for the 1st column c_1,

$$c_1=\begin{bmatrix}1\\0\end{bmatrix}. \tag{29}$$

Here

$$Y=\begin{bmatrix}1 & y_{12} & \cdots & y_{1j} & \cdots & y_{1v}\\ 0 & y_{22} & \cdots & y_{2j} & \cdots & y_{2v}\end{bmatrix}. \tag{30}$$

As the inner product of the first and jth column is an integer λ_{1j} we have $(c_1, c_j)=y_{1j}=\lambda_{1j}$ is integral and as $y_{1j}+y_{2j}=u_j$ is an integer then y_{2j} is also an integer. Hence in (30) Y is an integral completion for X making

$$W=\begin{bmatrix}X\\Y\end{bmatrix}$$

a full integral normal completion of X. This proves our theorem for $s=2$.

Case $s=3$. $0\leq u_j \leq 3$. If $u_j=0$, $c_j=[0,0,0]^{\mathrm{T}}$, while if $u_j=3$, $c_j=[1,1,1]^{\mathrm{T}}$. Since the last three rows are not identical there must be some $u_j=1$ or some $u_j=2$. Suppose some $u_j=1$, and we may take $u_1=1$ so that with an appropriate U

$$c_1=\begin{bmatrix}1\\0\\0\end{bmatrix}. \tag{31}$$

Hence for the 1st and jth columns $\lambda_{ij}=(c_1,c_j)=y_{1j}$ and so the first row of Y is integral. Adjoining this integral row to X we are reduced to the case $s=2$ and our result holds. Suppose that no $u_j=1$ but that some $u_j=2$. We may take $u_1=2$ and with an appropriate U the column c_1 becomes

$$c_1=\begin{bmatrix}1\\1\\0\end{bmatrix}. \tag{32}$$

Here if $c_j=[y_{1j}, y_{2j}, y_{3j}]^{\mathrm{T}}$ then

$$\lambda_{ij}=(c_1,c_j)=y_{1j}+y_{2j}, \qquad u_j=y_{1j}+y_{2j}+y_{3j}. \tag{33}$$

It now follows that y_{3j} is an integer for $j=1,\ldots,v$. Adjoining this integral row to X we are reduced to the case $s=2$ already proved.

This completes the proof for $s=3$.

Case $s=4$. $0\leqslant u_j\leqslant 4$. If $u_j=0$, $c_j=[0,0,0,0]^T$. If $u_j=4$, $c_j=[1,1,1,1]^T$. If some $u_j=1$ take $u_1=1$, $c_1=[1,0,0,0]^T$. Then $\lambda_{1j}=(c_1,c_j)=y_{1j}$ is an integer and adjoining this integral row to X we are reduced to the case $s=3$. If some $u_j=3$ take $u_1=3$, $c_1=[1,1,1,0]^T$. Then with $c_j=[y_{1j}, y_{2j}, y_{3j}, y_{4j}]^T$ we have

$$\lambda_{1j}=(c_1,c_j)=y_{1j}+y_{2j}+y_{3j}$$

and

$$u_j=y_{1j}+y_{2j}+y_{3j}+y_{4j}.$$

Hence y_{4j} is integral for $j=1,\ldots,v$. Adjoining this integral row to X we are reduced to the case $s=3$.

Thus we need consider only cases where u_j takes on only the values 0, 2, 4. There must be some cases with $u_j=2$. Take $u_1=2$ and $c_1=[1, 1, 0, 0]^T$. If a further column has $u_j=2$ then $c_j=[a, b, c, d]^T$ with $a+b+c+d=a^2+b^2+c^2+d^2=2$ and $\lambda_{1j}=a+b$. If $\lambda_{1j}\geqslant 3$ then $a^2+b^2\geqslant(\frac{3}{2})^2+(\frac{3}{2})^2=\frac{9}{2}$, a conflict. If $\lambda_{1j}\leqslant -1$ then $c+d\geqslant 3$ and $c^2+d^2>\frac{9}{2}$ a conflict. Hence $\lambda_{1j}=0$, 1, or 2. If $\lambda_{1j}=2$ then $c_j=[1, 1, 0, 0]^T$ and is identical with c_1. As Y is of rank 4 there must be 4 linearly independent columns. Thus there must be at least three columns with $u=2$, and besides $u_1=2$ two others with $u_j=2$ and $\lambda_{1j}=0$ or 1.

Suppose first that there are three columns with $u=2$ and $\lambda_{ij}=1$ in all three cases. Then Y, taking $u_1=2$, $u_2=2$, $u_3=2$, $\lambda_{12}=1$, $\lambda_{13}=1$, $\lambda_{23}=1$ will have the shape

$$Y=\begin{bmatrix} 1 & 1 & 0 & \cdots & y_{1j} & \cdots \\ 1 & 0 & 1 & \cdots & y_{2j} & \cdots \\ 0 & 1 & 1 & & y_{3j} & \cdots \\ 0 & 0 & 0 & \cdots & y_{4j} & \cdots \end{bmatrix}. \tag{34}$$

If $u_j=4$ then $c_j=[1,1,1,1]^T$ which is integral. If $u_j=0$, $c_j=[0,0,0,0]^T$. If $u_j=2$ then

$$\begin{aligned} &\lambda_{1j}=y_{1j}+y_{2j}, \qquad \lambda_{2j}=y_{1j}+y_{3j}, \qquad \lambda_{3j}=y_{2j}+y_{3j}, \\ &y_{1j}+y_{2j}+y_{3j}+y_{4j}=2. \end{aligned} \tag{35}$$

Hence

$$\lambda_{1j}+\lambda_{2j}+\lambda_{3j}=2(y_{1j}+y_{2j}+y_{3j}). \tag{36}$$

If $\lambda_{1j}+\lambda_{2j}+\lambda_{3j}$ is odd, then each of y_{1j}, y_{2j}, y_{3j} and so also y_{4j} is half an odd integer. But with odd integers a, b, c, d then

$$\left(\frac{a}{2}\right)^2+\left(\frac{b}{2}\right)^2+\left(\frac{c}{2}\right)^2+\left(\frac{d}{2}\right)^2=2, \qquad a^2+b^2+c^2+d^2=8, \tag{37}$$

which is impossible with odd a, b, c, d. Hence $\lambda_{1j}+\lambda_{2j}+\lambda_{3j}$ is even and y_{1j}, y_{2j}, y_{3j} and y_{4j} are all integers. Thus in this case Y is integral and our theorem holds.

There remains to be considered the cases in which there are two u's equal to 2 and the $\lambda_{ij}=0$. Here Y has the shape

$$Y=\begin{bmatrix} 1 & 0 & \cdots & a \\ 1 & 0 & \cdots & b \\ 0 & 1 & \cdots & c \\ 0 & 1 & \cdots & d \end{bmatrix}. \qquad \text{(column } (j) \text{ is the last shown)} \tag{38}$$

For a column c_j with $u_j=2$ different from c_1 and c_2 we have $\lambda_{1j}=a+b$, $\lambda_{2j}=c+d$, $\lambda_{1j}+\lambda_{2j}=u_j=2$. With $\lambda_{1j}=0$ or 1 and $\lambda_{2j}=0$ or 1 the only possibility is $\lambda_{1j}=1$, $\lambda_{2j}=1$ and taking $j=3$ we may put Y in the shape

$$Y=\begin{bmatrix} 1 & 0 & 1 & \cdots & a & \cdots \\ 1 & 0 & 0 & \cdots & b & \cdots \\ 0 & 1 & 1 & \cdots & c & \cdots \\ 0 & 1 & 0 & \cdots & d & \cdots \end{bmatrix}. \tag{39}$$

If $u_j=4$ then $c_j[1, 1, 1, 1]^{\mathrm{T}}=c_1+c_2$ so that a further independent column c_j must have $u_j=2$ and $\lambda_{1j}=1$, $\lambda_{2j}=1$, $\lambda_{3j}=0$ or 1. Thus with this $j=4$, Y has the shape

$$Y=\begin{bmatrix} 1 & 0 & 1 & 0 & \cdots & a & \cdots \\ 1 & 0 & 0 & 1 & \cdots & b & \cdots \\ 0 & 1 & 1 & 0 & \cdots & c & \cdots \\ 0 & 1 & 0 & 1 & \cdots & d & \cdots \end{bmatrix}, \qquad \lambda_{34}=0 \tag{40}$$

or

$$Y=\begin{bmatrix} 1 & 0 & 1 & 1 & \cdots & a & \cdots \\ 1 & 0 & 0 & 0 & \cdots & b & \cdots \\ 0 & 1 & 1 & 0 & \cdots & c & \cdots \\ 0 & 1 & 0 & 1 & \cdots & d & \cdots \end{bmatrix}, \qquad \lambda_{34}=1. \tag{41}$$

With $\lambda_{34}=0$ we have $y_{14}+y_{34}=0$, $y_{24}+y_{44}=2$ and so $y_{24}^2+y_{44}^2\geqslant 2$ but as $u_4=2$ this gives $c_4=[0, 1, 0, 1]^{\mathrm{T}}$ as in (40) above. But here $c_1+c_2=c_3+c_4$ and so c_4 is not an independent column and we must have Y in the shape (41) with $\lambda_{34}=1$. Here for a further column c_j we have $c_j=[0, 0, 0, 0]^{\mathrm{T}}$ if $u_j=0$ and $c_j=[1, 1, 1, 1]^{\mathrm{T}}$ if $u_j=4$. For any further column $c_j=[a, b, c, d]^{\mathrm{T}}$ we will have $u_j=2$, $\lambda_{1j}=a+b=1$, $\lambda_{2j}=c+d=1$, $\lambda_{3j}=a+c$ and $\lambda_{4j}=a+d$. Hence $\lambda_{3j}+\lambda_{4j}-\lambda_{2j}=2a$ is an integer. If a is half an odd integer, so is b, c, and d and as in (37) this leads to a conflict. Thus a is an integer and so are b, c, and d.

This covers the final possibility with $s=4$ and we conclude that Y can be taken as integral in this case.

For $s=8$ in the case $v=11$, $k=5$, $\lambda=2$ the initial rows

$$\begin{bmatrix} 1&1&1&1&1&0&0&0&0&0&0\\ 1&1&0&0&0&1&1&1&0&0&0\\ 1&1&0&0&0&0&0&0&1&1&1 \end{bmatrix} \tag{42}$$

cannot be integrally completed because an integral completion would be a symmetric 11, 5, 2 design and any two columns would also have inner product 2, which is not possible here for the first two columns. But there is a rational normal completion of denominator 2.

$$W=\frac{1}{2}\begin{bmatrix} 2&2&2&2&2&0&0&0&0&0&0\\ 2&2&0&0&0&2&2&2&0&0&0\\ 2&2&0&0&0&0&0&0&2&2&2\\ 2&-1&2&0&1&2&0&1&2&0&1\\ 2&-1&0&2&1&0&2&1&0&2&1\\ 0&1&2&2&-1&0&1&2&1&0&2\\ 0&1&2&0&1&2&1&0&-1&2&2\\ 0&1&2&0&1&0&2&1&2&2&-1\\ 0&1&0&2&1&2&-1&2&1&2&0\\ 0&1&0&2&1&2&2&-1&2&0&1\\ 0&1&0&0&3&0&1&2&1&0&2 \end{bmatrix}$$

References

[1] R.H. Bruck, Finite Nets II—Uniqueness and Imbedding, Pacific J. Math. 13 (1963) 421–457.

[2] R.H. Bruck and H.J. Ryser, The nonexistence of certain finite projective planes, Can. J. Math. 1 (1949) 88–93.

[3] S. Chowla and H.J. Ryser, Combinatorial problems, Can. J. Math. 2 (1950) 93–99.

[4] M. Hall Jr., Distinct representatives of subsets, Bull. Am. Math. Soc. 54 (1948) 922–926.

[5] M. Hall Jr., Combinatorial Theory (Wiley, New York, 1967).

[6] M. Hall Jr., Integral matrices A for which $AA^T=mI$ (Academic Press, New York, to appear).

[7] M. Hall Jr., Matrices satisfying the incidence equation, Proc. Fifth Hungarian Conference on Combinatorics.

[8] M. Hall and H.J. Ryser, Normal Completions of incidence matrices, Am. J. Math. 76 (1954) 581–589.

[9] J.J. Hsia, Two theorems on integral matrices, J. Linear Multilinear Algebra, to appear.

[10] H.J. Ryser, Matrices with integer elements in combinatorial investigations, Am. J. Math. 74 (1952) 769–773.

[11] E. Verheiden, Integral and rational completions of combinatorial matrices, J. Combinatorial Theory (A), to appear.

Annals of Discrete Mathematics 3 (1978) 125–127.

A CLASS OF REGULARISABLE GRAPHS

F. JAEGER and C. PAYAN

I.R.M.A., B.P. 53, 38041, Grenoble, Cédex, France

1. Introduction

In this volume Berge [1] gives a generalization to hypergraphs of the following result on line graphs: the line-graph of a graph with no pendent vertex is regularisable.

In this note we characterise regularisable graphs without an induced subgraph isomorphic to $K_{1,3}$, a star with three edges. Since no line-graph contains an induced $K_{1,3}$, this is another extension of the result above.

Throughout the note we use the notation of [1].

2. A class of non-regularisable graphs

Definition 2.1. A graph will be said to be of class C whenever it can be obtained by the following two operations:

(i) take an elementary cycle of even length together with a (proper) colouring of its vertices with colours α and β;

(ii) add some edges (at least one) whose endvertices are coloured α and are at distance 2 on the cycle.

For example, omitting an edge of K_4 we obtain a graph of class C. Clearly, a graph of class C has no induced subgraph isomorphic to $K_{1,3}$.

Lemma 2.2. *A graph G of class C is not regularisable.*

Proof. G is connected, it is not bipartite and the set S of vertices of G coloured β is a non-empty independent set of G such that $|\Gamma S| = |S|$. The result now follows from the [1 Theorem 3.1].

3. A class of regularisable graphs

Theorem 3.1. *A connected graph G with no pendent vertex and with no induced subgraph isomorphic to $K_{1,3}$ is regularisable if and only if it is not of class C.*

Proof. By Lemma 2.2 it is sufficient to prove the "if" part of this theorem.

Let then G be a connected graph with no pendent vertex, with no induced subgraph isomorphic to $K_{1,3}$, which is not regularisable. We shall show that G is of class C.

G is not bipartite (otherwise G would be an elementary cycle of even length). Hence by [1, Theorem 3.1], there exists a non-empty stable S of G such that $|\Gamma S| \leqslant |S|$.

Every vertex of S is adjacent to at least 2 vertices of ΓS since G has no pendent vertex. Moreover, every vertex of ΓS is adjacent to at most 2 vertices of S since G has no induced subgraph isomorphic to $K_{1,3}$.

Let m be the number of edges connecting a vertex of S to a vertex of ΓS. It follows from the above that

$$2\,|S| \leqslant m \leqslant 2\,|\Gamma S| \leqslant 2\,|S|.$$

Hence $|S| = |\Gamma S|$ and every vertex of S (respectively ΓS) is adjacent to exactly 2 vertices of ΓS (respectively S). Furthermore, there is no edge connecting a vertex of ΓS to a vertex of $V(G) - (S \cup \Gamma S)$ (otherwise we could find an induced subgraph of G isomorphic to $K_{1,3}$); since G is connected, $V(G) = S \cup \Gamma S$.

Consider now the subgraph $G[\Gamma S]$ of G induced by ΓS; $G[\Gamma S]$ has at least one edge (since G is not bipartite). Moreover, for every edge $e = \{x, y\}$ of $G[\Gamma S]$ there exists $z \in S$ such that $\{z, x\} \in E(G)$ and $\{z, y\} \in E(G)$ otherwise the subgraph induced by y, x and the 2 vertices of S adjacent to x would be isomorphic to $K_{1,3}$.

This, together with the fact that G is connected, shows that G is of class C.

4. An application to line graphs of graphs

Definition 4.1. A graph will be said to be of class K whenever it can be obtained by the following three operations:

(i) take an elementary cycle of length $l \geqslant 3$;

(ii) choose a non-empty subset S of the vertices such that if $|S| = 1$, then l is odd, and if $|S| \geqslant 2$, there is no chain of even length with its two endvertices in S and its other vertices outside S;

(iii) for every $s \in S$ add an edge joining s to a (new) pendent vertex.

Lemma 4.2. *A graph is of class C if and only if it is isomorphic to the line-graph of a graph of class K.*

We leave the proof to the reader.

Theorem 4.3. *The line-graph of a connected graph which is not of class K and with no edge adjacent to exactly one other edge is regularisable.*

Proof. Let G be a connected graph which is not of class K and with no edge adjacent to exactly one other edge. By a theorem of Whitney [2] $L(G)$ is not isomorphic to the line-graph of a graph of class K; by Lemma 4.2, $L(G)$ is not of class C. Moreover, $L(G)$ has no induced subgraph isomorphic to $K_{1,3}$ and no pendent vertex. The result now follows from Theorem 3.1.

References

[1] C. Berge, Regularisable graphs, Ann. Discrete Math. 3 (1978).
[2] H. Whitney, Congruent graphs and the connectivity of graphs, Am. J. Math. 54 (1932) 150–168.

Annals of Discrete Mathematics 3 (1978) 129–144.

ON MAXIMAL CIRCUITS IN FINITE GRAPHS

H.A. JUNG
Fachbereich Mathematik, Technische Universität Berlin, 1 Berlin 12, W. Germany

1. Introduction

If the valency $d(v)$ of each vertex v of a graph G is at least $\frac{1}{2}n(G)$, where $n(G)$ is the number of vertices of G and $n(G)>2$, then G allows a hamiltonian circuit, i.e. a circuit which contains every vertex of G. This result is due to Dirac [5]. It stimulated investigations of many authors (e.g. [2, 3, 7, 8, 9 and 10]).

In this paper we concentrate on the following refinement by Ore [9]:

If $d(v)+d(w)\geqslant n(G)>2$ for any two different, non-adjacent vertices v, w of G, then G contains a hamiltonian circuit.

It turns out that the main obstacle to a further improvement in this direction is given by the graphs in class $\mathcal{K}$. We say that G belongs to class $\mathcal{K}$ if for some $k\geqslant 1$, there exist vertices $x_1,\ldots,x_k$ such that $G-x_1-\cdots-x_k$ has at least $k+1$ components. Obviously no $G\in\mathcal{K}$ allows a hamiltonian circuit (cf. [4]).

The results of this investigation have the following main application:

Let $n(G)\geqslant 11$ and $G\notin\mathcal{K}$. If $d(v)+d(w)\geqslant n(G)-4$ for any two different, non-adjacent vertices v, w of G then G contains a hamiltonian circuit.

We note that there exist infinitely many graphs $G\notin\mathcal{K}$ without hamiltonian circuits such that $d(v)+d(w)\geqslant n(G)-5$ for any two different, non-adjacent vertices v, w of G.

For graphs with constant valency related results were obtained by Erdös and Hobbs [6] and by Bollobás and Hobbs [1].

2. Notations and auxiliary results

Given a subgraph H of G, let $V(H)$ denote the set of vertices of H and $G-V(H)$ the maximal subgraph of G with vertex set $V(G)-V(H)$.

Circuits and paths in a graph G are considered as subgraphs of G. A circuit C in G is called *maximal* if there exists no circuit C' such that $V(C')\supset V(C)$.

Given a circuit C together with a direction of traversing let $C[a,b]$ ($C^-[a,b]$) denote the path obtained by running on C from a to b in the given (resp. opposite) direction of C. Similarly $C(a,b]$, $C^-(a,b]$, ($C(a,b)$ and $C^-(a,b)$) are obtained from $C[a,b]$, $C^-[a,b]$ respectively by deletion of a (and b).

In Section 2 and 3 a maximal circuit C with a direction of transversing and a component H of $G-(C)$ are fixed. An *H-avoiding arc* or simply arc is a path Q joining vertices v, $v' \in V(C)$ such that

$$V(Q) \cap (V(C) \cup V(H)) = \{v, v'\}.$$

Given two subgraphs H_1, H_2 of G, let $N(H_1, H_2)$ denote the set of $v_2 \in V(H_2)$ such that some edge has terminal vertices $v_1 \in V(H_1)$ and v_2.

Let a maximal circuit C and a component H of $G-C$ be fixed. Then $m = m(H, C)$ denotes the minimum number s such that there exists a path $P = P[a, b]$ with $N(H, C) \subseteq V(P)$, $V(P) \cap V(H) = \emptyset$, $|V(C) - V(P)| = s$, $a \neq b$ and $a, b \in N(H, C)$. A path P of this form can be enlarged to a circuit by adding a path from b to a with inner vertices in H. We note

Lemma 2.1. *$m(H, C) \geq 1$ and $n(C[a, b]) \geq m(H, C) + 2$ for all $a, b \in N(H, C)$ such that $a \neq b$.*

In the following we construct paths P by putting together arcs and subpaths of C. For a vertex x on C and a subpath L of C let $N_\alpha(x, L)$ denote the set of $v \in V(L) - \{x\}$ such that there exists an H-avoiding arc $Q = Q[x, v]$. Note that $N(x, L) \subseteq N_\alpha(x, L)$. We denote

$$\bar{N}_\alpha(x, L) = V(L) - \{x\} - N_\alpha(x, L).$$

Lemma 2.2. *Let $L_i = C(a_i, b_i)$ $(i = 1, 2)$ be different components of $C - N(H, C)$ and Q an arc joining z_1 on $C[a_1, b_1]$ to z_2 on $C[a_2, b_2]$. If $z_1 \neq a_1$ and $z_2 \neq a_2$ then*

$$n(C(a_1, z_1)) + n(C(a_2, z_2)) \geq m. \tag{1}$$

If $z_1 \neq b_1$ and $z_2 \neq b_2$ then

$$n(C(z_1, b_1)) + n(C(z_2, b_2)) \geqq m. \tag{1'}$$

If $z_1 \in V(L_1)$ and $z_2 \in V(L_2)$ then

$$n(L_1) + n(L_2) \geq 2m + 2. \tag{1''}$$

Proof. If $z_1 \neq a_1$ and $z_2 \neq a_2$ then $C^-[a_2, z_1]$, Q and $C[z_2, a_1]$ define a path P. Now a_1, $a_2 \in N(H, C)$ implies $|V(C) - V(P)| \geq m(H, C)$ and hence (1). Similarly (1') is obtained. (1'') follows from (1) and (1').

In the following lemma L_i, z_i $(i = 1, 2)$ and Q are given as in Lemma 2.2.

Lemma 2.3. *Let Q_1 be an arc joining x_1 on $C[a_1, z_1)$ to y_1 on $C(z_1, b_1]$. If $x_1 \neq a_1$ and $z_2 \neq a_2$ then*

$$n(C(a_1, x_1)) + n(C(z_1, y_1)) + n(C(a_2, z_2)) \geq m. \tag{2}$$

If $y_1 \neq b_1$ *and* $z_2 \neq b_2$ *then*

$$n(C(x_1, z_1)) + n(C(y_1, b_1)) + n(C(z_2, b_2)) \geq m. \tag{2'}$$

If $x_1, y_1 \in V(L_1)$ *and* $z_2 \in V(L_2)$ *then*

$$n(L_1) + n(L_2) \geq 2m + 4. \tag{2''}$$

Proof. If $V(Q) \cap V(Q_1) \neq \emptyset$ then $Q \cup Q_1$ contains an arc Q' from z_2 to x_1. In this case (2) follows from (1) with respect to Q'. If $V(Q) \cap V(Q_1) = \emptyset$ and $x_1 \neq a_1$, $z_2 \neq a_2$ then we obtain a path P by running through $C^-[a_2, y_1]$, Q_1, $C[x_1, z_1]$, Q and $C(z_2, a_1]$. Now (2) is equivalent to $|V(C) - V(P)| \geq m$. (2″) follows from (2) and (2′).

Let $L_i = C[a_i, b_i]$, z_i $(i = 1, 2)$ and Q be given as in Lemma 2.2.

Lemma 2.4. *Let* Q_i *be an arc joining* x_i *on* $C[a_i, z_i)$ *to* y_i *on* $C(z_i, b_i]$ $(i = 1, 2)$. *If* $x_1 \neq a_1$ *and* $x_2 \neq a_2$ *then*

$$n(C(a_1, x_1)) + n(C(z_1, y_1)) + n(C(a_2, x_2)) + n(C(z_2, y_2)) \geq m. \tag{3}$$

If $y_1 \neq b_1$ *and* $y_2 \neq b_2$ *then*

$$n(C(x_1, z_1)) + n(C(y_1, b_1)) + n(C(x_2, z_2)) + n(C(y_2, b_2)) \geq m. \tag{3'}$$

If $x_1, y_1 \in V(L_1)$ *and* $x_2, y_2 \in V(L_2)$ *then*

$$n(L_1) + n(L_2) \geq 2m + 6. \tag{3''}$$

Proof. If $V(Q_2) \cap V(Q) \neq \emptyset$ then there exists an arc Q' from x_2 to z_1. In this case (3) follows from (2) with respect to Q', Q_1. The case $V(Q) \cap V(Q_1) \neq \emptyset$ is symmetric. If $V(Q_1) \cap V(Q_2) \neq \emptyset$ is symmetric. If $V(Q_1) \cap V(Q_2) \neq \emptyset$ then there exists an arc Q' from x_1 to x_2 in which case (3) follows from (1). Now let $x_1 \neq a_1$, $x_2 \neq a_2$ and Q, Q_1, Q_2 pairwise disjoint. We obtain a path P by running through $C^-[a_2, y_1]$, Q_1, $C[x_1, z_1]$, Q, $C^-[z_2, x_2]$, Q_2 and $C[y_2, a_1]$. In this case $|V(C) - V(P)| \geq m$ yields (3). Finally (3″) is a consequence of (3) and (3′).

Lemma 2.5. *Let* $a_i \in N(H, C)$ *and* u_i *successor of* a_i *on* C $(i = 1, 2;\ a_1 \neq a_2)$. *Then* $\bar{N}_\alpha(u_1, C)$ *contains the first* m *successors of* a_2. *For* $z \in N_\alpha(u_1, C(a_2, a_1))$ *the set* $\bar{N}_\alpha(u_2, C)$ *contains the first* m *successors of* z.

Proof. The first assertion is a special case of (1). Now let Q_1 be an arc joining u_1 to z on $C(a_2, a_1)$ and Q_2 an arc joining u_2 to z' on $C(z, a_1]$. Then $V(Q_1) \cap V(Q_2) = \emptyset$ since $u_2 \in \bar{N}_\alpha(u_1, C)$. We obtain a path by running through $C^-[a_2, u_1]$, Q_1, $C^-[z, u_2]$, Q_2 and $C[z', a_1]$. Hence $n(C(z, z')) \geq m$. Since $\bar{N}_\alpha(u_2, C)$ contains the first m successors of u_1, the second assertion follows.

Analogous statements can be made for the predecessors w_i of $a_i \in N(H, C)$ $(i = 1, 2)$.

The following observation is often used in the proofs to come.

Lemma 2.6. *Let $a, b \in N(H, C)$ such that $a \neq b$ and $C(a, b) = C[u, w]$. Let Q, Q' be arcs joining u to z and w to z' respectively where z, z' are different vertices of some component of $C - N(H, C) - C(a, b)$. Then z, z' have distance at least m on $C[b, a]$ or $V(Q) \cap V(Q') - \{u, w\} \neq \emptyset$. In either case z, z' are not neighbours on C.*

Proof. If Q, Q' have inner vertices in common then $Q \cup Q'$ contains a non-trivial arc from z to z'. In this case z, z' are not neighbours on C since C is maximal. Now let $V(Q) \cap V(Q') \subseteq \{u, w\}$. If z is on $C(z', a]$ then $C^-[a, z]$, Q, $C[u, w]$, Q' and $C^-[z', b]$ define a path, while if z' on $C(z, a]$ then $C^-[a, z']$, Q', $C[w, u]$, Q and $C^-[z, b]$ define a path.

3. Maximal circuits

As in the previous section, a maximal circuit C of G and a component H of $G - V(C)$ are fixed throughout Section 3. We use the abbreviations $m = m(H, C)$ and $k = |N(H, C)|$.

Proposition 3.1. *Let $L_i = C[u_i, w_i]$ $(i = 1, 2)$ be different components of $C - N(H, C)$ and $w_1 \in N_\alpha(u_2, L_1)$. Further let*

$$|\bar{N}_\alpha(v, C)| \leq (k-1)m + 3 \quad \text{for all } v \in V(C) - N(H, C).$$

Then $n(C) \leq k(m+1) + 3$ and $w_2 \in \bar{N}_\alpha(u_1, L_2)$.

Moreover there exists different vertices x, $x' \in V(C) - N(H, C)$ such that $x' \in \bar{N}_\alpha(x, C)$ and

$$|\bar{N}_\alpha(x, C)| + |\bar{N}_\alpha(x', C)| \geq n(C) + (k-2)(m-1) + \min(2, k-2).$$

Proof. Let

$$L_i = C(a_i, b_i) = C[u_i, w_i] \quad (1 \leq i \leq k)$$

be all components of $C - N(H, C)$ and let Q be an arc joining w_1 to u_2. Then $n(L_i) \geq m + 1$ $(i = 1, 2)$ by (1) and (1′).

(I) Let $a_2 \neq b_1$. For w_i on $C[b_2, a_1]$ the set $\bar{N}_\alpha(w_i, C)$ contains b_1, w_1 and m predecessors of w_1 by Lemma 2.6, moreover a_2 and m predecessors of b_j if $j \neq i$ (see remark after Lemma 2.5). On the other hand

$$|\bar{N}_\alpha(w_i, C)| \leq (k-1)m + 3$$

by hypothesis. Therefore $u_2 \in N_\alpha(w_i, b_2)$ which in turn yields $b_i \in \bar{N}_\alpha(w_2, C)$ by Lemma 2.6. Now by a similar argument $\bar{N}_\alpha(w_2, C)$ contains b_1, a_2, $m+1$ vertices on L_1 and m vertices on L_j for $j \neq 2$ contradicting the assumption

$$|\bar{N}_\alpha(w_2, C)| \leq (k-1)m + 3.$$

Hence $b_2 = a_1$.

(II) Let $n(L_i) \geq m+1$ and u_i be on $C(b_1, a_2)$ for some i. Then $u_i \in N_\alpha(w_2, L_i)$ and $a_2 = b_i$ according to (I). Now $\bar{N}_\alpha(w_i, C)$ contains a_i and $m+2$ vertices on $C(a_2, b_2]$. Hence $u_1 \in N_\alpha(w_i, L_1)$ and $b_1 = a_i$. This means $i = k = 3$. If $n(L_1) \geq m+2$ then $u_1 \in N_\alpha(w_2, L_1)$ and by Lemmas 2.5 and 2.6 $\bar{N}_\alpha(u_2, C)$ would contain the first $m+2$ vertices of $C[a_j, b_j)$ $(j = 1, 3)$. Hence $n(L_1) = m+1$ and, by symmetry, $n(L_2) = n(L_3) = m+1$.

This implies

$$w_2 \in \bar{N}_\alpha(u_1, L_2), \qquad w_3 \in \bar{N}_\alpha(u_2, L_3), \qquad w_1 \in \bar{N}_\alpha(u_3, L_1).$$

If $p_1 \in N_\alpha(a_1, C(u_1, b_1))$ then (2′) implies $n(C(p_1, b_1)) \geq m$ since $u_1 \in N_\alpha(w_3, L_1)$. Similarly $p_2 \in N_\alpha(b_2, C(a_2, w_2))$ yields a contradiction. In view of $u_3 \in N_\alpha(w_2, L_3)$ and $w_3 \in N_\alpha(u_1, L_2)$ we infer $V(L_3) \subseteq \bar{N}_\alpha(a_1, C)$. If $a_3 \in N_\alpha(a_1, C)$ then we would obtain a path by running through an arc from a_3 to a_1, $C[a_1, w_1]$, Q, $C[u_2, w_2]$, an arc from w_2 to u_3 and $C[u_3, a_2]$. We have shown $N_\alpha(a_1, C) = \{u_1, w_2\}$. By symmetry $N_\alpha(a_2, C) = \{u_2, w_3\}$ and $N_\alpha(a_3, C) = \{u_3, w_1\}$. If $p_1 \in N_\alpha(p_2, C(u_1, w_1))$ for some p_2 on $C[a_2, b_2]$ then p_2 on $C(u_2, w_2)$ by Lemma 2.6 and $q_2 \in N_\alpha(w_2, L_2)$ for the predecessor q_2 of p_2. Then $n(C(p_1, b_1)) \geq m$ by (2′), a contradiction. Hence

$$|\bar{N}_\alpha(p_1, C)| \geq 2(m+1)+3$$

for p_1 on $C(u_1, w_1)$ which yields $m = 1$. By symmetry, $n(L_j) = 2$ $(1 \leq j \leq 3)$.

If H' is a component of $G - V(C) - V(H)$ and $|N(H', C)| \geq 2$ then $N(H', C)$ is one of the sets $\{w_1, u_2\}$, $\{w_2, u_3\}$, $\{w_3, u_1\}$.

If $V(G) - V(C) = V(H)$ then the contraction of H to a single vertex yields the Petersen graph.

(III) Let $a_2 \neq b_1$. By (II) we may assume $n(L_i) = m$ for all a_i on $C[b_1, a_2)$. Let a_i on $C[b_1, a_2)$. Then $\bar{N}_\alpha(u_i, C)$ contains $2m+2$ vertices on $L_1 \cup L_2$. If $a_i \neq b_1$, then $a_i \in N_\alpha(u_1, C)$. Since $b_1 \in N_\alpha(u_i, C)$ or $a_1 \in N_\alpha(u_i, C)$ we would obtain a path by running through $C^-[a_2, u_i]$, an arc from u_i to b_1, $C[b_1, a_i]$, an arc from a_i to u_1, $C[u_1, w_1]$, Q, $C[u_2, a_1]$ or by running through $C[b_1, a_i]$, an arc from a_i to u_1, $C[u_1, w_1]$, Q, $C[u_2, a_1]$, an arc from a_1 to u_i and $C[u_i, a_2]$. We infer $k = 3$. If $n(L_1) \geq m+2$ then $u_1 \in N_\alpha(w_2, L_1)$ and $\bar{N}_\alpha(u_3, C)$ would contain $m+3$ vertices of $C[a_1, b_1)$, contrary to the hypothesis. Hence $n(L_1) = n(L_2) = m+1$. If $p_2 \in N_\alpha(p_1, L_2)$ for some p_1 on $C(a_1, w_1)$, then as in the proof of Lemma 2.6 we obtain inequalities

$$n(C(p_1, w_1)) + n(C(p_2, b_2)) \geq m$$

and

$$n(C(u_2, p_2)) + n(C(a_1, p_1)) \geq m,$$

contrary to $n(L_1) + n(L_2) \leq 2m+2$. By Lemmas 2.2 and 2.3 there exists no arc from $C[b_1, a_2]$ to $C(a_1, w_1) \cup C(u_2, b_1)$. Hence $|N_\alpha(x, C)| \leq m+1$ for all

$$x \in V(L_1 \cup L_2 \cup L_3) - \{w_1, u_2\}.$$

(IV) Let $a_2 = b_1$ and $k \geq 3$. If $n(L_2) \geq m+2$ and u_i on $C(b_2, a_1)$ then $\bar{N}_\alpha(u_i, C)$ contains by (II) and (III) $m+1$ vertices on L_1, and by Lemma 2.6 the first $m+2$ vertices of $C[a_2, b_2)$. Hence $n(L_2) \geq m+2$ implies $w_2 \in N_\alpha(u_i, L_2)$, and $\bar{N}_\alpha(u_1, C)$ would contain $m+4$ vertices on $C[a_2, b_2]$. Therefore $n(L_2) = m+1$ and, by symmetry, $n(L_1) = m+1$. Let $a_3 = b_2$. Since $\bar{N}_\alpha(u_3, C)$ contains all v on $C(a_1, b_2)$ and all w_i on $C[b_3, a_1]$ we infer $n(L_i) = m$ for all a_i on $C[b_3, a_1)$. Hence, by symmetry, $k \geq 4$ implies $n(L_i) = m$ for $i > 2$. If Q' is an arc joining $p_2 \in V(L_2)$ to $p_i \in V(L_i)$ and $i > 2$ then $n(L_i) \geq m+1$ by (1″) and hence $i = k = 3$. Further $p_3 \neq u_3$ since $n(L_2) = m+1$. There exists an arc Q_1 from u_3 to the successor q_3 of p_3 and therefore $n(C(a_2, p_2)) \geq m$ by Lemma 2.3. We would obtain a path P by running through b_2, Q', $C^-[p_3, u_3]$, Q_1, $C[q_3, w_1]$, Q and a_2 contrary to $n(L_2) = m+1$. This shows that there exists no arc from $C(a_1, b_1) \cup C(a_2, b_2)$ to $\bigcup_{i>2} L_i$. In particular $n(L_3) \leq m+1$. By Lemma 2.3 and the argument in the proof of Lemma 2.6 it follows that there exist no arcs from L_1 to L_2 except for arcs from w_1 to u_2. Also

$$N_\alpha(a_2, C) \subseteq \{w_1, u_2\} \cup N(H, C).$$

Hence $G - \{b_2, \ldots, b_k\}$ has $k-1$ components one of which has cut vertices a_2, w_1 and u_2.

(V) Let $k = 2$ and $u_1 \in \bar{N}_\alpha(w_2, L_1)$. If $x_1 \in N_\alpha(w_2, L_1)$, then $\bar{N}_\alpha(w_2, C)$ contains u_1 and the last $m+2$ vertices of $C(a_1, b_1]$. In this case $p_1 \in N_\alpha(w_2, L_1)$ for the successor p_1 of u_1 and hence $\bar{N}_\alpha(u_1, C)$ contains w_1 and the predecessor q_1 of w_1, a contradiction. Hence

$$N_\alpha(w_2, L_1) = N_\alpha(u_1, L_2) = \emptyset, \qquad n(L_i) \leq m+2 \quad (i = 1, 2).$$

Case 1. There exists $p_2 \in N_\alpha(w_1, L_2) - \{u_2\}$. Then p_2 is the successor of u_2 since $n(C(p_2, b_2)) \geq m$. Further $u_2 \in \bar{N}_\alpha(w_2, L_2)$ and hence $n(L_1) = m+1 = n(L_2) - 1$. There exists no arc from x_2 to $C(a_1, w_1) \cup C(p_2, a_1)$ $(x_2 = u_2, a_2)$. If there exists an arc from $C(a_1, w_1)$ to $C(u_2, a_1)$ then as in the proof of Lemma 2.6 $n(L_1) + n(L_2) \geq 2m+4$, a contradiction. Hence $G - \{a_1\}$ has cut vertices a_2, w_1, p_2.

Case 2. $|N_\alpha(w_1, L_2)| = |N_\alpha(u_2, L_1)| = 1$. If $x_1 \in N_\alpha(x_2, L_1)$ and $x_2 \in V(L_2) - \{u_2\}$, then $n(L_1) + n(L_2) \geq 2m+4$, hence $w_i \in N_\alpha(u_i, L_i)$, $(i = 1, 2)$ contrary to (3″) of Lemma 2.4. If $N_\alpha(a_2, C) \subseteq \{a_1, w_1, u_2\}$, then $G - \{a_1\}$ has cut vertices a_2, w_1, u_2. Therefore let $q_1 \in N_\alpha(a_2, L_1) - \{w_1\}$. Then $n(L_1) = m+2$ and q_1 is predecessor of w_1.

If Q_1 is an arc from w_1 to x_1 on $C(a_1, q_1)$, then we obtain a path by running through an arc from b_1 to q_1, $C^-[q_1, x_1]$, Q_1, Q and $C[u_2, b_2]$, contrary to

$n(L_1)=m+2$. In particular $w_1 \in \bar{N}_\alpha(u_1, L_1)$ and hence $n(L_2)=m+1$. We conclude that $G-\{a_1\}$ has cut vertices q_1, b_1, u_2.

(VI) In the remaining case $k=2$, $u_1 \in N_\alpha(w_2, L_1)$, we have $n(L_i) \geq m+2$ $(i=1,2)$.

Case 1. There exists an arc from w_2 to x_1 on $C(u_1, w_1)$. If x_1 is the successor of u_1, then $\bar{N}_\alpha(u_1, C)$ contains b_1, w_1 and the predecessor of w_1 on C contrary to $|\bar{N}_\alpha(u_1, L_2)| \geq m+1$. Since $\bar{N}_\alpha(u_2, C)$ contains a_1, u_1, the predecessor of x_1 and m vertices on $C(x_1, w_1)$, we have $n(C(u_1, x_1))=1$ and $x_1 \in N_\alpha(u_2, L_1)$. Using symmetry we obtain that L_1 has vertices u_1, p_1, x_1, q_1, w_1. Further $q_1 \in \bar{N}_\alpha(u_1, L_1)$, $p_1 \in \bar{N}_\alpha(w_1, L_1)$ and hence $|\bar{N}_\alpha(u_1, L_2)| \leq m+1$ which yields that L_2 has vertices u_2, p_2, w_2. From

$$x_1 \in N_\alpha(u_1, L_1) \cap N_\alpha(u_2, L_1)$$

follows $q_1 \in \bar{N}_\alpha(p_1, L_1)$. From $x_1 \in N_\alpha(u_2, L_1)$ and $w_2 \in N_\alpha(u_1, L_2)$ follows $a_1 \in \bar{N}_\alpha(p_1, C)$. From $u_2 \in N_\alpha(w_1, L_2)$ and $w_2 \in N_\alpha(u_1, L_2)$ follows $b_1 \in \bar{N}_\alpha(p_1, C)$ and $p_2 \in \bar{N}_\alpha(p_1, L_2)$. Hence

$$\bar{N}_\alpha(p_1, C)=7=m+6$$

contrary to hypothesis. In view of the symmetry the following case remains.

Case 2. Each of the sets $N_\alpha(u_2, L_1)$, $N_\alpha(w_2, L_1)$, $N_\alpha(u_1, L_2)$ and $N_\alpha(w_1, L_2)$ has cardinality 1. Then $n(L_i) \leq m+3$ $(i=1,2)$. If $x_1 \in N_\alpha(a_1,, L_1)$, then $n(L_1)=m+3$ and $x_1=p_1$. In this subcase $\bar{N}_\alpha(u_1, L_1)$ would contain w_1, hence not q_1 and we would obtain a path by running through an arc from a_1 to p_1, $C[p_1, q_1]$, an arc from q_1 to u_1, an arc from u_1 to w_2, $C^-[w_2, u_2]$, Q and b_1. If Q' is an arc joining x_1 on $C(u_1, w_1)$ to x_2 on $C(u_2, w_2)$, then we obtain a path by running through a_1, an arc from u_1 to w_2, $C^-[w_2, x_2]$, Q', $C[x_1, w_1]$, Q and b_1. We infer

$$n(C(u_1, x_1))+n(C(u_2, x_2)) \geq m$$

and, by symmetry,

$$n(C(x_1, w_1))+n(C(x_2, w_2)) \geq m.$$

We deduce $n(L_1)+n(L_2) \geq 2m+6$ and hence $n(L_1)=n(L_2)=m+3$, which in turn implies $y_i \in N_\alpha(u_i, L_i)$ for the successor y_i of x_i $(i=1,2)$. This is, in view of (3) of Lemma 2.4, impossible. Hence in Case 2 we have shown $|\bar{N}_\alpha(x, C)| \geq m+4$ for all x on $C(u_1, w_1) \cup C(u_2, w_2)$ contrary to the hypotheses.

Proposition 3.2. *Let $L_i=C[u_i, w_i]$ $(1 \leq i \leq k)$ be all components of $C-N(H, C)$. Let $N_\alpha(u_2, L_1) \cup N_\alpha(w_2, L_1) \neq \emptyset$, $u_1 \in \bar{N}_\alpha(w_2, L_1)$, $w_1 \in \bar{N}_\alpha(u_2, L_1)$ and $|\bar{N}_\alpha(v, C) \leq$ $(k-1)m+3$ for all $v \in \{u_1, u_2, w_1, w_2\}$.*

If moreover $G \notin \mathcal{K}$, *then* $m = 1$ *and*
(i) $|\bar{N}_\alpha(x, C)| \geq k+3$ *for some* $x \in V(C) - N(H, C)$, *or*
(ii) $k = 2$, $n(C) = 8$, $u_1 \in \bar{N}_\alpha(w_1, C)$ *and* $|\bar{N}_\alpha(u_1, C)| = |\bar{N}_\alpha(w_1, C)| = 4$.

Proof. The proof of Proposition 3.2 is divided into Cases A, B, C and D.

Case A. There exists some vertex $z_1 \in N_\alpha(u_2, L_1) \cup N_\alpha(w_2, L_1)$, such that $n(C(a_1, z_1)) > m$ and $n(C(z_1, b_1)) > m$.

If $z_1 \in N_\alpha(u_2, L_1)$, then $\bar{N}_\alpha(w_2, C)$ contains u_1, the predecessor p_1 of z_1, the successor q_1 of z_1 and the last m vertices of L_1. This implies b_1, $z_1 \in N_\alpha(w_2, L_1)$ and $\bar{N}_\alpha(u_2, L_1)$ contains p_1, q_1, the last m and the first m vertices of L_1. We infer

$$z_1 \in N_\alpha(u_2, L_1) \cap N_\alpha(w_2, L_1)$$

and $m = 1$.

(I) We first assume $n(C(a_1, z_1)) > 2$. Since $N_\alpha(u_2, L_1) \cap N_\alpha(w_2, L_1)$ contains all vertices of $C(u_1, p_1)$, the path $C(a_1, z_1)$ has vertices u_1, x_1, p_1. Then p_1, $q_1 \in \bar{N}_\alpha(u_1, L_1)$ and p_1, $u_1 \in \bar{N}_\alpha(w_1, L_1)$, yielding $n(L_i) = 1$ for all $i \geq 2$ and $z_1 \in N_\alpha(u_1, L_1)$. If $n(C(z_1, b_1)) > 2$ then $C(z_1, b_1)$ has vertices q_1, y_1, w_1, where $y_1 \in N_\alpha(u_2, L_1)$ and $q_1 \in \bar{N}_\alpha(w_1, L_1)$. If Q is an arc joining q_1 to p_1, then $C^-[a_2, q_1]$, Q, an arc from z_1 to u_1, an arc from x_1 to u_2 and $C[u_2, a_1]$ defines a path P such that $V(P) \supseteq V(C)$. Therefore $p_1 \in \bar{N}_\alpha(q_1, L_1)$.

In the subcase $n(L_1) = 7$ the graph $G - N(H, C) - \{x_1, z_1, y_1\}$ has at least $k+4$ components. In the subcase $n(L_1) = 6$ the graph $G - N(H, C) - \{x_1, z_1\}$ has at least $k+3$ components.

(II) Since the subcase $n(C(z_1, b_1)) > 2$ is symmetric we assume that L_1 has vertices u_1, p_1, z_1, q_1, w_1. Then $q_1 \in \bar{N}_\alpha(u_1, L_1)$, $p_1 \in \bar{N}_\alpha(w_1, L_1)$ and $n(L_i) = 1$ for all $i > 2$.

(III) Let $u_1 \in \bar{N}_\alpha(w_1, L_1)$. Then $n(L_2) \leq 2$, since otherwise $u_2 \in N_\alpha(w_2, L_2)$ and $q_2 \in N_\alpha(u_1, L_1)$ for the predecessor q_2 of w_2, a contradiction by Lemma 2.3. If $q_1 \in \bar{N}_\alpha(p_1, L_1)$, then $G - N(H, C) - \{z_1\}$ has at least $k+2$ components. Let $q_1 \in N_\alpha(p_1, L_1)$. Then $z_1 \in \bar{N}_\alpha(w_1, L_1)$, since otherwise $C[b_2, p_1]$, an arc from p_1 to q_1, an arc from w_1 to z_1, an arc from z_1 to w_2 and $C^-[w_2, b_1]$ would define a path P such that $V(P) \supseteq V(C)$. Similarly $z_1 \in \bar{N}_\alpha(u_1, L_1)$. If $u_3 \in N_\alpha(z_1, L_3)$ and u_3 on $C(b_1, a_2)$, then $C^-[a_3, b_1]$, an arc from b_1 to u_1, an arc from p_1 to q_1, an arc from w_1 to a_2, $C^-[a_2, u_3]$, an arc from u_3 to z_1, an arc from z_1 to u_2 and $C[u_2, a_1]$ define a path P such that $V(P) \supseteq V(C)$, a contradiction. Since the case u_3 on $C(b_2, a_1)$ is symmetric, we infer that $G - N(H, C)$ has k components one of which has cut vertices p_1, q_1, z_1. Note that

$$|\bar{N}_\alpha(u_i, L_1)| \geq (k-1)m + 4 \quad \text{for all } i > 2.$$

(IV) Let $u_1 \in N_\alpha(w_1, L_1)$. Then $p_1 \in \bar{N}_\alpha(q_1, L_1)$, since otherwise we would obtain a path P by running through $C^-[a_2, w_1]$, an arc from w_1 to u_1, an arc from p_1 to q_1, an arc from z_1 to u_2 and $C[u_2, a_1]$. If $b_1 \in N_\alpha(p_1, L_1)$, then we could replace part of P by $C^-[a_2, b_1]$, an arc from b_1 to p_1, an arc from u_1 to w_1 and q_1. A similar argument yields $a_1 \in \bar{N}_\alpha(p_1, C)$. Since $w_i \in \bar{N}_\alpha(p_1, L_1)$ for all i we have shown $|\bar{N}_\alpha(p_1, C)| \geq km+3$. By symmetry also $|\bar{N}_\alpha(q_1, C)| \geq k+3$.

Case B. $|N_\alpha(u_2, L_1)| \geq 2$. Let $x_1, y_1 \in N_\alpha(u_2, L_1)$, where y_1 is a vertex of $C(x_1, b_1)$. If $n(C(y_1, b_1)) < m$, then $\bar{N}_\alpha(w_2, C)$ would contain at least $m+4$ vertices on $C(a_1, b_1]$. In view of Case A we assume

$$n(C(a_1, x_1)) = n(C(y_1, b_1)) = m$$

and

$$N_\alpha(v_2, L_1) \subseteq \{x_1, y_1\} \quad \text{for } v_2 = u_2, w_2.$$

Now $\bar{N}_\alpha(w_2, L_1)$ contains the first m and the last m vertices of L_1, the successor of x_1 and the predecessor of y_1. Hence $m \geq 2$ implies $n(C(x_1, y_1) = 1$ and $|\bar{N}_\alpha(u_1, L_1)| \geq m+2 \geq 4$, contrary to the hypothesis.

(I) Let $n(L_1) \geq 6$. Then L_1 has vertices $u_1, x_1, p_1, q_1, y_1, w_1$ and $u_1, p_1, q_1, w_1 \in \bar{N}_\alpha(w_2, L_1)$. Therefore $x_1, y_1 \in N_\alpha(w_2, L_1)$ and $w_1, p_1 \in \bar{N}_\alpha(u_1, L_1)$, $q_1 \in \bar{N}_\alpha(w_1, L_1)$. We first assume $q_1 \in N_\alpha(u_1, L_1)$. If $p_1 \in N_\alpha(w_1, L_1)$, then we get a path P by running through $C[b_2, u_1]$, an arc from u_1 to q_1, y_1, an arc from w_1 to p_1, an arc from x_1 to w_2 and $C^-[w_2, b_1]$. If $x_1 \in N_\alpha(w_1, L_1)$, then we replace $P(q_1, w_2]$ by p_1, an arc from x_1 to w_1, and an arc from y_1 to w_2. We have shown that $q_1 \in N_\alpha(u_1, L_1)$ implies $|\bar{N}_\alpha(w_1, L_1)| \geq 4$. Therefore $q_1 \in \bar{N}_\alpha(u_1, L_1)$ and, by symmetry, $p_1 \in \bar{N}_\alpha(w_1, L_1)$. Since $|\bar{N}_\alpha(u_1, L_1)| \geq 3$, we have $n(L_i) = 1$ for $i > 1$, and $G - N(H, C) - \{x_1, y_1\}$ has at least $k+3$ components.

(II) Let L_1 have vertices u_1, x_1, p_1, y_1, w_1. Then $p_1, w_1 \in \bar{N}_\alpha(u_1, L_1)$. If $p_1 \in N_\alpha(w_1, L_1)$, then $y_1 \in \bar{N}_\alpha(w_2, L_1)$ and hence $u_2 \neq w_2$, which yields $y_1 \in N_\alpha(u_1, L_1)$. We would obtain a path by running through $C[b_2, u_1]$, an arc from u_1 to y_1, an arc from w_1 to p_1, an arc from x_1 to w_2 and $C^-[w_2, b_1]$. Therefore $p_1 \in \bar{N}_\alpha(w_1, L_1)$. If $n(L_i) \geq 2$ for some $i > 1$, then $y_1 \in N_\alpha(u_1, L_1)$ and $x_1 \in N_\alpha(w_1, L_1)$, which in turn implies $u_j, w_j \in \bar{N}_\alpha(p_1, L_j)$ for all $j \geq 1$. Therefore in the subcase "$n(L_i) \leq 2$ for all $i > 1$" the graph $G - N(H, C) - \{x_1, y_1\}$ has at least $k+3$ components. Now let $n(L_j) \geq 3$ for some $j > 1$. Then $u_1, w_1 \in N_\alpha(x_j, L_1)$ for all vertices x_j of $C(u_j, w_j)$ and hence $n(L_j) = 3$, $u_j \in \bar{N}_\alpha(w_j, L_j)$. In this case $n(L_i) = 1$ for all $i \neq 1$, j and $G - N(H, C) - \{x_1, y_1, x_j\}$ has at least $k+4$ components.

(III) Let $n(L_1) = 4$. Then $w_1, y_1 \in \bar{N}_\alpha(u_1, L_1)$ and $u_2 \neq w_2$, hence $b_1 \in N_\alpha(u_1, L_1)$. If $x_1 \in N_\alpha(w_1, L_1)$, then we would obtain a path by running through $C^-[a_2, b_1]$, an arc from b_1 to u_1, an arc from x_1 to w_1, an arc from y_1 to u_2 and $C[u_2, a_1]$. Hence

$x_1 \in \bar{N}_\alpha(w_1, L_1)$, which yields $n(L_i) = 1$ for $i > 2$. If successor x_2 of u_2 belongs to $N_\alpha(u_1, C)$, then $C^-[a_2, y_1]$, an arc from y_1 to u_2, an arc from u_2 to x_1, an arc from u_1 to x_2 and $C[x_2, a_1]$ would define a path. We infer $n(L_2) = 2$. If $N_\alpha(w_3, L_1) \neq \emptyset$, then w_3 on $C(a_1, a_2)$ and $y_1 \in N_\alpha(w_3, L_1)$, $w_2 \in N_\alpha(a_2, L_2)$. We would obtain a path defined by $C^-[a_3, y_1]$, an arc from y_1 to u_3, $C[u_3, a_2]$, an arc from a_2 to w_2, an arc from u_2 to x_1 and $C^-[x_1, b_2]$. Hence $G - N(H, C)$ has k components, one of which has cut vertices x_1, y_1, u_2. Note that

$$|\bar{N}_\alpha(u_i, L_1)| \geq (k-1)m + 4 \quad \text{for all } i > 2.$$

Case C. There exists some vertex $z_1 \in N_\alpha(u_2, L_1) - N_\alpha(w_2, L_1)$. We assume

$$|N_\alpha(v_2, L_1)| \leq 1 \quad (v_2 = u_2, w_2),$$

according to cases A and B.

(I) Let $n(C(z_1, b_1)) < m$. Then $z_1 \neq w_1$ and $\bar{N}_\alpha(w_2, C)$ contains u_1, a_2, all vertices of $C[z_1, b_1]$ and m predecessors of z_1. Hence $n(C(z_1, b_1)) = 1$, $n(L_1) = m+2$ and $a_2 = b_1$. Moreover $\bar{N}_\alpha(w_1, L_1)$ contains all vertices of $C(a_1, z_1)$. If $n(L_3) \geq m+1$, then $u_3 \in N_\alpha(w_2, L_1)$ and hence $n(L_2) \geq m+1$ and w_2, $b_2 \in \bar{N}_\alpha(u_1, C)$. In this subcase $\bar{N}_\alpha(u_1, C)$ would contain $m+4$ vertices on $C[z_1, b_2]$. We have shown $n(L_i) = m$ for $i > 2$. If $n(L_2) \geq m+2$, then $p_2 \in N_\alpha(w_1, L_2)$, where p_2 is the successor of u_2, and hence $u_2 \in \bar{N}_\alpha(w_2, L_2)$, a contradiction. Therefore $n(L_2) \leq m+1$. For $x_j \in N_\alpha(x_1, L_j)$ such that $j \neq 1$ and $x_1 \in \bigcup_{i \neq j} V(L_i)$, we have $x_1 \in V(L_1)$. If $x_1 \neq z_1$, then $x_j \neq u_2$ and therefore

$$n(C(x_j, b_j)) + n(C(x_1, z_1)) \geq m.$$

This is impossible since also

$$n(C(a_j, x_j)) + n(C(a_1, x_1)) \leq m.$$

We have shown that $G - N(H, C) - \{z_1\}$ has at least $k+2$ components.

(II) Let $m \geq 2$. According to (I) we assume $n(C(z_1, b_1)) \geq m$. Then $\bar{N}_\alpha(w_2, L_1)$ contains u_1, w_1, z_1, m predecessors and m successors of z_1. Hence L_1 has vertices u_1, p_1, z_1, q_1, w_1. From $|\bar{N}_\alpha(u_2, L_1)| = 4 = m+2$ we deduce $n(L_i) = 2$ for $i > 2$. Let Q be an arc from $x_1 \in \{u_1, p_1\}$ to $y_1 \in \{w_1, q_1\}$. Then $x_1 = p_1$, and we obtain a path P by running through $C[b_1, a_2]$, an arc from a_2 to w_2, $C^-[w_2, u_2]$, an arc from u_2 to z_1, $C[z_1, y_1]$, Q^- and $C^-[p_1, b_2]$. This is impossible since $|V(C) - V(P)| \leq 1$. In particular $|\bar{N}_\alpha(v_1, L_1)| \geq 2$ $(v_1 = u_1, w_1)$. If $n(L_2) \geq m+2$, then $p_2 \in N_\alpha(w_1, L_2)$, where p_2 is the successor of u_2. But then $\bar{N}_\alpha(u_1, C)$ would contain m successors of p_2 and q_1, w_1, a_2, u_2, p_2. Hence $n(L_2) \leq 3$. Let Q be an arc from $x_j \in V(L_j) \neq V(L_1)$ to $x_1 \in V(L_1) - \{z_1\}$. If x_j on $C(a_2, b_1)$, we get a contradiction as at the end of (I). If x_j on $C(b_1, a_2)$ and for instance x_1 on $C[z_1, b_1]$, then we would obtain a path by running through $C^-[a_j, x_1]$, Q^-, $C[x_j, a_2]$, an arc from a_2 to w_2,

$C^-[w_2, u_2]$, an arc from u_2 to z_1 and $C^-[z_1, b_2]$. We would obtain

$$n(C(a_j, x_j)) + n(C(z_1, x_1)) \geq 2$$

and hence $x_j = w_j$, $x_1 = w_1$, a contradiction. The case x_1 on $C(a_1, z_1)$ is treated similarly. We have shown that there exists no Q and therefore that $G - N(H, C) - \{z_1\}$ has at least $k+2$ components.

(III) In the remaining subcase we have $m = 1$ and $n(L_1) \leq 5$. If $n(L_1) = 5$, then $N_\alpha(w_2, L_1) = \{y_1\}$. We first assume $n(L_1) = 5$. Then $|\bar{N}_\alpha(v_2, L_1)| = m+3$ ($v_2 = u_2, w_2$) and hence $n(L_i) = 1$ for $i > 2$. Note that L_1 has vertices u_1, y_1, p_1, z_1, w_1 or vertices u_1, z_1, p_1, y_1, w_1. If $p_1 \in N_\alpha(w_1, L_1)$, then z_1 is the successor of p_1 and we would obtain a path by running through $C^-[a_2, w_1]$, an arc from w_1 to p_1, an arc from z_1 to u_2, $C[u_2, w_2]$, an arc from w_2 to y_1 and $C^-[y_1, b_2]$. Hence $p_1 \in \bar{N}_\alpha(w_1, L_1)$ and, by symmetry, $p_1 \in \bar{N}_\alpha(u_1, L_1)$. Similarly it follows that $u_1 \in \bar{N}_\alpha(w_1, L_1)$. If $n(L_2) \geq 3$ then $p_2 \in N_\alpha(w_1, L_1)$ for the successor p_2 of u_2 and hence $u_2 \in \bar{N}_\alpha(w_2, L_2)$, a contradiction. Consequently $G - N(H, C) - \{y_1, z_1\}$ has at least $k+3$ components.

(IV) Let $n(L_1) = 4$. Then $m = 1$ and $\bar{N}_\alpha(w_2, L_1) = V(L_1)$. Moreover $n(L_i) = 1$ for $i > 2$ since $|\bar{N}_\alpha(u_2, L_1))| = m+2$ by assumption. Let Q be an arc joining x_1 on $C(a_1, z_1)$ to y_1 on $C(z_1, b_1)$. Then

$$n(C(a_1, x_1)) + n(C(z_1, y_1)) \geq 1$$

by (2) of Lemma 2.3. We obtain a path P by running through $C[b_2, x_1]$, Q, $C^-[y_1, z_1]$, an arc from z_1 to u_2, $C[u_2, w_2]$, an arc from w_2 to a_2 and $C^-[a_2, b_1]$. But then

$$|V(C) - V(P)| = n(C(x_1, z_1)) + n(C(y_1, b_1)) \geq 1$$

and hence $n(L_1) \geq 5$ contrary to the hypothesis. This disproves the existence of Q.

If $n(L_2) \geq 3$, then $u_2 \in N_\alpha(w_2, L_2)$ and hence $\bar{N}_\alpha(u_1, L_2)$ contains u_2, w_2 and the predecessor of w_2; $\bar{N}_\alpha(w_1, L_2)$ contains u_2, w_2 and the successor of u_2. Since

$$|\bar{N}_\alpha(v_1, L_1)| \geq 2 \quad (v_1 = u_1 \quad \text{or} \quad v_1 = w_1),$$

this is impossible. Therefore $n(L_2) \leq 2$.

If $x_1 \in N_\alpha(u_j, L_1)$ and $j > 1$, $x_1 \neq z_1$, then u_j on $C(b_1, a_2)$ and x_1 on $C(z_1, b_1)$. We obtain a path by running through $C[b_2, z_1]$, an arc from z_1 to u_2, $C[u_2, w_2]$, an arc from w_2 to a_2, $C^-[a_j, u_j]$, an arc from u_j to x_1 and $C[x_1, a_j]$. Hence x_1 is not the immediate successor of z_1, which yields $x_1 = w_1$, a contradiction. We have shown that $G - N(H, C) - \{z_1\}$ has at least $k+2$ components.

Case D. $N_\alpha(u_2, L_1) = N_\alpha(w_2, L_1) = \{z_1\}$. According to Case A we may assume $n(C(a_1, z_1)) = m$. Then $n(C(z_1, b_1)) \leq 3$, and $\bar{N}_\alpha(u_1, L_1)$ contains w_1 and m successors of z_1.

(I) Let Q be an arc joining x_1 on $C(a_1, z_1)$ to y_1 on $C(z_1, b_1)$. Then

$$n(C(a_1, x_1)) + n(C(z_1, y_1)) \geq m$$

and

$$n(C(x_1, z_1)) + n(C(y_1, b_1)) \geq m$$

according to Lemma 2.3. Hence $n(C(z_1, b_1)) \geq m+2$.

We infer $m=1$ and that L_1 has vertices u_1, z_1, p_1, y_1, w_1. Further $n(L_i)=1$ for $i>2$ and $p_1 \in \bar{N}_\alpha(w_1, L_1)$ (cf. (I) in Case B). Therefore $|\bar{N}_\alpha(v_1, L_1)| \geq 2$ for $v_1 = u_1, w_1$. If $n(L_2) \geq 3$, then $q_2 \in N_\alpha(u_1, L_2)$ for the predecessor q_2 of w_2 and hence $w_2 \in \bar{N}_\alpha(u_2, L_2)$, a contradiction. Hence $n(L_2) \leq 2$, which yields that $G - N(H, C) - \{z_1, y_1\}$ has at least $k+3$ components. For the remainder of Case D we assume that there exists no arc from $C(a_1, z_1)$ to $C(z_1, b_1)$.

(II) Let $x_j \in N_\alpha(u_1, L_j)$ for some $j>1$. If $n(C(x_j, b_j)) < m$, then $\bar{N}_\alpha(w_1, C)$ contains a_1 and $m+2$ vertices on $C(a_j, b_j]$, which would imply $m=1$, $x_j = w_j$ and $b_j = a_1$ contrary to $z_1 \in N_\alpha(u_2, L_1)$. Hence $n(C(x_j, b_j)) \geq m$, $|\bar{N}_\alpha(v_1, L_j)| \geq 2m = 2$ $(v_1 = u_1, w_1)$ and $n(L_1) \leq 4$. If $n(L_j) \geq 4$, then $n(C(z_1, b_1)) = 1$, since otherwise $N_\alpha(u_1, L_j) = V(L_j) - \{u_j, w_j\}$, which yields $N_\alpha(w_1, L_j) = V(L_j)$, a contradiction. Hence $n(C(z_1, b_1)) \geq 2$ yields

$$V(L_j) = \{u_j, x_j, w_j\}, \qquad u_j \in \bar{N}_\alpha(w_j, L_j)$$

and

$$n(L_i) = 1 \quad \text{for all } i \neq 1, j.$$

If Q' is an arc joining different components of $C - N(H, C) - \{z_1, x_j\}$ and $n(L_1) = 4$, then Q' has one terminal vertex y_1 on $C(z_1, b_1)$ and the other y_j on L_j. But then $y_j = w_j$, w_j on $C(b_1, a_2)$ and $z_1 \in \bar{N}_\alpha(w_1, L_1)$, hence $x_j \in N_\alpha(w_1, L_j)$ and we would obtain a path by running through $C[b_1, x_j]$, an arc from x_j to w_1, Q', $C[w_j, w_2]$, an arc from w_2 to z_1 and $C^-[z_1, b_2]$. We have shown that in the subcase $n(C(z_1, b_1)) \geq 2$ the graph $G - N(H, C) - \{z_1, x_j\}$ has at least $k+3$ components.

(III) Let $n(C(z_1, b_1)) \geq 2$. According to (II) we may assume $N_\alpha(u_1, C) \subseteq N(H, C) \cup V(L_1)$. Then

$$\sum_{i>1} n(L_i) \leq (k-1)m + 1.$$

Hence every arc Q' joining different components of $C - N(H, C) - \{z_1\}$ has one terminal vertex y_1 on L_1 and the other y_j on L_j for some $j>1$. If $u_j = y_j$, then y_1 lies on $C(z_1, b_1)$, $n(C(z_1, y_1)) \geq m$ and $w_1 \neq y_1$, since $y_1 = w_1$ implies $b_1 \in \bar{N}_\alpha(u_1, C)$ and $n(L_j) \geq m+1$, a contradiction. Hence $u_j = y_j$ yields that L_1 has vertices u_1, z_1, p_1, y_1, w_1. In this subcase $n(L_i) = 1$ for all $i>1$, and $G - N(H, C) - \{y_1, z_1\}$ has at least $k+3$ components. If $y_j = w_j$ and $n(L_j) \geq 2$, then

y_1 on $C(z_1, b_1)$ and $n(C(y_1, b_1)) \geq m$. In this subcase $n(C(z_1, b_1)) = 2$, since otherwise $n(L_j) = m = 2$ and hence $n(C(z_1, y_1)) \geq m - 1 = 1$, a contradiction. Hence $n(C(z_1, b_1)) \geq 2$ in conjunction with $y_j = w_j$ and $w_j \neq u_j$ imply $m = 1$, w_j on $C(b_1, a_2)$ and $z_1 \in \bar{N}_\alpha(w_1, L_1)$, which in turn imply $b_1 \in N_\alpha(u_1, C)$ and $a_1 \in N_\alpha(w_1, C)$. We would obtain a path running through $C[b_j, u_2]$, an arc from u_2 to z_1, an arc from u_1 to b_1, $C[b_1, w_j]$, Q', an arc from w_1 to a_1 and $C^-[a_1, b_2]$. We have shown that the case $y_j = w_j \neq u_j$ cannot occur. In the remaining subcase $y_j \neq u_j$, w_j we infer $m \geq 2$, $n(L_j) = 3$ and $n(C(z_1, b_1)) = m$. Then y_1 has distance at least $m - 1$ from z_1 on C since $u_2, w_2 \in N_\alpha(z_1, C)$. In this subcase $y_1 = u_1$ or $y_1 = w_1$, a contradiction by Lemma 2.2.

(IV) For the remainder of case D we assume $n(L_1) = 3 = 3m$. If $n(L_j) \geq 4$ for some $j > 1$, then $N_\alpha(v_1, L_j) \neq \emptyset$ and $|\bar{N}_\alpha(v_1, L_j)| \geq 3$ $(v_1 = u_1, w_1)$. Note that there cannot exist neighboring vertices on C in $V(L_j) \cap N_\alpha(v_1, L_j)$. Hence $n(L_i) = 1$ for $i \neq 1, j$ and the path L_i has vertices u_j, p_j, q_j, w_j or u_j, x_j, p_j, y_j, w_j. In the former case

$$N_\alpha(u_1, L_j) = N_\alpha(w_1, L_j) = \{z_j\} \quad (z_j = p_j \quad \text{or} \quad z_j = q_j),$$

in the latter case

$$N_\alpha(u_1, L_j) = N_\alpha(w_1, L_j) = \{x_j, y_j\}.$$

Then $G - N(H, C) - \{z_1, z_j\}$ has at least $k + 3$ components or $G - N(H, C) - \{z_1, x_j, y_j\}$ has at least $k + 4$ components.

(V) Let $V(L_j) = \{u_j, x_j, w_j\}$ and $j > 1$. If $V(L_{j'}) = \{u_{j'}, x_{j'}, w_{j'}\}$ for some $j' \neq 1, j$, then

$$x_j, x_{j'} \in N_\alpha(u_1, C) \cap N_\alpha(w_1, C)$$

and $G - N(H, C) - \{z_1, x_j, x_{j'}\}$ has at least $k + 4$ components. Let $n(L_i) \leq 2$ for all $i \neq 1, j$. If $u_j \in \bar{N}_\alpha(w_j, C)$, then $G - N(H, C) - \{z_1, x_j\}$ has at least $k + 3$ components. If $u_j \in N_\alpha(w_j, L_j)$ then

$$x_j \in \bar{N}_\alpha(u_i, L_j) \cap \bar{N}_\alpha(w_i, L_j) \quad \text{for all } i \neq j$$

and hence $G - N(H, C) - \{z_1\}$ has at least $k + 2$ components.

(VI) Let $n(L_i) \leq 2$ for all $i \geq 2$. As shown in (II) there exists no arc from u_1 to $\bigcup_{i>1} V(L_i)$. By symmetry the same is true for w_1 instead of u_1. An arc Q' joining different components of $G - N(H, C) - \{z_1\}$ has terminal vertices u_j, $w_{j'}$. In this subcase $n(L_j) = n(L_{j'}) = 2$ and $n(L_i) = 1$ for $i \neq 1, j, j'$. Moreover $w_{j'}$ lies on $C(b_j, a_1)$ since otherwise $a_j \in \bar{N}_\alpha(w_1, C)$. If u_2 on $C(b_j, b_{j'})$ then $a_j, w_j, b_{j'} \in \bar{N}_\alpha(u_2, C)$, a contradiction. In particular $L_2 \neq L_{j'}$ and by symmetry $L_2 \neq L_j$. Note that $\bar{N}_\alpha(w_j, C)$ contains a_j, $u_{j'}$, $b_{j'}$ and u_1, which yields

$$|\bar{N}_\alpha(w_j, C)| \geq (k - 1)m + 4.$$

Since Cases A, B, C, D cover all possibilities Proposition 3.2 is proved.

Theorem 3.3. *Let C be a maximal circuit of G and let H be some component of $G-V(C)$. Assume $n(G)\geq 11$ and $G\notin\mathcal{K}$. Then there exist nonadjacent different vertices x, y in G such that*

$$d(x)+d(y)\leq n(G)-(k-2)(m-1)-5,$$

where $k=|N(H, C)|$ and $m=m(H, C)$.

Proof. Let $G\notin\mathcal{K}$ and assume

$$d(x)+d(y)\geq n(G)-(k-2)(m-1)-4$$

for any two non-adjacent and different vertices $x, y\in V(G)$. We adopt the notations of Proposition 3.2. Let

Case 1. Let

$$N_\alpha\Big(u_i, \bigcup_{j\neq i} L_j\Big)\cup N_\alpha\Big(w_i, \bigcup_{j\neq i} L_j\Big)\neq\emptyset \quad \text{for some } i\geq 1.$$

Without loss of generality we may assume $N_\alpha(u_2, L_1)\neq\emptyset$. For arbitrary $x\in V(C)-N(H, C)$ we put

$$|\bar{N}_\alpha(x, C)|=(k-1)m+\delta$$

and

$$\varepsilon=n(G-C-H)-|N(x, G-C)|.$$

Then

$$\begin{aligned} d(v)+d(x)&\leq n(H)-1+k+n(C)-1-(k-1)m-\delta+|N(x, G-C)|\\ &=n(G)-(k-2)(m-1)-(\delta+\varepsilon+m). \end{aligned}$$

Hence $\delta+m+\varepsilon\leq 4$, and by Propositions 3.1 and 3.2 we can find $x, x'\in V(C)-N(H, C)$ such that $x'\in\bar{N}_\alpha(x, C)$ and

$$|\bar{N}_\alpha(x, C)|+|\bar{N}_\alpha(x', C)|\geq n(C)+(k-2)(m-1)+\min(2, k-2).$$

Then also $\delta'+m+\varepsilon'\leq 4$ and $\delta+\delta'\geq n(C)-k(m+1)+\min(4, k)$, where δ', ε' have the meaning for x' as δ, ε have for x.

From $N(x, G-C)\cap N(x', G-C)=\emptyset$ we deduce $\varepsilon+\varepsilon'\geq n(G-C-H)$. Therefore

$$(n(C)-k(m+1)+\min(4, k))+2m+n(G-H-C)\leq 8,$$

which simplifies to

$$\begin{aligned} n(G-H-C)+2m+4+\min(2, k-2)&\leq n(G-H)-(k-2)(m+1)\\ &\quad+\min(2, k-2)\leq 8. \qquad (*) \end{aligned}$$

The first part of $(*)$ is equivalent to $n(C)\geq k(m+1)+2$, which in turn is an application of Lemma 2.2, since $G\notin\mathcal{K}$ implies that $G-N(H,C)$ has at most k components. On the other hand we have

$$\begin{aligned} d(x)+d(x')&\leq 2n(C)-2-|\bar{N}_\alpha(x,C)|-|\bar{N}_\alpha(x',C)|+|N(x,G-C)|+N(x',G-C)| \\ &\leq n(C)-(k-2)(m-1)-\min(4,k)+2n(G-C-H)-\varepsilon-\varepsilon' \\ &\leq n(G)-(k-2)(m-1)-n(H)-\min(4,k). \end{aligned}$$

Hence by hypothesis $n(H)+\min(4,k)\leq 4$. We infer $k\leq 3$ and $n(H)+k\leq 4$. Combining the last inequality with the second part of $(*)$ we obtain

$$n(G)-(k-2)(m+1)+k-2+k\leq 12.$$

In the subcase $m=1$ this means $n(G)\leq 10$. If $m\geq 2$ then $m=2$ and $k=2$ by $(*)$ and again $n(G)\leq 10$.

For the remainder of the proof we assume

$$\bar{N}_\alpha(u_i,L_j)\cup\bar{N}_\alpha(w_i,L_j)=\emptyset \quad \text{for all } i\neq j.$$

We call a vertex z on some L_i a good vertex if $n(C(a_i,z))=n(C(z,b_i))\neq 0$ and moreover there exists no arc joining $C(a_i,z)$ to $C(z,b_i)$.

Case 2. Let Q be an arc joining different components of $C-N(H,C)$ such that none of the terminal vertices is a good vertex. Without loss of generality we may assume that Q joins $z_1\in V(L_1)$ to $z_2\in V(L_2)$.

Case 2.1. For $i=1$ and $i=2$ there exist arcs from $C(a_i,z_i)$ to $C(z_i,b_i)$. Then $n(L_1)+n(L_2)\geq 2m+6$ by (3″) of Lemma 2.4 and hence $n(L_1)=n(L_2)=m+3$ by hypothesis. Moreover $q_i\in N_\alpha(u_i,L_i)$, where q_i is the successor of z_i on L_i ($i=1,2$). But this subcase cannot occur (see (3) of Lemma 2.4).

For the remainder of Case 2 we assume that there exist no arcs from $C(a_1,z_1)$ to $C(z_1,b_1)$.

Case 2.2. Let $n(C(z_1,b_1))\geq 3$. Then $n(C(z_1,b_1))=3$ and $3\leq n(L_2)=m$, since $|\bar{N}_\alpha(u_1,L_1)|\geq n(C(z_1,b_1))$. We infer $n(C(a_1,z_1))\leq m-1$, since $n(L_1)\leq m+3$. By Lemma 2.3 we have $q_2\in\bar{N}_\alpha(u_2,L_2)$, which in turn implies $n(L_1)\leq m+2$ and $n(C(a_1,z_1))\leq m-2$. From this we deduce $n(C(a_2,z_2))\geq 2$ by Lemma 2.2. Moreover there exists no arc from $C(a_2,z_2)$ to $C(z_2,b_2)$, since $n(L_1)+n(L_2)\leq 2m+2$.

Therefore $|\bar{N}_\alpha(w_2,L_2)|\geq 2$, which implies $n(L_1)\leq m+1$, a contradiction.

Case 2.3. Let $n(C(z_1,b_1))=2$. In view of Case 2.2 we obtain $n(C(a_1,z_1))=1$, hence $n(C(a_2,z_2))\geq m-1$ by Lemma 2.2. On the other hand $n(L_2)\leq m+1$, since $|\bar{N}_\alpha(u_1,L_1)|\geq 2$. We infer $3\leq n(L_2)=m+1$. From $n(C(a_1,z_1))=1<m$ we deduce $q_2\in\bar{N}_\alpha(u_2,L_2)$ by Lemma 2.3. Since z_2 is not a good vertex this implies $n(L_2)\geq 4$,

hence $m \geq 3$. On the other hand $n(L_1) = 4 \geq m+1$ by Lemma 2.2. Consequently

$$n(L_1) = n(L_2) = 4 = m+1$$

and, by Lemma 2.3, there exists no arc from $C(a_2, z_2)$ to $C(z_2, b_2)$. Now $n(L_i) = m$ for all $i > 2$. An arc joining different components of $C - N(H, C) - \{z_1\}$ would join q_1 to the predecessor of z_2 which, by the construction in Lemma 2.6, is impossible. If such an arc does not exist, then $G - N(H, C) - \{z_1\}$ has at least $k+2$ components contrary to $G \notin \mathcal{K}$.

Since z_1 is not a good vertex the discussion of Case 2 is exhaustive and leads to a contradiction in any subcase.

Case 3. Each arc joining different components of $C - N(H, C)$ has a good terminal vertex. Let S be the set of all good vertices. By assumption $G - N(H, C) - S$ has at least $k + |S| + 1$ components contrary to $G \notin \mathcal{K}$.

This completes the proof of the theorem.

Corollary. *Let G be graph without hamiltonian circuit and having at least* 11 *vertices. Then*

(i) *there exist non-adjacent vertices* x, y *such that* $d(x) + d(y) \leq n(G) - 5$ *or*

(ii) *there exist for some* $t \geq 1$ *vertices* $x_1, x_2, \ldots, x_t$ *such that* $G - x_1 - \cdots - x_t$ *has at least* $t+1$ *components.*

Proof. If $n(G) \geq 3$ and if G has not property (ii) then G is 2-connected and we can find a maximal circuit C in G. Also $|N(H, C)| \geq 2$ for each component H of $G - V(C)$. In this case (i) by Theorem 3.3.

References

[1] B. Bollobás and A.M. Hobbs, Hamiltonian cycles in regular graphs, Ann. Discrete Math. 3 (1978). 29–34.

[2] J.A. Bondy, Properties of graphs with constraints on degrees, Studia Sci. Math. Hungr. 4 (1969) 473–475.

[3] V. Chvátal, On Hamilton's ideals, J. Combinatorial Theory B12 (1972) 163–168.

[4] V. Chvátal, Tough graphs and hamiltonian circuits, Discrete Math. 5 (1973) 215–228.

[5] G.A. Dirac, Some theorems on abstract graphs, Proc. London Math. Soc. 2 (1952) 69–81.

[6] P. Erdös and A.M. Hobbs, A class of Hamiltonian regular graphs, to appear.

[7] A. Ghouila-Houri, Une condition suffisante d'existence d'un circuit hamiltonien, Comp. Rend. Acad. Sci. Paris 251 (1960) 495–497.

[8] M. Las Vergnas, Sur l'existence des cycles hamiltoniens dans un graphe, Comp. Rend. Acad. Sci. Paris Sér. A (1970) 1361–1364.

[9] O. Ore, Arc coverings of graphs, Ann. Mat. Pura Appl. 55 (1961) 315–321.

[10] L. Pósa, A theorem concerning Hamilton lines, Magyar Tud. Akad. Mat. Kutató Intézetének Közleményei 7 (1962) 225–226.

Annals of Discrete Mathematics 3 (1978) 145–164.

A REDUCTION METHOD FOR EDGE-CONNECTIVITY IN GRAPHS

W. MADER

Freie Universität Berlin, Fachbereich Mathematik, 1000 Berlin 33, W. Germany

Let $V(G)$ be the vertices and $E(G)$ the edges of the multigraph $G = (V(G), E(G))$. (In a multigraph parallel edges are allowed, but not loops.) We denote the set of edges between the vertices x and y of G by $[x, y]_G$, sometimes without the index G. Let $\lambda(x, y; G)$ be the maximal number of edge-disjoint paths between x and y in G.

Let $h \in [z, x]_G$ and $k \in [z, y]_G$ with $x \neq y$, and denote by G^{hk} the multigraph which arises from $G - \{h, k\} = (V(G), E(G) - \{h, k\})$ by the addition of exactly one new edge between x and y. The multigraph G^{hk} is called a *lifting of G at z*, arising from the *lifting of h and k at z*. For pairs $x \neq y$ in $V(G) - \{z\}$ it is obvious that $\lambda(x, y; G^{hk}) \leq \lambda(x, y; G)$. If for all such pairs $\lambda(x, y; G^{hk}) = \lambda(x, y; G)$ holds, we call the lifting *admissible*. In [5], Lovász proved that at each vertex of an eulerian multigraph there is an admissible lifting. In his talk at the conference on graph theory in Prague in June 1974 he also announced the result that at each non-separating vertex z of even degree in a finite multigraph G, there are edges h and k such that

$$\min\{\lambda(x, y; G^{hk}) : \{x, y\} \subseteq V(G) - \{z\}\} = \min\{\lambda(x, y; G) : \{x, y\} \subseteq V(G) - \{z\}\}$$

holds. Furthermore he advanced the conjecture (see [6]) that at each non-separating vertex of even degree in a finite multigraph there is an admissible lifting.

As the main result of the present work we will prove the following somewhat more general result, namely that if z is a non-separating vertex of degree at least 4 in the multigraph G then there exists an admissible lifting of G at z.

With the help of this result we will describe a simple construction procedure for all n-fold edge connected multigraphs, thereby proving Conjecture 2 of Simmons [11] in a somewhat modified form.

In a directed multigraph $\vec{G}$ let $\lambda(x, y; \vec{G})$ denote the maximal number of (continuously directed) edge-disjoint paths from x to y. An orientation $\vec{G}$ of the multigraph G is called *admissible* if for each pair $x \neq y$ of vertices of G $\lambda(x, y; \vec{G}) \geq [\frac{1}{2}\lambda(x, y; G)]$ holds, where $[r]$ denotes the integer part of the real number r. As suggested by Lovász, we will deduce from our main result a theorem of Nash-Williams [9], stating that every finite multigraph possesses an admissible orientation.

We mention a few more definitions and symbols. In place of $x \in V(G)$, resp. $k \in E(G)$, we mostly write $x \in G$, resp. $k \in G$. For $k \in E(G)$, say $k \in [x, y]_G$, let $V_G(k) = \{x, y\}$ (usually written without the index); when $\|[x, y]\| = 1$ we sometimes regard $[x, y]$ simply as the edge between x and y. Let $A \subseteq V(G)$ and $B \subseteq V(G)$ with $A \cap B = \emptyset$. Then let $\bar{A} = V(G) - A$ and let $G(A)$ denote the submultigraph induced by A. Further let $G - A = G(\bar{A})$ and for $E' \subseteq E(G)$ let $G - E' = (V(G), E(G) - E')$. Furthermore let

$$E(A, B; G) = \{k \in E(G) : V(k) \cap A \neq \emptyset \quad \text{and} \quad V(k) \cap B \neq \emptyset\},$$

$$d(A, B; G) = |E(A, B; G)|$$

and

$$E(A; G) = E(A, \bar{A}; G), \qquad d(A; G) = |E(A; G|.$$

In the case $A = \{a\}$ we write simply a in the symbols, and the same convention is used if $E' = \{k\}$; thus for instance $d(a; G)$ is the degree of the vertex a in G. Let $\delta(G) = \min_{x \in V(G)} d(x; G)$ and let $|G| = |V(G)|$. A multigraph G is called n-regular if $d(x; G) = n$ holds for all $x \in V(G)$. For $x \in G$ let

$$N(x; G) = \{y \in V(G) : [x, y]_G \neq \emptyset\}$$

and for a submultigraph $H \subseteq G$ and $k \in E(G)$ with $V_G(k) \subseteq V(H)$ let

$$H \cup k = (V(H), E(H) \cup \{k\})$$

with $V_{H \cup k}(k) = V_G(k)$. A path P with endvertices x and y is called an x, y-path, and for $x', y' \in P$ we denote by $P[x', y']$ the x', y'-path contained in P. In general the multigraphs we consider are finite; only in a few places will infinite multigraphs be allowed, and these will be manifest from the context. For a finite multigraph G we put $\lambda(x, x; G) = \infty$ for $x \in G$ and $\lambda(G) = \min \lambda(x, y; G)$, so in particular $\lambda(G) = \infty$ in the case $|G| = 1$. A multigraph G with $\lambda(G) \geqslant n$ is called *n-fold edge connected.* By Menger's theorem in edge form (see for instance [12, 9.3] and for infinite multigraphs compare [2, p. 45]) we know that

$$\lambda(x, \bar{x}; G) = \min \{d(X; G) : X \subset V(G) \quad \text{with} \quad x \in X \quad \text{and} \quad \bar{x} \in \bar{X}\} \quad \text{for} \quad x \neq \bar{x} \text{ in } G.$$

We will frequently make use of this without explicit reference. Thus for $x \neq \bar{x}$ in G, the set

$$T(x, \bar{x}; G) = \{X \subseteq V(G) : x \in X, \bar{x} \in \bar{X} \quad \text{and} \quad d(X; G) = \lambda(x, \bar{x}; G)\}$$

is non-empty. A set $E' \subseteq E(G)$ with $|E'| = \lambda(x, \bar{x}; G)$ with the property that x and $\bar{x}$ lie in different components of $G - E'$ we call a *smallest separating edge set* for x and $\bar{x}$ in G.

We shall use most of the above definitions analogously for directed multigraphs (or multidigraphs). The set of edges k from x to y in the multidigraph G we denote by $(x, y)_G$; x is called the initial vertex of k and y the terminal vertex. An orientation of the multigraph G to a multidigraph $\vec{G}$ is equivalent to a decomposition of $[x, y]_G$ into two disjoint sets $(x, y)_{\vec{G}}$ and $(y, x)_{\vec{G}}$ for all $x \neq y$ in G. For a multidigraph G let $E(A, B; G) = \{k \in E(G)$: the initial vertex of k belongs to A, the terminal vertex to $B\}$. In contrast to the undirected case we do not now put $E(A; G) = E(A, \bar{A}; G)$ but we define $E^+(A; G) = E(A, \bar{A}; G), E^-(A; G) = E^+(\bar{A}; G)$ and $E(A; G) = E^+(A; G) \cup E^-(A; G)$. We define $d(A, B; G)$, $d^{\pm}(A; G)$ and $d(A; G)$ correspondingly. For the directed case of Menger's theorem and for the definition of $T(x, \bar{x}; G)$ one has merely to substitute $d^+(X; G)$ for $d(X; G)$. In a directed multigraph we understand by a path, resp. cycle, a continuously directed path, resp. cycle, and an x, y-path is a directed path from x to y.

Let A be a set of vertices of the multigraph G with $\emptyset \neq A \neq V(G)$ and let $a \in A$. The multigraph G_a has vertex set $\bar{A} \cup \{a\}$ and $E(G_a)$ is defined by $G_a - a = G(\bar{A})$ and $|[x, a]_{G_a}| = d(x, A; G)$ for all $x \in \bar{A}$. We say that the multigraph G_a arises from G by *identification of A to a*. (In directed multigraphs the identification of A to a is defined analogously.) For all $X \subseteq \bar{A}$ it is clear that $d(X; G_a) = d(X; G)$ and $d(X \cup \{a\}; G_a) = d(X \cup A; G)$ hold. Furthermore there exists a bijective function $i: E(a; G_a) \to E(A; G)$ with $i([x, a]_{G_a}) = E(x, A; G)$ for all $x \in \bar{A}$; we call i an *associative bijection*.

The following lemma is an immediate consequence of Menger's theorem.

Lemma 1. *For any three distinct vertices a, b, c we have*

$$\lambda(a, c; G) \geqslant \min\{\lambda(a, b; G), \lambda(b, c; G)\},$$

where G is a (possibly infinite) multigraph or directed multigraph.

The next lemma may be compared with [1, Lemma 3.1 in Ch. IV].

Lemma 2. *In the multigraph G let $A \in T(a, \bar{a}; G)$ for certain a, $\bar{a}$ in G. G_a arises from G by identification of A to a. Then $\lambda(x, y; G_a) = \lambda(x, y; G)$ holds for all $x \neq y$ in $V(G_a)$.*

Proof. For $x \neq y$ in G_a, each system of n edge-disjoint x, y-paths in G clearly yields a system of n edge-disjoint x, y-paths in G_a; hence $\lambda(x, y; G_a) \geqslant \lambda(x, y; G)$. Since $d(A; G) = \lambda(a, \bar{a}; G)$ one can associate with each $k \in E(A; G)$ an $a, \bar{a}$-path P_k in G with $k \in P_k$, so that $E(P_k) \cap E(P_{k'}) = \emptyset$ for $k \neq k'$. Let $i: E(a; G_a) \to E(A; G)$ be an associative bijection. Given a system of n edge-disjoint x, y-paths in G_a we may replace each edge $k \in E(a; G_a)$, say $k \in [z, a]_{G_a}$, by $P_{i(k)}[z, a]$, and so obtain n edge-disjoint x, y-paths in G. Thus $\lambda(x, y; G) \geqslant \lambda(x, y; G_a)$ also holds, and thereby Lemma 2 is proved.

Lemma 3. *In the multigraph G let $A \in T(a, \bar{a}; G)$ for certain a, $\bar{a}$ in G. Let G_a arise from G by identification of A to a. Let i' be an associative bijection, which we continue through the identity on $E(G - A)$ to a map $i: E(G_a) \to E(G)$. Let G_a^{hk} be* *an admissible lifting of G_a at $z \in \bar{A}-\{\bar{a}\}$. Then $G^{i(h)i(k)}$ is an admissible lifting of G at z.*

Proof. Let $n=\lambda(a, \bar{a}; G)$ and let $G_{\bar{a}}$ arise from G by identification of $\bar{A}$ to $\bar{a}$. As the lifting G_a^{hk} is admissible and $z \notin \{a, \bar{a}\}$, it follows that $\lambda(\bar{a}, a; G_a^{hk})= \lambda(\bar{a}, a; G_a)=n$ by Lemma 2. Since $n=d(A; G^{i(h)i(k)})=d(a; G_a^{hk})$, we can extend n edge-disjoint $\bar{a}$, a-paths in G_a^{hk} to n edge-disjoint $\bar{a}$, a-paths in $G^{i(h)i(k)}$ as in the proof of Lemma 2. Thus

$$\lambda(a, \bar{a}; G^{i(h)i(k)})=\lambda(a, \bar{a}; G_a^{hk})=\lambda(a, \bar{a}; G_a)=\lambda(a, \bar{a}; G)$$

and

$$A \in T(a, \bar{a}; G^{i(h)i(k)}).$$

If $x, y \in (\bar{A}-\{z\}) \cup \{a\}$, by Lemma 2,

$$\lambda(x, y; G^{i(h)i(k)})=\lambda(x, y; G_a^{hk})=\lambda(x, y; G_a)=\lambda(x, y; G).$$

Since G_a arises from $G^{i(h)i(k)}$ by identification of $\bar{A}$ to $\bar{a}$, we have

$$\lambda(x, y; G^{i(h)i(k)})=\lambda(x, y; G_{\bar{a}})=\lambda(x, y; G) \quad \text{for all} \quad x, y \in A \cup \{\bar{a}\}.$$

If now $x \in A$ and $y \in \bar{A}-\{z\}$, we have $\lambda(x, y; G) \leq d(A; G)=n$. Hence by Lemma 1 and the above equalities.

$$\begin{aligned}\lambda(x, y; G^{i(h)i(k)}) &\geq \min\{\lambda(x, a; G^{i(h)i(k)}), \lambda(a, y; G^{i(h)i(k)})\}\\ &= \min\{\lambda(x, a; G), \lambda(a, y; G)\}\\ &\geq \min\{\lambda(a, y; G), \lambda(x, y; G)\}.\end{aligned}$$

But

$$\lambda(a, y; G)=\lambda(a, y; G_a) \geq \lambda(x, y; G),$$

and so

$$\lambda(x, y; G^{i(h)i(k)}) \geq \lambda(x, y; G).$$

Thus

$$\lambda(x, y; G^{i(h)i(k)})=\lambda(x, y; G) \quad \text{for all } x, y \in V(G)-\{z\},$$

and Lemma 3 is proved.

Lemma 4. *In the multigraph G let $A \subseteq V(G)$ with $d(A; G) \leq \lambda(a, \bar{a}; G)+1$ for certain vertices $a \in A$ and $\bar{a} \in \bar{A}$. Then $d(X; G(A)) \geq d(X, \bar{A}; G)-1$ holds for all $X \subseteq A-\{a\}$.*

Proof. There exist edge-disjoint a, $\bar{a}$-paths $P_1, \ldots, P_n$ in G with $n=\lambda(a, \bar{a}; G)$. Then since $d(A; G) \leq n+1$ each path P_ν contains exactly one edge of $E(A; G)$, so $V(P_\nu) \cap X \neq \emptyset$ holds for at least $d(X, \bar{A}; G)-1$ many ν. Since $a \notin X$ the lemma follows.

Lemma 5. *Let $d(a; G) \geq n$ and suppose that for all $\{x, y\} \subseteq V(G-a)$, $\lambda(x, y; G) \geq n$ holds. Then $\lambda(G) \geq n$.*

Proof. If there existed an $A \subset V(G)$ with $a \in A$ and $d(A; G) < n$, then $|A| \geq 2$ since $d(a; G) \geq n$, and we would have $\lambda(x, y; G) < n$ for $x \in A - \{a\} \neq \emptyset$ and $y \in \bar{A} \neq \emptyset$.

The multigraph G is called *irreducible relative to the vertices* $a \neq b$ if, for each smallest separating edge set E' for a and b, either $E' = E(a; G)$ or $E' = E(b; G)$ holds. In particular it then holds that

$$\lambda(a, b; G) = \min\{d(a; G), d(b; G)\}.$$

We call the multigraph G *irreducible* if it is irreducible relative to all pairs $a \neq b$ of vertices of G, and G is *z-irreducible* if it is irreducible relative to all pairs $a \neq b$ of $G - z$, where $z \in V(G)$.

Lemma 6. *Let G be a z-irreducible multigraph and let $k \in [x, y]_G$. Suppose that for the vertices $a \neq \bar{a}$ in $G - z$, $\lambda(a, \bar{a}; G - k) < \lambda(a, \bar{a}; G)$ holds. Then $\{a, \bar{a}\} \cap \{x, y\} \neq \emptyset$. If $a \in \{x, y\}$, but $\bar{a} \notin \{x, y\}$, then $d(a; G) \leq d(\bar{a}; G)$ holds.*

Proof. There is an $A \in T(a, \bar{a}; G - k)$. Since

$$d(A; G - k) = \lambda(a, \bar{a}; G - k) < \lambda(a, \bar{a}; G)$$

we have $k \in E(A; G)$ and by the z-irreducibility of G we have also $E(A; G) = E(a; G)$ or $E(A; G) = E(\bar{a}; G)$. Therefore k is incident with a or $\bar{a}$. We now assume that $a \in \{x, y\}$ and $\bar{a} \notin \{x, y\}$. Then $E(A; G) = E(a; G)$, and so

$$d(a; G) = d(A; G) = \lambda(a, \bar{a}; G) \leq d(\bar{a}; G)$$

holds.

Lemma 7. *Let $G^{kk'}$ be a lifting of the multigraph G at z and let there be $a \neq \bar{a}$ in $G - z$ with $\lambda(a, \bar{a}; G^{kk'}) < \lambda(a, \bar{a}; G)$. Furthermore let $A \in T(a, \bar{a}; G^{kk'})$. Then $\{k, k'\} \subseteq E(A; G)$. In the case $3 \leq d(z; G) \leq 4$ we have also $E(z; G) \cap E(A; G) = \{k, k'\}$. If G is z-irreducible then $\lambda(a, \bar{a}; G^{kk'}) = \lambda(a, \bar{a}; G) - 1$.*

Proof. Let say $k \in [z, x]$ and $k' \in [z, x']$. Since $d(A; G^{kk'}) < \lambda(a, \bar{a}; G) \leq d(A; G)$, we have $\{x, x'\} \subseteq A$ or $\{x, x'\} \subseteq \bar{A}$, and so $\{k, k'\} \subseteq E(A; G)$. In particular it follows that

$$d(A; G) = d(A; G^{kk'}) + 2 \leq \lambda(a, \bar{a}; G) + 1 = n + 1,$$

say. The second statement then follows by Lemma 4. Since $\{k, k'\} \subseteq E(A; G) = E_0$, say, neither $E_0 = E(a; G)$ nor $E_0 = E(\bar{a}; G)$ holds. If G is z-irreducible, it follows then that $|E_0| > n$ and so

$$\lambda(a, \bar{a}; G^{kk'}) = d(A; G^{kk'}) = n - 1.$$

Remark 8. Let G be an eulerian, z-irreducible multigraph and let $G^{kk'}$ be a lifting of G at z. Since in a multigraph H in which all vertices have even degree

$d(A;H)$ is even for all $A\subseteq V(H)$, there cannot be vertices $a\neq \bar{a}$ with $\lambda(a,\bar{a};G^{kk'})=\lambda(a,\bar{a};G)-1$. By Lemma 7 then the lifting is admissible. Hence by Lemma 3 the existence of an admissible lifting at any given vertex z of an eulerian multigraph can be proved at once by induction [5, Theorem 1].

If $|N(z;G)|=2$, say $N(z;G)=\{x,y\}$, then G^{hk} is an admissible lifting of G at z for all $h\in[z,x]$ and $k\in[z,y]$. We need then only consider the case $|N(z;G)|\geq 3$. The following example shows that, in general, if $d(z;G)=3$ then no admissible lifting of G at z can be found. Let G be an n-fold edge connected graph and let a, b, c be three distinct vertices of degree n in G. The graph G' is constructed by adding a new vertex z and exactly one edge between each of z and a, z and b and z and c. Then

$$\lambda(a,b;G')=\lambda(a,c;G')=\lambda(b,c;G')=n+1,$$

but any lifting of G' at z will reduce the connectivity between a, b and c.

Let the graph G have components $C_1,\ldots,C_n$ and let $c_\nu\in V(C_\nu)$ for $\nu=1,\ldots,n$. The graph G' arises from G by the addition of one further vertex z and exactly one edge between z and c_ν for $\nu=1,\ldots,n$. In the case $n\geq 3$ there is no admissible lifting of G' at z. If however we assume that z is not separating, there is always, in the case $d(z;G)\geq 4$, an admissible lifting of G at z. We prove this by induction on the degree of z; to start the induction we need the following lemma.

Lemma 9. *Let z be a non-separating vertex of the finite multigraph G with $d(z;G)=4$ and $|N(z;G)|\geq 2$. Then there exists an admissible lifting of G at z.*

Proof. We suppose that Lemma 9 is already proved for all multigraphs G' with $|G'|+|E(G')|<|G|+|E(G)|$. Suppose further that there exists for certain $a\neq\bar{a}$ in $G-z$ a set $A\in T(a,\bar{a};G)$ with $|A|\geq 2$ and $|\bar{A}|\geq 2$. Let $z\in\bar{A}$, say, and let G_a arise from G by identification of A to a. Then z doesn't separate G_a either and since $E(z;G)\not\subseteq E(A;G)$ we have also $|N(z;G_a)|\geq 2$. By the inductive hypothesis there exists an admissible lifting of G_a at z, and by Lemma 3 an admissible lifting of G at z also. We may therefore suppose that G is z-irreducible. The case $|N(z;G)|=2$ has already been dealt with. Suppose $|N(z;G)|=3$, say $N(z;G)=\{x,x',x''\}$. Let say $\|[z,x]\|=2$ and let $k\in[z,x]$ and $k'\in[z,x']$. Then $G^{kk'}$ is an admissible lifting of G at z. For if this were not the case, there would exist a set $A\subseteq V(G)$ with $E(z;G)\cap E(A;G)=\{k,k'\}$, by Lemma 7. But this is impossible, since $k\in E(A;G)$ implies $[z,x]\subseteq E(A;G)$. Thus we may assume $|N(z;G)|=4$.

If an $a\in G-z$ with $|N(a;G)|=1$ exists, then $z\notin N(a;G)$, since z is not a cutvertex, and an admissible lifting of $G-a$ at z obviously yields one such for G. It is assumed then that $|N(x;G)|\geq 2$ for all $x\in G$. Since G is z-irreducible, it follows that $\lambda(G)\geq 2$ by Lemma 5.

Suppose now that there exists a vertex $a\in G-z$ with $d(a;G)=2$. Let say $E(a;G)=\{k,k'\}$. We consider the multigraph $H=G^{kk'}-a$. Let say $K(H)-K(G)=\{k_0\}$. By the induction hypothesis there is an admissible lifting

$H^{hh'}$ of H at z. If $F^{k_1k_2}$ is an admissible lifting of a multigraph F at a vertex z with $d(z; F) = |N(z; F)| = 4$, then obviously $F^{k_3k_4}$ for $\{k_3, k_4\} = E(z; F) - \{k_1, k_2\}$ is also an admissible lifting of F at z. Since $|N(z; H)| = 4$ or there is an edge in H parallel to k_0, we may assume $k_0 \notin \{h, h'\}$. Hence $\{h, h'\} \subseteq E(G)$ and $G^{hh'}$ results from $H^{hh'}$ by subdividing the edge k_0 by the vertex a. For all $x \neq y$ in $G - \{a, z\}$ it then follows that

$$\lambda(x, y; G^{hh'}) = \lambda(x, y; H^{hh'}) = \lambda(x, y; H) = \lambda(x, y; G).$$

Since $\lambda(G) \geq 2$ it follows from Lemma 5 that $\lambda(H^{hh'}) \geq 2$ and so also that $\lambda(G^{hh'}) \geq 2$. Thus $G^{hh'}$ is an admissible lifting of G at z.

We may now assume that $\delta(G) \geq 3$. By the z-irreducibility of G it follows in particular that $\lambda(G) \geq 3$, by Lemma 5. We show that there is no $k \in E(G)$ for which the vertex z is a cutvertex of $G - k$. For otherwise, since $\lambda(G) \geq 3$, $(G - k) - z$ would have exactly two components, C_1 and C_2, and $d(V(C_1); G) = d(V(C_2); G) = 3$ would hold. By the z-irreducibility of G we would then have $|C_1| = |C_2| = 1$, contradicting $|N(z; G)| = 4$.

Let $a_0 \in G - z$ with

$$d(a_0; G) = \min_{x \in G - z} d(x; G) = d_0,$$

say, and let $a \in N(a_0; G - z) \neq \emptyset$ with

$$d(a; G) = \min_{x \in N(a_0;\, G - z)} d(x; G) = d,$$

say. Further let $k_0 \in [a_0, a] \neq \emptyset$ and let $H = G - k_0$. Then by the preceding paragraph z is not a cutvertex of H, and so by the induction hypothesis there exists an admissible lifting $H^{hh'}$ of H at z. We consider the multigraph $G_0 = G^{hh'} = H^{hh'} \cup k_0$, and we assume that G_0 is not an admissible lifting of G at z.

There exist then $a_1 \neq \bar{a}$ in $G - z$ with $\lambda(a_1, \bar{a}; G_0) < \lambda(a_1, \bar{a}; G)$. Since

$$\lambda(x, y; G - k_0) - \lambda(x, y; H^{hh'}) \leq \lambda(x, y; G_0) \quad \text{for all} \quad x \neq y \text{ in } G - z,$$

it follows that $\lambda(a_1, \bar{a}; G - k_0) < \lambda(a_1, \bar{a}; G)$. Hence $\{a_1, \bar{a}\} \cap \{a_0, a\} \neq \emptyset$ by Lemma 6. Since

$$\lambda(a_0, a; H^{hh'}) = \lambda(a_0, a; G - k_0) = \lambda(a_0, a; G) - 1,$$

and so $\lambda(a_0, a; G_0) = \lambda(a_0, a; G) = d_0$, we have $\{a_1, \bar{a}\} \neq \{a_0, a\}$. Let say $\{a_1, \bar{a}\} \cap \{a_0, a\} = \{a_1\}$. We cannot have

$$\lambda(a, \bar{a}; G_0) = \lambda(a, \bar{a}; G) = \min\{d(a; G), d(\bar{a}; G)\} \geq d_0,$$

for then by Lemma 1 we would also have $\lambda(a_0, \bar{a}; G_0) \geq d_0 = \lambda(a_0, \bar{a}; G)$, since $\lambda(a_0, a; G_0) = d_0$. Thus we have $\lambda(a, \bar{a}; G_0) < \lambda(a, \bar{a}; G)$ and we may assume $a_1 = a$. By Lemma 6, $d(a; G) \leq d(\bar{a}; G)$ holds and by Lemma 7

$$\lambda(a, \bar{a}; G_0) = \lambda(a, \bar{a}; G) - 1 = d - 1.$$

We first prove the following statement.

(∗) *If* $d(\{a_0, a\}; G-z)=d-1$ *and* $\{a_0, a\}\subseteq N(z; G)$ *then there is an admissible lifting of* G *at* z.

Let $v=|[a_0, a]_G|\geq 1$. By Lemma 6,

$$\lambda(a, \bar{a}; G-[a_0, z])=\lambda(a, \bar{a}; G)=d$$

holds. Since

$$d(\{a_0, a\}; G-[a_0, z])=d=d(a; G)$$

we have $d(a_0; G-\{a, z\})=v$, in particular $d(a_0; G)=2v+1$. Since $v\geq 1$ there is a $k_2\in E(a_0; G-\{a, z\})$, say $k_2\in[a_0, u_2]$. Let $E(z, G-\{a_0, a\})=\{k, k_1\}$, say $k\in[z, u]$ and $k_1\in[z, u_1]$. Since $u\neq u_1$ we may assume $u_1\neq u_2$. We now show that if $k'\in[a, z]$ then $G^{kk'}$ is an admissible lifting of G at z.

We show first of all $\lambda(x, \bar{x}; G^{kk'})=\lambda(x, \bar{x}; G)$ for all $\{x, \bar{x}\}\subseteq V(G)-\{a_0, z, u_1\}$. Let $G_1=G-k_1$. By Lemma 6, $\lambda(x, \bar{x}; G_1)=\lambda(x, \bar{x}; G)$ holds. It thus suffices to derive a contradiction from the assumption $\lambda(x, \bar{x}; G_1^{kk'})<\lambda(x, \bar{x}; G_1)$. Let $X\in T(x, \bar{x}; G_1^{kk'})$ and let say $z\in X$. By Lemma 7 $E(z; G_1)\cap E(X; G_1)=\{k, k'\}$, and thus $\{a, u\}\subseteq\bar{X}$ and $a_0\in X$. It follows therefore that $[a_0, a]\subseteq E(X; G_1)$ and thereby $d(\{a_0, z\}, \bar{X}; G_1)\geq v+2$. Since $d(X; G_1)\leq\lambda(x, \bar{x}; G_1)+1$ and $x\notin\{a_0, z\}$ we deduce from Lemma 4 that $d(\{a_0, z\}; G_1(X))\geq v+1$, in contradiction to $E(\{a_0, z\}; G_1(X))\subseteq E(a_0; G-\{a, z\})$.

We show now that $\lambda(x, \bar{x}; G^{kk'})=\lambda(x, \bar{x}; G)$ for all $\{x, \bar{x}\}$ in $V(G)-\{a_0, z, u_2\}$. First we consider $G_2=G-k_2$. By Lemma 6 we have $\lambda(x, \bar{x}; G_2)=\lambda(x, \bar{x}; G)$. We again derive a contradiction from the assumption $\lambda(x, \bar{x}; G_2^{kk'})<\lambda(x, \bar{x}; G_2)$. Let $X\in T(x, \bar{x}; G_2^{kk'})$ and let say $z\in X$. By Lemma 7 it follows as before that $\{a, u\}\subseteq\bar{X}$ and $\{a_0, u_1\}\subseteq X$, and furthermore $d(\{a_0, z\}, \bar{X}; G_2)\geq v+2$. Once again Lemma 4 yields a contradiction to $d(\{a_0, z\}; G_2(X))\leq v$.

We now derive a contradiction from the assumption that $\lambda(u_1, u_2; G^{kk'})<\lambda(u_1, u_2; G)$. Let $U\in T(u_1, u_2; G^{kk'})$. By Lemma 7, $E(z; G)\cap E(U; G)=\{k, k'\}$, so $z\in U$ and hence both $a_0\in U$ and $\{a, u\}\subseteq\bar{U}$. Since $[a_0, a]\cup\{k, k', k_2\}\subseteq E(U; G)$ Lemma 4 once again yields a contradiction to $d(\{a_0, z\}; G(U))\leq v$.

In the last three paragraphs we have proved that $\lambda(x, \bar{x}; G^{kk'})=\lambda(x, \bar{x}; G)$ for all $\{x, \bar{x}\}\subseteq V(G)-\{a_0, z\}$. Since

$$\lambda(x, \bar{x}; G)=\min\{d(x; G), d(\bar{x}; G)\}\geq d_0 \quad \text{for all } \{x, \bar{x}\}\subseteq V(G-z),$$

it follows by Lemma 5 (since $d(z; G^{kk'})=2$) that $\lambda(a_0, x; G^{kk'})\geq d_0$ for all $x\in G-\{a_0, z\}$. Hence the statement (∗) is proved.

Since $\lambda(a, \bar{a}; G_0)=d-1$ there exists an $A\subseteq V(G)$ with $a\in A$, $\bar{a}\in\bar{A}$ and $d(A; G_0)=d-1$. By Lemma 7, $\{h, h'\}\subseteq E(A; G)$ and $d(A; G)=d+1$. Since $\lambda(a, \bar{a}; H^{hh'})=d-1$ also holds, $k_0\notin E(A; G)$. Thus $a_0\in A$. Since for each

$x \in A - \{a_0, a, z\}$ we have

$$\lambda(x, \bar{a}; G_0) = \lambda(x, \bar{a}; G) = \min\{d(x; G), d(\bar{a}; G)\},$$

we must have $N(a_0; G) \cap A \subseteq \{a, z\}$. Otherwise there would be an $x_0 \in A \cap N(a_0; G - \{a, z\})$, so $d(x_0; G) \geq d$, by choice of a, and so $\lambda(x_0, \bar{a}; G_0) \geq d$ in contradiction to $d(A; G_0) = d - 1$. We distinguish two cases.

Case (i). $[a_0, z]_G = \emptyset$.

By the above, $E(a_0, G - a) \subseteq E(A; G)$. Since

$$\lambda(a, \bar{a}; G_0 - k_0) = d(A; G_0) = d - 1 = d(a; G_0 - k_0)$$

it follows that

$$d(a_0; G - a) = |[a_0, a]_G| - 1 = v - 1,$$

say. The inequality $d(a_0; G - a) < |[a_0, a]_G|$ stands, however, in contradiction to $\lambda(a, \bar{a}; G) = d(a; G)$.

Case (ii). $[a_0, z]_G \neq \emptyset$.

If $z \in A$ let $B = A - \{z\}$ and if $z \notin A$ let $B = A$. By Lemma 7, $E(z; G) \cap E(A; G) = \{h, h'\}$. Thus both $|E(z; G) \cap E(B; G)| = 2$ and $d(B; G) = d + 1$ hold. Then by Lemma 6

$$\lambda(a, \bar{a}; G - [a_0, z]) = \lambda(a, \bar{a}; G) = d$$

holds, hence $d(a_0; G - \{a, z\}) = v$ since $d(B; G - [a_0, z]) = d = d(a; G)$ and since $N(a_0; G - a) \subseteq \bar{B}$. It follows however that $d(B - \{a_0\}; G) = d$ and, by the z-irreducibility of G, $B = \{a_0, a\}$. But then $a \in N(z, G)$ and by (*) there is an admissible lifting of G at z.

Theorem 10. *Let G be a finite multigraph and let z be a non-separating vertex of G with $d(z; G) \geq 4$ and $|N(z; G)| \geq 2$. Then there exists an admissible lifting of G at z.*

Proof. We suppose that Theorem 10 is already proved for all multigraphs G' with $|G'| + |E(G')| < |G| + |E(G)|$. By Lemma 3 we can thus assume that G is z-irreducible. By Lemma 9 we may assume that $d(z; G) \geq 5$. Furthermore we may obviously assume that $|N(x; G)| \geq 2$ for all $x \in G$ and $|N(z; G)| \geq 3$. We can choose $a \in N(z; G)$ with

$$d(a; G) = \min_{x \in N(z; G)} d(x; G) = d,$$

say, and let $k \in [a, z]$. Furthermore let $H = G - k$. Since z fulfils the conditions of Theorem 10 in H also, there is by the induction hypothesis an admissible lifting

$H^{hh'}$ at z. We suppose that $G^{hh'}$ is not an admissible lifting. Since by Lemma 6

$$\lambda(x, y; G^{hh'}) \geq \lambda(x, y; H^{hh'}) = \lambda(x, y; H) = \lambda(x, y; G)$$

holds for all $\{x, y\} \subseteq V(G) - \{a, z\}$, there is an $\bar{a} \in G - \{a, z\}$ with

$$\lambda(a, \bar{a}; G^{hh'}) = \lambda(a, \bar{a}; G) - 1 = d - 1 \text{ by Lemma 6 and Lemma 7.}$$

Let $A \in T(a, \bar{a}; G^{hh'})$. Since

$$\lambda(a, \bar{a}; H^{hh'}) = \lambda(a, \bar{a}; H) = d - 1$$

also holds, we have $k \notin E(A; G)$, so $z \in A$. We may now assume that $N(z; G) \cap A = \{a\}$. Of course, if there were an $x \in N(z; G - a) \cap A$, we would have $d(x; G) \geq d$ by choice of a, and so by Lemma 6

$$\lambda(x, \bar{a}; G^{hh'}) \geq \lambda(x, \bar{a}; H^{hh'}) = \lambda(x, \bar{a}; H) = \lambda(x, \bar{a}; G) = \min\{d(x; G), d(\bar{a}; G)\} \geq d,$$

in contradiction to $d(A; G^{hh'}) = d - 1$. Thus $E(z; G - a) \cup \{h, h'\} \subseteq E(A; G)$ by Lemma 7. Since

$$\lambda(a, \bar{a}; H^{hh'}) = d - 1 = d(a; H^{hh'}) = d(A; H^{hh'})$$

we then have $d(z; H^{hh'} - a) = |[a, z]_G| - 1 = v - 1$, say, and so $d(z; G - a) = v + 1$, in particular $d(z; G) = 2v + 1$. But then $d(A - \{z\}; G) = d$, whence $A = \{a, z\}$ by the z-irreducibilty of G.

Let $l \neq k'$ be elements of $E(z; G - a)$, where say $l \in [z, u]$. We now show $\lambda(x, \bar{x}; G^{kk'}) = \lambda(x, \bar{x}; G)$ for all $\{x, \bar{x}\} \subseteq E(G) - \{z, u\}$. We consider the multigraph $L = G - l$. By Lemma 6 $\lambda(x, \bar{x}; L) = \lambda(x, \bar{x}; G)$ holds. We derive a contradiction from the assumption that $\lambda(x, \bar{x}; L^{kk'}) < \lambda(x, \bar{x}; L)$. Let $X \in T(x, \bar{x}; L^{kk'})$ and let say $z \in X$. By Lemma 7, $\{k, k'\} \subseteq E(X; L)$, so $a \in \bar{X}$ and $[z, a]_G \subseteq E(X; L)$. Hence $d(z, \bar{X}; L) \geq v + 1$, which by Lemma 4 yields a contradiction to $d(z; L) = 2v$.

Since $|N(z; G)| \geq 3$ there exists $u \neq u'$ in $N(z; G - a)$; let $l \in [z, u]$ and $l' \in [z, u']$. Since $d(z; G) \geq 5$, hence $d(z; G - a) \geq 3$, there exists $k' \in E(z; G - a) - \{l, l'\}$. We show that $G^{kk'}$ is an admissible lifting of G at z. By the previous paragraph $\lambda(x, \bar{x}; G^{kk'}) = \lambda(x, \bar{x}; G)$ holds for all $\{x, \bar{x}\} \subseteq E(G - z)$ with $\{x, \bar{x}\} \neq \{u, \bar{u}\}$. We need then only derive a contradiction from the assumption $\lambda(u, u'; G^{kk'}) < \lambda(u, u'; G)$. Let $U \in T(u, u'; G^{kk'})$, and let say $z \in U$. Since $\{k, k'\} \subseteq E(U; G)$ by Lemma 7, $a \in \bar{U}$ and thus $[z, a] \subseteq E(U; G)$. Then since $u' \in \bar{U}$, $l' \in E(U; G)$ also, and $d(z, \bar{U}; G) \geq v + 2$ holds, which by Lemma 4 yields a contradiction to $d(z; G) = 2v + 1$. Hence Theorem 10 is proved.

Theorem 10 may be extended immediately to infinite multigraphs.

Corollary 11. *Let z be a non-separating vertex of the infinite multigraph G with $d(z; G) \geq 4$, $|N(z; G)| \geq 2$, and $d(z; G)$ finite. Then there is an admissible lifting of G at z.*

Proof. There exists a finite subgraph $G_0 \subseteq G$ with $d(z; G_0) = d(z; G)$ such that $G_0 - z$ is connected. Let

$$\mathcal{R} = \{E' \subseteq E(G) : E' \supseteq E(G_0) \quad \text{and} \quad E' \text{ finite}\},$$

and denote by P the set of all two element subsets of $E(z; G)$. We define a function $f : \mathcal{R} \to P$. Let $E' \in \mathcal{R}$. The graph $H = (V(G), E')$ has by Theorem 10 an admissible lifting $H^{h'k'}$ at z. We set $f(E') = \{h', k'\}$. Then there exists an element $\{h, k\} \in P$ with the property that to each $E \in \mathcal{R}$ there is an $E' \in \mathcal{R}$ with $E \subseteq E'$ and $f(E') = \{h, k\}$. For otherwise we could associate with each $p \in P$ an $E(p)$ with the property that for each $E' \in \mathcal{R}$ with $E' \supseteq E(p)$, $f(E') \neq p$ holds. This is impossible, though, since $f(\bigcup_{p \in P} E(p)) \in P$. However G^{hk} is then an admissible lifting of G at z. For given $x \neq y$ in G with $\lambda(x, y; G)$ finite there is an $E \in \mathcal{R}$ with $\lambda(x, y; (V(G), E)) = \lambda(x, y; G)$ and by choice of $\{h, k\}$ there is an $E' \in \mathcal{R}$ with $E \subseteq E'$ and $f(E') = \{h, k\}$.

Remark 12. Let z be a cutvertex of the connected multigraph G, such that no $k \in E(z; G)$ is a separating edge of G. Let $C_1 \neq C_2$ be components of $G - z$ and $k_i \in E(z, V(C_i); G)$ for $i = 1, 2$. Then $\lambda(x, y; G^{k_1k_2}) = \lambda(x, y; G)$ holds for all $x \neq y$ in G. If $\{x, y\} \subseteq V(C) \cup \{z\}$ for some component C of $G - z$, this is clear since $G - k_i$ is connected. If x and y belong to different components of $G - z$, by Lemma 1

$$\begin{aligned} \lambda(x, y; G^{k_1k_2}) &\geq \min\{\lambda(x, z; G^{k_1k_2}), \lambda(z, y; G^{k_1k_2})\} \\ &= \min\{\lambda(x, z; G), \lambda(z, y; G)\} = \lambda(x, y; G). \end{aligned}$$

We will now describe operations by which all n-fold edge-connected finite multigraphs may be successively constructed. It is somewhat more convenient to consider here pseudographs (that is, to allow loops). The extension of our terminology to include pseudographs need cause no problems. (A loop adds 2 to the degree of the vertex. No loops arise as a result of identifications. For any vertex a, $a \notin N(a; G)$ always holds.) A pseudograph G is called n-minimal if $\lambda(G) \geq n$ but for any $k \in E(G)$, $\lambda(G - k) < n$ holds. A pseudograph G with $\lambda(G) \geq n$ is obviously n-minimal exactly when G has no loops and for each pair x, y of adjacent vertices, $\lambda(x, y; G) = n$ holds.

We need also the result indicated in [7, Theorem 5b].

Lemma 13. *Each finite, n-minimal pseudograph G with $|G| \geq 2$ has at least two vertices of degree n.*

Proof. We use induction on the number of vertices. We can assume that an element $k \in E(G)$, say $k \in [a, \bar{a}]$, exists with $d(a; G) > n$ and $d(\bar{a}; G) > n$. Since $\lambda(a, \bar{a}; G - k) < n$ there exists $A \subseteq V(G)$ with $a \in A$, $\bar{a} \in \bar{A}$ and $d(A; G) = n$. The multigraph G_a, resp. $G_{\bar{a}}$, arises from G by identification of A to a, resp. $\bar{A}$ to $\bar{a}$. By Lemma 2 G_a is n-minimal, since $\lambda(x, a; G) = n$ for all $x \in \bar{A}$ (or since $d(a; G_a) = n$). Since $|A| \geq 2$ (as $d(a; G) > n$) there are two vertices of degree n in

G_a, by the induction hypothesis. The same is true of $G_{\bar{a}}$ and so G also has two vertices of degree n.

Let G be finite pseudograph and z a non-separating vertex of G with $d(z; G) = 2m$ such that z has no loops attached. By Theorem 10, G admits successive admissible liftings at z so as to get a pseudograph G' with $|N(z; G')| \leq 1$. (After the lifting z remains a non-separating vertex.) The pseudograph G^z arises from $G' - z$ by addition of $\frac{1}{2} d(z; G')$ loops at the (possibly non-existent) vertex of $N(z; G')$. We say that G^z arises from G by admissible splitting of z, and call G^z an admissible splitting of G at z. For all $x \neq y$ in G^z, $\lambda(x, y; G^z) = \lambda(x, y; G)$. We now consider the following three operations on pseudographs.

O_m. Choose m different edges $k_1, \ldots, k_m$ of G, divide k_i by a vertex $z_i \notin G$ and identify $\{z_1, \ldots, z_m\}$ to a vertex $z \notin G$.

O_m^+. Act as in O_m, then choose an $a \in V(G)$ and add an edge between z and a.

$O_m^{(2)}$. Act as in O_m, thereby constructing G'. Now choose m different edges $h_1, \ldots, h_m$ of G' with $\{h_1, \ldots, h_m\} \not\subseteq E(z; G')$, divide h_i by a vertex $v_i \notin G'$, identify $\{v_1, \ldots, v_m\}$ to a vertex $v \notin G'$ and add exactly one (further) edge between v and z.

(Note that in O_m and O_m^+ and $O_m^{(2)}$ the chosen k_i and h_i may also be loops. For $m = 0$ the operation O_m is simply the addition of a new vertex z.)

If the pseudograph G' arises by O_m from a pseudograph G with $\lambda(G) \geq 2m$, then $\lambda(G') \geq 2m$ also, by Lemma 5. We will now show that conversely each $(2m)$-fold edge-connected pseudograph may be obtained from the graph G with $|G| = 1$ by successive addition of edges and repeated application of O_m. The next corollary was announced by Lovász in Prague.

Corollary 14. *Let G be a finite, $(2m)$-minimal multigraph with $|G| \geq 2$. Then there is a pseudograph G' with $\lambda(G') \geq 2m$ and $|G'| = |G| - 1$, from which G arises by O_m.*

Proof. By Lemma 13 there is a $z \in G$ with $d(z; G) = 2m$; z cannot be a cutvertex. Thus there is an admissible splitting G^z at z. Further $\lambda(G^z) \geq 2m$ holds. As O_m is the opposite operation to splitting, G^z is the pseudograph we are looking for.

If G is a $(2m)$-regular multigraph with $\lambda(G) \geq 2m$ and $|G| \geq 3$ then the pseudograph which exists by Corollary 14 is also $(2m)$-regular and so likewise has no loops. These successive constructions of all $(2m)$-fold edge connected $(2m)$-regular multigraphs were displayed by Kotzig in [3] and cited [4, Theorem 8], (See also [11, Conjecture 1]).

To obtain a similar construction of all $(2m+1)$-fold edge-connected pseudographs, we need the following result.

Lemma 15. *Let G be a $(2m+1)$-minimal, finite multigraph with $|G| \geq 3$, and let a_0 be a non-separating vertex of G. Then there exists a $k \in E(G - a_0)$, say $k \in [a, \bar{a}]$, with $d(a; G) = 2m+1$, such that $\lambda(x, y; G - k) \geq 2m+1$ for all $x \neq y$ in $G - \{a, \bar{a}\}$, and furthermore for all $x \neq y$ in $G - a$ in the case $d(\bar{a}; G) > 2m+1$.*

Proof. We assume that Lemma 15 is already proven for all multigraphs G' with $|G'|<|G|$.

By Lemma 13 there is a $b\neq a_0$ with $d(b;G)=2m+1$. Since a_0 doesn't separate the multigraph G, $N(b;G)\neq\{a_0\}$. Thus there exists $h\in E(b;G-a_0)$, say $h\in[b,\bar{b}]$. We can assume that $u\neq\bar{u}$ exist in $G-b$ with $\lambda(u,\bar{u};G-h)\leq 2m$, where further $\bar{b}\notin\{u,\bar{u}\}$ if $d(\bar{b};G)=2m+1$. Then there exists a $U\subseteq V(G)$ with $u\in U$, $\bar{u}\in\bar{U}$, $d(U;G)=2m+1$ and $h\in E(U;G)$; let say $b\in U$ and $\bar{b}\in\bar{U}$. Since $u\neq b$, $|U|\geq 2$. Since either $\bar{u}\neq\bar{b}$ or $d(\bar{b};G)>2m+1$, $|\bar{U}|\geq 2$ also holds. In the following we make use only of $|U|\geq 2$, $|\bar{U}|\geq 2$, and $d(U;G)=2m+1$, so we may assume say $a_0\in U$. Let G_u arise from G by identification of U to u. By Lemma 1, G_u is also $(2m+1)$-minimal and $3\leq|G_u|<|G|$ holds. Since $G_u-u=G(\bar{U})$ is connected, there is by the induction hypothesis a $k\in E(G_u-u)$, say $k\in[a,\bar{a}]_G$, with the properties (in G_u) described in the statement of Lemma 15.

We now show that k also has these properties in G. Clearly for all $x\in\bar{U}$, $d(x;G_u)=d(x;G)$ holds, in particular $d(a;G)=2m+1$. We observe that there is a $u'\in\bar{U}$ with $\lambda(u,u';G_u-k)=2m+1$. The case $|G_u|\geq 4$ each $u'\in U-\{a,\bar{a}\}\neq\emptyset$ will do. In the case $|G|=3$, $d(\bar{a};G_u)>2m+1$ holds, since not all three vertices of G_u can have degree $2m+1$. Then $d(u;G_u-k)=2m+1$ and $d(\bar{a};G_u-k)\geq 2m+1$, and since $|G_u|=3$ we have $\lambda(\bar{a},u;G_u-k)=2m+1$. Now for each such $u'\in\bar{U}$ with $\lambda(u,u';G_u-k)=2m+1$ it is obvious that $\lambda(u',x;G-k)=2m+1$ for all $x\in U$. By Lemma 2, $\lambda(x,y;G-k)=\lambda(x,y;G_u-k)$ for all $\{x,y\}\subseteq\bar{U}$. Since identification of $\bar{U}$ to $\bar{u}$ in $G-k$ yields the same graph as does the same operation in G, Lemma 2 also implies that $\lambda(x,y;G-k)=\lambda(x,y;G)$ for all $\{x,y\}\subseteq U$. Hence k has the desired properties in G.

Remark 16. A multigraph G with $\lambda(G)\geq 2m$ need by no means possess an edge $k\in G$ with the property that for all $x\neq y$ in $G-V(k)$, $\lambda(x,y;G-k)\geq 2m$ holds. For consider any graph G with $\lambda(G)\geq 2$ and subdivide each edge by exactly three vertices. If we then replace each edge by a set of m parallel edges we obtain a multigraph G' with $\lambda(G')\geq 2m$, but no edge in G' has the above property.

We now consider a pseudograph G with $\lambda(G)=2m+1$. For a pseudograph G' which arises from G by O_m^+, we have $\lambda(G')\geq 2m+1$ by Lemma 5. Analogously it is easily seen that $\lambda(G')\geq 2m+1$ if G' arises from G by $O_m^{(2)}$. We will now see that conversely each finite pseudograph G with $\lambda(G)\geq 2m+1$ may be obtained from a vertex by addition of edges and by use of O_m^+ and $O_m^{(2)}$.

Corollary 17. *Let G be a finite, $(2m+1)$-minimal multigraph with $|G|\geq 2$. Then either there is a pseudograph G_1 with $\lambda(G_1)\geq 2m+1$ and $|G_1|=|G|-1$, from which G arises by O_m^+, or else there is a pseudograph G_2 with $\lambda(G_2)\geq 2m+1$ and $|G_2|=|G|-2$, from which G arises by $O_m^{(2)}$.*

Proof. If $|G|=2$, G arises from a vertex with m loops by O_m^+. Let thus $|G|\geq 3$. There exists a non-separating vertex a_0 of G. Let $k\in[a,\bar{a}]\subseteq E(G-a_0)$ with $d(a;G)=2m+1$, where k is the edge described in Lemma 15. Let $H=G-k$. Since $d(a;G)=2m+1$, $H-a=G-a$ is connected, and so there is an admissible splitting H^a of H at a.

Case (i). $d(\bar{a};G)>2m+1$.

Then $\lambda(x,y;H)\geq 2m+1$ holds for all $x\neq y$ in $G-a$. Hence $\lambda(H^a)\geq 2m+1$ and G arises from H^a by O_m^+.

Case (ii). $d(\bar{a};G)=2m+1$.

This can only occur for $m>0$. Now $d(\bar{a};H^a)=2m$ and $\lambda(\bar{a},x;H^a)=2m$ for $x\in H^a-\bar{a}\neq\emptyset$. Thus H^a has no loops at $\bar{a}$ and since $m\geq 1$ $H^a-\bar{a}$ is connected. Hence there is an admissible splitting $(H^a)^{\bar{a}}$ at $\bar{a}$. But then $\lambda((H^a)^{\bar{a}})\geq 2m+1$ holds and G arises from $(H^a)^{\bar{a}}$ by $O_m^{(2)}$.

Let G' be a pseudograph with $\lambda(G')\geq 2m+1$ and $|G'|\geq 2$. If G arises from G' by O_m^+ then there is a vertex $x\in G$ with $d(x;G)>2m+1$. If G arises from G' by $O_m^{(2)}$ then there are two adjacent vertices in G of degree $2m+1$. Whilst the 1-minimal finite graphs may be constructed by O_0^+, alone, we cannot manage in Corollary 17 for $m>0$ with just one of the operations O_m^+ and $O_m^{(2)}$. (The complete bipartite graph $K_{2m+1,s}$ with $s>2m+1$ is $(2m+1)$-minimal, and contains no two adjacent vertices of degree $2m+1$.)

The situation is different, though, if we restrict ourselves to the class $\mathscr{k}_m (m\geq 1)$ of all $(2m+1)$-fold edge connected finite pseudographs whose vertices all have degree $2m+1$, with the possible exception of one vertex of degree $2m+2$. If $G\in\mathscr{k}_m$ has loops, then G is a single vertex with $m+1$ loops. Let $G\in\mathscr{k}_m$ with $|G|$ even, $|G|\geq 4$. Then G is $(2m+1)$-regular and arises by $O_m^{(2)}$ from a pseudograph G' with $\lambda(G')\geq 2m+1$, by Corollary 17. Since G' is also $(2m+1)$-regular, it follows that $G'\in\mathscr{k}_m$. Now let $G\in\mathscr{k}_m$ with $|G|$ odd, $|G|\geq 3$. Then G has exactly one vertex b of degree $2m+2$. Since $m\geq 1$ b is not a cutvertex. Thus we can assume $a_0=b$, where a_0 is the vertex described in the proof of Corollary 17. Then case (ii) will occur and G arises by $O_m^{(2)}$ from a pseudograph G' with $\lambda(G')\geq 2m+1$. With the exception of $d(b;G')=2m+2$, all vertices of G' have degree $2m+1$ and so $G'\in\mathscr{k}_m$. Thus we see that the graphs $\mathscr{k}_m$ arise by repeated application of $O_m^{(2)}$ either from a vertex with $m+1$ loops or from the graph consisting of two vertices joined by exactly $2m+1$ edges.[1]

In particular this settles [11, Conjecture 2]. (This conjecture states that each $G\in\mathscr{k}_m$ with $|G|\geq 4$ may be obtained from an appropriate $G'\in\mathscr{k}_m$ by the following operations: choose $2m$ different edges $k_1,\ldots,k_{2m}$ of G', subdivide k_i by a vertex z_i, identify $\{z_1,\ldots,z_m\}$ to a vertex z and $\{z_{m+1},\ldots,z_{2m}\}$ to a vertex $z'\neq z$, and join z and z' by exactly one edge. As stated this construction is not possible, however, since it implies that each $G\in\mathscr{k}_m$ with $|G|\geq 4$ must contain vertices $x\neq y$ with $\|[x,y]_G|=1$, and this is only fulfilled when $m=1$ or $m=2$.

[1] This successive construction of all $G\in\mathscr{k}_m$ with $|G|$ even had already been found by Kotzig in [3].

Consider for $m \geq 3$ a polygonal prism where each edge of both polygons has been replaced by $[(2m+1)/3]$ parallel edges and each other edge by $(2m+1-2[(2m+1)/3)]$ parallel edges.) One might suppose that each $G \in \mathscr{k}_m$ with $|G|$ odd arises by O_m^+ from a $G' \in \mathscr{k}_m$ with $|G'| = |G| - 1$. The following graph shows that this is not true. Let C be a cycle of length 12 with vertices $a_1, \ldots, a_{12}$ (in cyclic order) and let $a \notin C$. Join a to the vertices a_2, a_5, a_8 and a_{11} by a single edge and add further the edges $[a_n, a_{n+2}]$ for $n = 1, 4, 7, 10$ — Naturally it is possible, however, to obtain each $G \in \mathscr{k}_m$ with $|G| \geq 3$ odd from a $G' \in \mathscr{k}_m$ by O_{m+1}, and conversely applying O_{m+1} to a pseudograph $G' \in \mathscr{k}_m$ with $|G'|$ even always results in another element of $\mathscr{k}_m$.

We now turn to the theorem of Nash-Williams mentioned at the start. A multidigraph G is said to be *strongly connected* if $\lambda(x, y; G) \geq 1$ and $\lambda(y, x; G) \geq 1$ for all $x \neq y$ in G. Robbins shows in [10] that each 2-fold edge-connected multigraph admits a strongly connected orientation. Let G be a multigraph and let $\emptyset \neq A \subset V(G)$. Let G_{a_1}, resp. G_{a_2} arise from G by identification of A to $a_1 \in A$, resp. $\bar{A}$ to $a_2 \in \bar{A}$. Furthermore let i_n be an associative bijection

$$i_n : E(a_n; G_{a_n}) \to E(A; G) \quad \text{for} \quad n = 1, 2$$

and let $\vec{G}_{a_n}$ be an orientation of G_{a_n}. We say that $\vec{G}_{a_1}$ and $\vec{G}_{a_2}$ are *compatibly oriented* if for all $k \in E(A; G)$, $i_1^{-1}(k) \in E^+(a_1; \vec{G}_{a_1})$ if and only if $i_2^{-1}(k) \in E^-(a_2; \vec{G}_{a_2})$. Then obviously for all $k \in E(A; G)$, $i_1^{-1}(k) \in E^-(a_1; \vec{G}_{a_1})$ if and only if $i_2^{-1}(k) \in E^+(a_2; \vec{G}_{a_2})$ and $\vec{G}_{a_1}$ and $\vec{G}_{a_2}$ yield in a natural manner an orientation of G. Let C be a (continuously directed) cycle in the multidigraph G. G' arises from G by reversing the direction of each edge of C and leaving the other edges of G unchanged. We say that G' arises from G by *reorientation* of C. The dual multidigraph G' arises from G by reversing the direction of every edge of G.

We shall need the following simple lemmata.

Lemma 18. *Let $\vec{G}$ be an admissible orientation of the multigraph G. Let C be a cycle of $\vec{G}$ and let $\overleftarrow{G}$ arise from $\vec{G}$ by reorientation of C. Then $\overleftarrow{G}$ is an admissible orientation of G.*

Proof. For all $A \subseteq V(G)$, $|E^+(A; \vec{G}) \cap E(C)| = |E^-(A; \vec{G}) \cap E(C)|$ holds. Thus $d^+(A; \vec{G}) = d^+(A; \overleftarrow{G})$ for all $A \subseteq V(G)$, and the lemma follows by Menger's Theorem.

Lemma 19. *Let a be a vertex of the multidigraph G with*

$$\lambda(a, \bar{a}; G) = \lambda(\bar{a}, a; G) = d^+(a; G) = d^-(a; G) \quad \textit{for a certain } \bar{a} \in G - a.$$

Then for each $E^+ \subseteq E^+(a; G)$ and each $E^- \subseteq E^-(a; G)$ with $|E^+| = |E^-| = m$ there exist edge-disjoint cycles $C_1, \ldots, C_m$ with $E^+ \cup E^- \subseteq \bigcup_{\mu=1}^m E(C_\mu)$.

(Since by the definition of a cylce no vertex is passed through twice it follows that $a \in C_\mu$ for $\mu = 1, \ldots, m$.)

Proof. Let $E_0 = E(a; G) - (E^+ \cup E^-)$ and let $G_0 = G - E_0$. We split a in G_0 into vertices a and $a' \notin G_0$ to obtain a multidigraph G' thus: add to $G_0 - E^-$ the vertex a' and let $(x, a')_{G'} = (x, a)_{G_0}$ for all $x \in G_0$ and let $(a', x) = \emptyset$ for all $x \in G'$. Since however, $\lambda(a, \bar{a}; G') = \lambda(\bar{a}, a'; G') = m$ holds, it follows that $\lambda(a, a'; G') = m$ by Lemma 1, and so the existence of the cycles C_μ follows also.

Lemma 20. *Let a be a vertex of the multidigraph G with the property that for all $E^+ \subseteq E^+(a; G)$ and $E^- \subseteq E^-(a; G)$ with $|E^+| = |E^-| = m$ there exist edge-disjoint cycles $C_1, \ldots, C_m$ with $E^+ \cup E^- \subseteq \bigcup_{\mu=1}^m E(C_\mu)$. Then for each $E_0 \subseteq E(a; G)$ with $|E_0| = d^+(a; G)$ there is a multigraph G', obtained by reorienting cycles, with $E^+(a; G') = E_0$.*

Proof. Let $E^+ = E^+(a; G) - E_0$ and $E^- = E_0 - E^+(a; G)$. Since $|E_0| = d^+(a; G)$ we have $|E^+| = |E^-| = m$. Thus edge-disjoint cycles $C_1, \ldots, C_m$ exist with $E^+ \cup E^- \subseteq \bigcup_{\mu=1}^m E(C_\mu)$, and reorienting $C_1, \ldots, C_m$ yields the desired G'.

Lemma 21. *Let G be a multigraph with $A \in T(a, \bar{a}; G)$ for certain $a, \bar{a} \in G$, and let $d(A; G) = \lambda(a, \bar{a}; G)$ be even. Let G_a, resp. $G_{\bar{a}}$, arise from G by identification of A to a, resp. $\bar{A}$ to $\bar{a}$. If G_a and $G_{\bar{a}}$ possess admissible orientations then so does G.*

Proof. Let i and $\bar{i}$ be associative bijections of G_a and $G_{\bar{a}}$ respectively, and let $\vec{G}_a$, resp. $\vec{G}_{\bar{a}}$ be an admissible orientation of G_a, resp. $G_{\bar{a}}$. Since $\lambda(a, \bar{a}; G) = 2n$ then

$$\lambda(a, \bar{a}; \vec{G}_a) = \lambda(\bar{a}, a; \vec{G}_a) = n = \lambda(a, \bar{a}; \vec{G}_{\bar{a}}) = \lambda(\bar{a}, a; \vec{G}_{\bar{a}})$$

by Lemma 2. Consider say $\vec{G}_a$. By Lemma 19 the conditions of Lemma 20 for the vertex a in $\vec{G}_a$ are fulfilled. We choose $E_0 = i^{-1}(\bar{i}(E^-(\bar{a}; \vec{G}_{\bar{a}})))$. Since $|E_0| = n$ we obtain by Lemma 20 an orientation $\overleftarrow{G}_a$ with $E^+(a; \overleftarrow{G}_a) = E_0$. Then $\overleftarrow{G}_a$ and $\vec{G}_{\bar{a}}$ are compatibly oriented and by Lemma 18 $\overleftarrow{G}_a$ is also an admissible orientation. The orientations $\overleftarrow{G}_a$ and $\vec{G}_{\bar{a}}$ yield in a natural way an orientation $\vec{G}$ of G. We now show that $\vec{G}$ is an admissible orientation of G. First of all $\lambda(x, y; \vec{G}) = \lambda(x, y; \vec{G}_{\bar{a}})$ for all $\{x, y\} \subseteq A$ by Lemma 19 (this lemma implies that x, y-paths in $\vec{G}_{\bar{a}}$ can be transferred to x, y-paths in $\vec{G}$), and so

$$\lambda(x, y; \vec{G}) \geq [\tfrac{1}{2}\lambda(x, y; G_{\bar{a}})] = [\tfrac{1}{2}\lambda(x, y; G)]$$

by Lemma 2. Similarly $\lambda(x, y; \vec{G}) \geq [\frac{1}{2}\lambda(x, y; G)]$ for $x, y \in \bar{A}$. By the above we have $\lambda(a, \bar{a}; \vec{G}) = d^+(A; \vec{G}) = n = \lambda(\bar{a}, a; \vec{G})$. Let now, say, $x \in A$ and $y \in \bar{A}$ and let $m = [\frac{1}{2}\lambda(x, y; G)]$. Then since $\lambda(x, y; G) \leq \lambda(x, \bar{a}; G_{\bar{a}})$ we have $\lambda(x, \bar{a}; G_{\bar{a}}) \geq 2m$, similarly $\lambda(a, y; G_a) \geq 2m$, and so $\lambda(x, \bar{a}; \vec{G}_{\bar{a}}) \geq m$ and $\lambda(a, y; \overleftarrow{G}_a) \geq m$. Hence $\lambda(x, \bar{a}; \vec{G}) \geq m$ and $\lambda(a, y; \vec{G}) \geq m$. Since $E(A; G)$ separates x from y we have $2n = \lambda(a, \bar{a}; G) \geq 2m$, and so

$$\lambda(x, y; \vec{G}) \geq \min\{\lambda(x, \bar{a}; \vec{G}), \lambda(\bar{a}, a; \vec{G}), \lambda(a, y; \vec{G})\} \geq m,$$

as required.

Theorem 22 (Nash-Williams). *Each finite multigraph has an admissible orientation.*

Proof. We suppose that Theorem 22 is false. Let G be a counter-example to Theorem 22 with $|G|+|E(G)|$ as small as possible. It is easily seen that G has no cutvertex and $\lambda(G)\geq 2$. Obviously $|G|\geq 3$, so $|N(x, G)|\geq 2$ for all $x\in G$. Further it is easily seen that $\delta(G)\geq 3$. By Lemma 21 we have the following result.

(α) If $A\in T(a, \bar{a}; G)$ and if $d(A; G)$ is even, then $|A|=1$ or $|\bar{A}|=1$.

Let $k\in E(G)$ and consider $G-k$. We cannot have $[\frac{1}{2}\lambda(x, y; G-k)]=[\frac{1}{2}\lambda(x, y; G)]$ for all $x, y\in G$, since otherwise an admissible orientation of $G-k$ yields one such for G. Hence there exist $a\neq\bar{a}$ in G with $\lambda(a, \bar{a}; G)$ even and $\lambda(a, \bar{a}; G-k)<\lambda(a, \bar{a}; G)$. Let $A\in T(a, \bar{a}; G-k)$. Then $k\in E(A; G)$ and $d(A; G)$ is even. Thus by (α) either $A=\{a\}$ or else $\bar{A}=\{\bar{a}\}$, and so either $k\in E(a; G)$ and $d(a; G)$ is even or $k\in E(\bar{a}; G)$ and $d(\bar{a}; G)$ is even. Hence we have the following.

(β) Every edge of G is incident with a vertex of even degree.

By the result of Robbins [10] (and anyway from (β)) there exist in G vertices of degree greater than 3; let

$$V_0=\{x\in V(G): d(x; G)>3\}$$

and let $d_0=\min_{x\in V_0} d(x; G)$. We now prove the following:

(γ) There is no $z\in V(G)$ with $d(z; G)=d_0$ and $\lambda(z, x; G)=d_0$ for all $x\in V_0-\{z\}$.

We assume there were such a z. By Theorem 10 we may obtain a multigraph H from G by successive admissible liftings at z, in which $d(z; H)=3$ or $|N(z; H)|\leq 1$. Since $|E(H)|<|E(G)|$ there is an admissible orientation $\vec{H}$ of H. Let $E(H)-E(G)=\{h_1, \ldots, h_s\}$. We divide the edges h_i in the directed multigraph $\vec{H}$ by a vertex z_i identify $\{z_1, \ldots, z_s\}$ to z, thus obtaining $\vec{G}$. Obviously $\vec{G}$ is an orientation of G; we now show that it is admissible. For $x\neq y$ in $G-z$ we have

$$\lambda(x, y; \vec{G})\geq\lambda(x, y; \vec{H})\geq[\tfrac{1}{2}\lambda(x, y; H)]=[\tfrac{1}{2}\lambda(x, y; G)].$$

In the case $d(z; H)\geq 2$ we have $\lambda(H)\geq 2$ by Lemma 5, and so also $\lambda(G)\geq 2$, and then $\vec{H}$ is strongly connected and $|d^+(z; \vec{H})-d^-(z; \vec{H})|\leq 1$. If $d(z; H)<2$ we have $\lambda(H-z)\geq 2$, and so $\vec{H}-z$ is strongly connected. Hence in both cases $\vec{G}$ is strongly connected and both $d^+(z; \vec{G})\geq m=[\frac{1}{2}d_0]$ and $d^-(z; \vec{G})\geq m$ hold. We have to show besides that $\lambda(z, x; \vec{G})\geq[\frac{1}{2}\lambda(z, x; G)]$ and $\lambda(x, z; \vec{G})\geq[\frac{1}{2}\lambda(x, z; G)]$ for all $x\in G-z$. For $x\in G-V_0$ this is on account of the strong connectivity of $\vec{G}$. By our assumptions and by Lemma 1 $\lambda(x, y; G)\geq d_0$ for all $\{x, y\}\subseteq V_0-\{z\}$, and hence $\lambda(x, y; \vec{G})\geq m$. We now deduce a contradiction, say, from the assumption $\lambda(z, \bar{z}; \vec{G})<m$ for some $\bar{z}\in V_0-\{z\}$. Let $Z\in T(z, \bar{z}; \vec{G})$ be minimal with respect to inclusion. Since $\lambda(x, y; \vec{G})\geq m$ for $\{x, y\}\subseteq V_0-\{z\}$, and $\bar{z}\in\bar{Z}$, we have $V_0-\{z\}\subseteq\bar{Z}$, and so $d(x; G)=3$ for all $x\in Z-\{z\}$. By (β) we have $E(G(Z))\subseteq E(z; G)$. Since $d^+(z; \vec{G})\geq m$ we have $|Z|\geq 2$; let $x\in Z-\{z\}$. Since Z is minimal we have $d^+(Z-\{x\}, \vec{G})>d^+(Z; \vec{G})$ and so $d(z, x; \vec{G})>d(x, \bar{Z}; \vec{G})$. If

$d(x, \bar{Z}; \vec{G})>0$ we have $|[z, x]_G|=2$. If $d(x, \bar{Z}; \vec{G})=0$ we have $d(x, z; \vec{G})>0$ by the strong connectivity of $\vec{G}$, so again $|[z, x]_G|\geq 2$. But $|[z, x]|\geq 2$ is in contradiction to $\lambda(z, \bar{z}; G)=d_0=d(z; G)$ since $d(x; G)=3$.

By (γ) the set

$$P=\{\{x, y\}\subseteq V(G):\lambda(x, y; G)<\min\{d(x; G), d(y; G)\}\}$$

is non-empty. Let $\{a, \bar{a}\}\in P$ with

$$\lambda(a, \bar{a}; G)=\min_{\{x, y\}\in P}\lambda(x, y; G).$$

By (α), $\lambda(a, \bar{a}; G)$ is odd, say $\lambda(a, \bar{a}; G)=2n+1$. Since $\lambda(G)\geq 2$ we have $n\geq 1$, and so $\{a, \bar{a}\}\subseteq V_0$. Let $A\in T(a, \bar{a}; G)$. The multigraph G_a, resp. $G_{\bar{a}}$, arises from G by identification of A to a, resp. $\bar{A}$ to $\bar{a}$. G_a and $G_{\bar{a}}$ have admissible orientations $\vec{G}_a$ and $\vec{G}_{\bar{a}}$ respectively. By Lemma 2 $\lambda(G_a)\geq 2$ holds and $\lambda(a, \bar{a}; G_a)=2n+1$. Thus $\vec{G}_a$ is strongly connected and both $\lambda(a, \bar{a}; \vec{G}_a)\geq n$ and $\lambda(\bar{a}, a; \vec{G}_a)\geq n$ hold; in particular $d^+(a; \vec{G}_a)\geq n$ and $d^-(a; \vec{G}_a)\geq n$. Similar considerations hold for $\vec{G}_{\bar{a}}$

(δ) Each (non-empty) set $X\subseteq V(G_a-a)$ with $d^+(X; \vec{G}_a)<n$ contains only vertices x with $d(x; G)=3$.
(δ) remains true if d^+ is replaced by d^- or a by $\bar{a}$.

Proof of (δ). For $x\in X$, $\lambda(x, a; \vec{G}_a)<n$. By Lemma 2 $\lambda(x, a; G)=\lambda(x, a; G_a)<2n$, since $\vec{G}_a$ is an admissible orientation of G_a. By choice of $\{a, \bar{a}\}$ we have $\{x, a\}\notin P$, and so

$$\lambda(x, a; G)=\min\{d(x; G), d(a; G)\}.$$

Hence $d(x; G)<2n$ since $\lambda(x, a; G)<2n<d(a; G)$. However, then $d(x; G)=3$, for if $d(x; G)>3$ we would have $d_0<2n$. In the case $d_0\leq 2n$ though we would have for all $z\in G$ with $d(z; G)=d_0$ and $y\in G-z$ that $\{z, y\}\notin P$, by choice of $\{a, \bar{a}\}$, and so

$$\lambda(z, y; G)=\min\{d(z; G), d(y; G)\}\quad \text{for all } y\in G-z,$$

in contradiction to (γ).

(ε) Let $E^+\subseteq E^+(a; \vec{G}_a)$ and $E^-\subseteq E^-(a; \vec{G}_a)$ with $|E^+|=|E^-|=m$. Then there exist edge-disjoint cycles $C_1, \ldots, C_m$ with $E^+\cup E^-\subseteq \bigcup_{\mu=1}^m E(C_\mu)$;
(ε) holds similarly with $\vec{G}_a$ replaced by $\vec{G}_{\bar{a}}$.

Proof of (ε). Let $H=\vec{G}_a-(E(a; \vec{G}_a)-(E^+\cup E^-))$. The multidigraph D arises in the following way from H: add to $H-E^-$ a vertex $a'\notin H$ and set $(x, a')_D=(x, a)_H$ for all $x\in H$ and let $d^+(a'; D)=0$. Let $d^-(a; \vec{G}_a)=n$; the case $d^+(a; \vec{G}_a)=n$ follows dually. Let $B\in T(a, a'; D)$ be minimal w.r.t. inclusion. It suffices to deduce a contradiction from the assumption $d^+(B; D)<m$. Since $d^+(a; D)=m$ we have $B-\{a\}\neq\emptyset$ and since $d^-(a; D)=0$ we have $d^+(B-\{a\}; D)<m$. Since

$$d^-(a; \vec{G}_a)=n=m+|E^-(a; \vec{G}_a)-E^-|,$$

$d^+(B-\{a\}; \vec{G}_a)<n$ follows. By (δ) then $d(b'; G)=3$ for all $b' \in B-\{a\}$ and by (β) $B-\{a\}$ is a set of independent vertices. Let $b \in B-\{a\} \neq \emptyset$. By choice of B, $d^+(B-\{b\}; D) > d^+(B; D)$, and so $d(a, b; D) > d(b, V(D)-B; D)$, and in particular $(a, b)_D \neq \emptyset$. Since $\vec{G}_a$ is strongly connected there is a $k \in E^+(b; \vec{G}_a)$. If $k \in E(D)$, then $k \in E(b, V(D)-B; D)$ since $d^-(a; D)=0$, so $|(a, b)_D| \geq 2$. If $k \notin E(D)$, then $k \in E(a; G_a)$, so again $|[a, b]_{G_a}| \geq 2$. Since $d(b; G)=3$, $b \neq \bar{a}$ and $|[a, b]_{G_a}| \geq 2$ contradicts $\lambda(a, \bar{a}; G_a)=2n+1=d(a; G_a)$.

We now conclude the proof of Theorem 22. We can suppose $d^+(a; \vec{G}_a)= d^-(\bar{a}; \vec{G}_{\bar{a}})$, for otherwise we may replace say $\vec{G}_a$ by its dual. Since by (ε) the conditions in Lemma 20 are fulfilled for $a \in \vec{G}_a$ we may suppose, in view of Lemma 20 and Lemma 18, that $\vec{G}_a$ and $\vec{G}_{\bar{a}}$ are compatibly (and admissibly) oriented. $\vec{G}$ arises naturally from $\vec{G}_a$ and $\vec{G}_{\bar{a}}$. We will now show that $\vec{G}$ is an admissible orientation of G. For $\{x, y\} \subseteq \bar{A}$, it follows easily by ($\varepsilon$) and Lemma 2 that

$$\lambda(x, y; \vec{G}) = \lambda(x, y; \vec{G}_a) \geq [\tfrac{1}{2}\lambda(x, y; G_a)] = [\tfrac{1}{2}\lambda(x, y; G)],$$

and similarly for $\{x, y\} \subseteq A$. Next we show that $\lambda(a, \bar{a}; \vec{G}) \geq n$ and $\lambda(\bar{a}, a; \vec{G}) \geq n$. We deduce a contradiction from the assumption that $\lambda(a, \bar{a}; \vec{G})<n$, say. Since $\lambda(a, \bar{a}; \vec{G}_{\bar{a}}) \geq n$ and $\lambda(a, \bar{a}; \vec{G}_a) \geq n$ we see immediately that if $d^+(A; \vec{G})=n$ then $\lambda(a, \bar{a}; \vec{G}) \geq n$. Thus $d^+(A; \vec{G})=n+1$ must hold. Let $B \in T(a, \bar{a}; \vec{G})$ with

$$|B \cap A| + |\bar{B} \cap \bar{A}| = \max\{|B' \cap A| + |\bar{B}' \cap \bar{A}| : B' \in T(a, \bar{a}; \vec{G})\}.$$

Since $\lambda(a, \bar{a}; \vec{G}_{\bar{a}}) \geq n$ we have $B \cap \bar{A} \neq \emptyset$ and since $\lambda(a, \bar{a}; \vec{G}_a) \geq n$ we have $\bar{B} \cap A \neq \emptyset$. Since

$$d^+(B \cap \bar{A}; \vec{G}) + d^-(\bar{B} \cap A; \vec{G}) \leq d^-(A; \vec{G}) + d^+(B; \vec{G}) \leq n+n-1,$$

it follows that $d^+(B \cap \bar{A}; \vec{G})<n$ or $d^-(\bar{B} \cap A; \vec{G})<n$. Let say $d^-(\bar{B} \cap A; \vec{G})<n$; the case $d^+(B \cap \bar{A}; \vec{G})<n$ leads similarly to a contradiction. Then $d^-(\bar{B} \cap A; \vec{G}_{\bar{a}})<n$ also and by (δ) we have $d(x; G)=3$ for all $x \in \bar{B} \cap A$. Let $b \in \bar{B} \cap A \neq \emptyset$. By choice of B it follows that $B \cup \{b\} \notin T(a, \bar{a}; \vec{G})$, so $d^+(B \cup \{b\}; \vec{G}) \geq d^+(B; \vec{G})$ and so by (β) then $d(b, \bar{B} \cap \bar{A}; \vec{G}) > d(B, b; \vec{G})$; in particular $d(b; \bar{B} \cap \bar{A}; \vec{G}) \geq 1$. Since $A \in T(a, \bar{a}; G)$, we have $d(b, \bar{A}; G) \leq 1$ since $d(b; G)=3$ and $b \neq a$. Therefore $d(b, \bar{B} \cap \bar{A}; \vec{G})=1$ and $d(\bar{A}, b; \vec{G})=0$. Thus $d(B, b; \vec{G})=0$, and by (β) $d^-(b; \vec{G})=0$, contradicting the strong connectivity of $\vec{G}_{\bar{a}}$.

Let now $x \in A$ and $y \in \bar{A}$, and let $m=[\tfrac{1}{2}\lambda(x, y; G)] \leq n$. Then $\lambda(x, \bar{a}; G) \geq 2m$, so by Lemma 1 $\lambda(y, \bar{a}; G) \geq 2m$ and so by the above, $\lambda(y, \bar{a}; \vec{G}) \geq m$ and $\lambda(\bar{a}, y; \vec{G}) \geq m$. Since similarly $\lambda(a, x; \vec{G}) \geq m$ and $\lambda(x, a; \vec{G}) \geq m$, we have by Lemma 1 that $\lambda(x, y; \vec{G}) \geq m$ and $\lambda(y, x; \vec{G}) \geq m$.

Note added in proof

Many thanks to Dr. Bollobás for translating my paper from German into English. The original German version may be obtained from the author upon request.

References

[1] L.R. Ford and D.R. Fulkerson, Flows in Networks (Princeton Univ. Press, Princeton, NJ, 1962).
[2] R. Halin, Zum Mengerschen Graphensatz, in Beiträge zur Graphentheorie (Teubner, Leibzig, 1968) 41–48.
[3] A. Kotzig, Súvilost a pravidelná súvilost konečných grafov, Dissertation, Karlsuniversität Prag, (VŠE Bratislava, 1956).
[4] A. Kotzig, Vertex splittings preserving the maximal edge connectedness of 4-regular multigraphs, to appear in J. Combinatorial Theory (B).
[5] L. Lovász, On some connectivity properties of eulerian graphs, Acta Math. Acad. Sci. Hung. 28 (1976) 129–138.
[6] L. Lovász, Proceedings of the Fifth British Combinatorial Conference 1975 (Utilitas Mathematica Publishing, Winnipeg, 1976) 684.
[7] W. Mader, Minimale n-fach kantenzusammenhängende Graphen, Math. Ann. 191 (1971) 21–28.
[8] L. Mirsky, Transversal Theory (Academic Press, New York, 1971).
[9] C.St.J.A. Nash-Williams, On orientations, connectivity and odd-vertex-pairings in finite graphs, Can. J. Math. 12 (1960) 555–567.
[10] H.E. Robbins, A theorem on graphs, with an application to a problem of traffic control, Am. Math. Monthly 46 (1939) 281–283.
[11] G.J. Simmons, Command graphs, Infinite and finite sets, (North-Holland, Amsterdam, 1975) 1277–1349.
[12] K. Wagner, Graphentheorie, Bibliographisches Institut, Mannheim (1970).

Annals of Discrete Mathematics 3 (1978) 165–179.

ANOTHER CRITERION FOR MARRIAGE IN DENUMERABLE SOCIETIES

C. St. J. A. NASH-WILLIAMS

Department of Mathematics, University of Reading, Whiteknights, Reading RG6 2AX, Great Britain

A society is an ordered triple (M, W, K) of sets such that M, W are disjoint and $K \subseteq M \times W$. An espousal of (M, W, K) is a subset of K of the form $\{(a, E(a)) : a \in M\}$ where $E(a_1) \neq E(a_2)$ whenever $a_1 \neq a_2$. For every transfinite sequence f of distinct elements of W, we define (in a somewhat complicated manner) a number $q(f)$. We prove that a necessary and, if M is countable, sufficient condition for (M, W, K) to have an espousal is that $q(f) \geq 0$ for every countable transfinite sequence f of distinct elements of W.

1. Introduction

We shall use the following set-theoretic conventions. A *relation* is a set of ordered pairs. Let R be a relation, A and B be sets and a be an element. Then $R\langle a\rangle$ denotes $\{y : (a, y) \in R\}$, $R(a)$ denotes the element of $R\langle a\rangle$ if $|R\langle a\rangle| = 1$, and $R[A]$ denotes $\bigcup_{a \in A} R\langle a\rangle$. The *domain* $\operatorname{dom} R$ and *range* $\operatorname{rge} R$ of R are $\{x : (x, t) \in R \text{ for some } t\}$ and $\{y : (t, y) \in R \text{ for some } t\}$ respectively. A *function* is a relation f such that $|f\langle x\rangle| = 1$ for every $x \in \operatorname{dom} f$. A *function f from A into B* or function $f : A \to B$ is a function f such that $\operatorname{dom} f = A$, $\operatorname{rge} f \subseteq B$. A *function from A onto B* is a function with domain A and range B. A function f is *injective* if there do not exist distinct elements x, y of $\operatorname{dom} f$ such that $f(x) = f(y)$. A set S is *countable* if $|S| \leq \aleph_0$. The axiom of choice will be assumed in this paper.

A *society* is an ordered triple (M, W, K) such that M, W are disjoint sets and K is a subset of $M \times W$. Elements of M are *men* of a society $\Gamma = (M, W, K)$ and elements of W are *women* of Γ, and a man m will be said to *know* a woman w if $(m, w) \in K$. An *espousal* of Γ is an injective function $E : M \to W$ such that $E \subseteq K$ (i.e., intuitively, a prescription for finding wives for all the men so that each man marries a woman whom he knows). A society is *espousable* if it has an espousal. A society (M, W, K) is *male-countable* if M is countable.

Damerell and Milner [1] proved a conjecture of [2] that a certain condition would be necessary and sufficient for a male-countable society to be espousable. In [3], I gave an alternative version of this proof and conjectured another necessary and sufficient condition for a male-countable society to be espousable. In fact, this conjecture is clearly false in the form suggested in [3], but it requires only a minor modification and the present paper proves the modified conjecture. This criterion for espousability of male-countable societies has perhaps a somewhat more direct character than the previous one (which involved a transfinite sequence of functions called "margin functions") and its proof seems a little shorter.

Further discussion of the background and history of this work may be found in [3].

We shall assume ordinals (ordinal numbers) to have been defined so that an ordinal α is the set of all ordinals less than α. In particular, the ordinal 0 is the empty set. An ordinal β is a *successor ordinal* if $\beta=\alpha+1$ for some ordinal α. A *limit ordinal* is an ordinal which is neither 0 nor a successor ordinal. We define a *transfinite sequence* to be a function whose domain is an ordinal number. If f is a transfinite sequence and λ is an ordinal less than or equal to dom f, then f_λ will denote the transfinite sequence $\{(\alpha, f(\alpha)): \alpha<\lambda\}$, i.e. the restriction of f to λ. We define a *queue* to be a countable injective transfinite sequence, i.e. an injective transfinite sequence whose domain is a countable ordinal. A *queue in* a set A is a queue whose range is a subset of A. The transfinite sequence whose domain is the ordinal 0 will be denoted by $\square$. In fact, since the ordinal 0 is the empty set, it follows that $\square$ is also the empty set, but we denote the empty set by $\square$ and not $\emptyset$ when it plays the rôle of a transfinite sequence.

Let $\mathcal{Q}$ denote a set whose elements are the integers and two further "numbers" ∞ and $-\infty$. Elements of $\mathcal{Q}$ will be called *quasi-integers.* The *size* $\|A\|$ of a set A is defined to be its cardinality $|A|$ if A is finite and to be ∞ if A is infinite: thus $\|A\|\in\mathcal{Q}$ for every set A. The *sum* $a_1+\cdots+a_n$ of n quasi-integers $a_1,\ldots,a_n$ has its usual meaning if the a_i are all integers, is defined to be ∞ if at least one a_i is ∞, and is defined to be $-\infty$ if no a_i is ∞ but at least one is $-\infty$. The *difference* $a-b$ of two quasi-integers is the sum of a and $-b$; and likewise the sum of the quasi-integers a, $-b$, c may be denoted by $a-b+c$, etc. For our purposes, the most important distinctive feature of these definitions is that $\infty-\infty$ is defined to be ∞, since we wish to think of $\infty-\infty$ as the *largest possible* value of $\|A\setminus B\|$ for sets A, B such that $B\subseteq A$ and $\|A\|=\|B\|=\infty$. *Inequalities* between quasi-integers are defined in the obvious way. The *infimum* inf $\mathcal{A}$ of a non-empty subset $\mathcal{A}$ of $\mathcal{Q}$ is the greatest quasi-integer q such that $q\leqslant a$ for every $a\in\mathcal{A}$, and the *supremum* sup $\mathcal{A}$ is analogously defined. If λ is a limit ordinal and $a_\theta\in\mathcal{Q}$ for every $\theta<\lambda$, we define $\liminf_{\theta\to\lambda} a_\theta$ to be $\sup\{i_\phi: \phi<\lambda\}$, where i_ϕ denotes $\inf\{a_\theta: \phi\leqslant\theta<\lambda\}$.

Throughout this paper, it will be understood that we are discussing a society $\Gamma=(M, W, K)$ and the symbols Γ, M, W, K should henceforward be interpreted accordingly.

The *demand-set* $D(X)$ of a subset X of W is $\{m\in M: K\langle m\rangle\subseteq X\}$, i.e., intuitively, the set of all men who demand wives from X when an espousal of Γ is sought. If f is a queue in W, then $\Delta(f)$ will denote $D(\text{rge } f)$.

With any queue f in W we shall associate a quasi-integer $q(f)$, called the *margin* of f, which is defined as follows. Define $q(\square)$ to be $-\|D(\emptyset)\|$. If now dom f is an ordinal $\lambda>0$ and $q(f')$ has been defined for every queue f' in W whose domain is less than λ, then define $q(f)$ to be

(i) $q(f_\kappa)+1-\|\Delta(f)\setminus\Delta(f_\kappa)\|$ if λ is a successor ordinal $\kappa+1$,

(ii) $\liminf_{\theta\to\lambda} q(f_\theta)-\|\Delta(f)\setminus\bigcup_{\theta<\lambda}\Delta(f_\theta)\|$ if λ is a limit ordinal.

Roughly speaking, $q(f)$ is the largest number of women whom we could hope to leave unmarried in rge f after working along the sequence f term by term, trying at each stage to ensure that wives have been found for all the men who demand them from amongst the set of women so far considered. Of course, if $D(\emptyset) \neq \emptyset$, then the men in $D(\emptyset)$ know no women at all and we encounter, so to speak, a shortfall of $\|D(\emptyset)\|$ women before we even start working along the sequence f: hence $q(\square)$ is defined to be $-\|D(\emptyset)\|$. Intuitively, (i) expresses the idea that, when the men in $\Delta(f_\kappa)$ have been married to women in rge f_κ with the maximum possible number $q(f_\kappa)$ of such women left unmarried, then adding $f(\kappa)$ to these unmarried women gives us $q(f_\kappa)+1$ women amongst whom to find wives for the men in $\Delta(f) \setminus \Delta(f_\kappa)$, which we might at best hope to achieve leaving $q(f_\kappa)+1-\|\Delta(f) \setminus \Delta(f_\kappa)\|$ of these women still unmarried. If, now, dom f is a limit ordinal λ as in (ii), and if, for $\theta < \lambda$, $q(f_\theta)$ is the maximum number of women who could be left unmarried in rge f_θ after wives have been found for the men in $\Delta(f_\theta)$, then we could at most hope, after working along the sequence f, to leave $l = \liminf_{\theta \to \lambda} q(f_\theta)$ women in rge f unmarried to men in $C = \bigcup_{\theta < \lambda} \Delta(f_\theta)$, and amongst these l women we must still find wives for the men in $\Delta(f) \setminus C$, which can leave at most $l - \|\Delta(f) \setminus C\|$ unmarried women in rge f. The proof of our first lemma translates this informal explanation into a more precise argument.

Lemma 1.1. *If E is an espousal of Γ and f is a queue in W then*

$$\|(\text{rge } f) \setminus E[\Delta(f)]\| \leqslant q(f). \tag{1}$$

Proof. The assumed existence of E implies that, for every $m \in M$, $E(m) \in K\langle m \rangle$ and therefore $K\langle m \rangle \neq \emptyset$ and therefore $m \notin D(\emptyset)$. Hence $0 = -\|D(\emptyset)\| = q(\square)$, so that (1) is true when $f = \square$, i.e. when dom $f = 0$.

To continue the proof by transfinite induction on dom f, suppose now that dom $f = \lambda > 0$ and assume the inductive hypothesis that $\|(\text{rge } g) \setminus E[\Delta(g)]\| \leqslant q(g)$ for every queue g in W with domain less than λ.

Suppose first that λ is a successor ordinal $\kappa + 1$. Then

$$E[\Delta(f_\kappa)] \subseteq K[\Delta(f_\kappa)] \subseteq \text{rge } f_\kappa = (\text{rge } f) \setminus \{f(\kappa)\}. \tag{2}$$

Moreover rge $f_\kappa \subseteq$ rge f and therefore

$$E[D(\text{rge } f_\kappa)] \subseteq E[D(\text{rge } f)] \subseteq \text{rge } f,$$

i.e.

$$E[\Delta(f_\kappa)] \subseteq E[\Delta(f)] \subseteq \text{rge } f. \tag{3}$$

Using the inductive hypothesis, (2) and (3), we see that

$$\begin{aligned} q(f) &= q(f_\kappa) + 1 - \|\Delta(f) \setminus \Delta(f_\kappa)\| \\ &\geqslant \|(\text{rge } f_\kappa) \setminus E[\Delta(f_\kappa)]\| + 1 - \|E[\Delta(f)] \setminus E[\Delta(f_\kappa)]\| \\ &= \|(\text{rge } f) \setminus E[\Delta(f_\kappa)]\| - \|E[\Delta(f)] \setminus E[\Delta(f_\kappa)]\| \\ &\geqslant \|((\text{rge } f) \setminus E[\Delta(f_\kappa)]) \setminus (E[\Delta(f)] \setminus E[\Delta(f_\kappa)])\| \\ &= \|(\text{rge } f) \setminus E[\Delta(f)]\|. \end{aligned}$$

Now suppose that λ is a limit ordinal. Let i_ϕ denote $\inf\{q(f_\theta)\colon \phi \leq \theta < \lambda\}$ for every $\phi < \lambda$, and let ∇ denote $\bigcup_{\theta<\lambda} \Delta(f_\theta)$. Since $\operatorname{rge} f_\theta \subseteq \operatorname{rge} f$ and consequently $\Delta(f_\theta) \subseteq \Delta(f)$ for every $\theta < \lambda$, it follows that $\nabla \subseteq \Delta(f)$. Therefore

$$E[\nabla] \subseteq E[\Delta(f)] \subseteq K[\Delta(f)] \subseteq \operatorname{rge} f. \tag{4}$$

For $\phi \leq \theta < \lambda$, the inductive hypothesis gives

$$q(f_\theta) \geq \|(\operatorname{rge} f_\theta) \setminus E[\Delta(f_\theta)]\| \geq \|(\operatorname{rge} f_\phi) \setminus E[\nabla]\|.$$

Therefore $i_\phi \geq \|(\operatorname{rge} f_\phi) \setminus E[\nabla]\|$ for every $\phi < \lambda$, and consequently

$$\begin{aligned} \liminf_{\theta \to \lambda} q(f_\theta) &= \sup\{i_\phi\colon \phi < \lambda\} \\ &\geq \sup\{\|(\operatorname{rge} f_\phi) \setminus E[\nabla]\|\colon \phi < \lambda\} = \|(\operatorname{rge} f) \setminus E[\nabla]\|. \end{aligned}$$

Hence

$$\begin{aligned} q(f) &= \liminf_{\theta \to \lambda} q(f_\theta) - \|\Delta(f) \setminus \nabla\| \\ &\geq \|(\operatorname{rge} f) \setminus E[\nabla]\| - \|E[\Delta(f)] \setminus E[\nabla]\| \\ &\geq \|((\operatorname{rge} f) \setminus E[\nabla]) \setminus (E[\Delta(f)] \setminus E[\nabla])\| \\ &= \|(\operatorname{rge} f) \setminus E[\Delta(f)]\| \end{aligned}$$

in view of (4).

We shall say that a society (M, W, K) is *good* if $q(f) \geq 0$ for every queue f in W, and is *bad* if $q(f) < 0$ for some queue f in W. Lemma 1.1 implies that every espousable society is good. The purpose of this paper is to establish the following theorem.

Theorem 1.2. *A male-countable society is espousable if and only if it is good.*

Thus, if a male-countable society (M, W, K) is not espousable, then there is a queue f in W with $q(f) < 0$. Informally speaking, f is a countable transfinite sequence of women which provides a fairly obvious obstruction to the existence of an espousal, because an impossibility is encountered when we work along this sequence term by term, trying at each stage to ensure that wives have been found for all the men who demand them from amongst the women so far considered.

In essence, Theorem 1.2 is a conjecture of [3]; but in [3] I mistakenly suggested taking $q(\square) = 0$ as the first step in the definition of $q(f)$.

2. Preliminary lemmas

Definition 2.1. If $f^1, f^2, \ldots, f^n$ are queues with disjoint ranges and $\operatorname{dom} f^i = \alpha_i$ $(i = 1, \ldots, n)$ then $f^1 * f^2 * \cdots * f^n$ will denote the queue h with domain $\alpha_1 + \alpha_2 + \cdots + \alpha_n$ such that

$$h(\alpha_1 + \alpha_2 + \cdots + \alpha_{i-1} + \theta) = f^i(\theta), \quad i = 1, 2, \ldots, n;\ \theta < \alpha_i$$

(where $\alpha_1+\alpha_2+\cdots+\alpha_{i-1}+\theta$ means θ if $i=1$). Informally, the terms of the queue $f^1*f^2*\cdots*f^n$ are the terms of f^1 followed by those of f^2 followed by those of f^3 etc. The queue $\{(0, x)\}$ will be denoted by $[x]$: in other words, $[x]$ is the queue f such that $\operatorname{dom} f=1=\{0\}$, $f(0)=x$.

If $A\subseteq M$, $U\subseteq W$ then $\Gamma[A, U]$ will denote the society $(A, U, K\cap(A\times U))$; and $\Gamma-A$, $\Gamma-U$, $\Gamma-A-U$ will denote $\Gamma[M\setminus A, W]$, $\Gamma[M, W\setminus U]$, $\Gamma[M\setminus A, W\setminus U]$ respectively. If $a\in M$, $u\in W$ then $\Gamma-a$, $\Gamma-u$, $\Gamma-a-u$ will denote $\Gamma-\{a\}$, $\Gamma-\{u\}$, $\Gamma-\{a\}-\{u\}$ respectively. If f is a queue in W then Γ/f will denote $\Gamma-\Delta(f)-\operatorname{rge} f$ and $\Gamma[f]$ will denote $\Gamma[\Delta(f), \operatorname{rge} f]$.

As previously stated, we shall throughout this paper be considering a society denoted by both Γ and (M, W, K); but we may also wish to consider other societies, such as $\Gamma-A$ where $A\subseteq M$. In such situations, the name of a society in which the symbols D, Δ, q are to be interpreted may be attached to these symbols as a suffix. However, if no suffix is attached, then D, Δ and q should be interpreted in the society denoted by Γ and by (M, W, K). For example, Lemma 2.2 below asserts that $D_{\Gamma-A}(X)=D_\Gamma(X)\setminus A$.

Lemma 2.2. *If $A\subseteq M$, $X\subseteq W$, then $D_{\Gamma-A}(X)=D(X)\setminus A$.*

Proof. Let $K_A=K\setminus(A\times W)$. Then $\Gamma-A=(M\setminus A, W, K_A)$. Moreover $K_A\langle m\rangle=K\langle m\rangle$ for each $m\in M\setminus A$. Hence, for any $m\in M\setminus A$, we have $m\in D_{\Gamma-A}(X)$ iff $K_A\langle m\rangle\subseteq X$, which occurs iff $K\langle m\rangle\subseteq X$, which occurs iff $m\in D(X)$. Therefore

$$D_{\Gamma-A}(X)=D(X)\cap(M\setminus A)=D(X)\setminus A.$$

Corollary 2.3. *If $A\subseteq M$ and f is a queue in W then $\Delta_{\Gamma-A}(f)=\Delta(f)\setminus A$.*

Lemma 2.4. *If U, X are disjoint subsets of W then $D_{\Gamma-U}(X)=D(X\cup U)$.*

Proof. Let $K^U=K\setminus(M\times U)$. Then $\Gamma-U=(M, W\setminus U, K^U)$. Moreover $K^U\langle m\rangle=K\langle m\rangle\setminus U$ for each $m\in M$. Hence, for any $m\in M$, we have $m\in D_{\Gamma-U}(X)$ iff $K^U\langle m\rangle\subseteq X$, which occurs iff $K\langle m\rangle\setminus U\subseteq X$, which occurs iff $K\langle m\rangle\subseteq X\cup U$, i.e. $m\in D(X\cup U)$.

Corollary 2.5. *If f, g are queues in W with disjoint ranges then $\Delta_{\Gamma-\operatorname{rge} g}(f)=\Delta(g*f)$.*

Proof. By Lemma 2.4,

$$D_{\Gamma-\operatorname{rge} g}(\operatorname{rge} f)=D((\operatorname{rge} g)\cup(\operatorname{rge} f))=D(\operatorname{rge}(g*f)).$$

Lemma 2.6. *If $A\subseteq M$ and U, X are disjoint subsets of W then $D_{\Gamma-A-U}(X)=D(X\cup U)\setminus A$.*

Proof. By Lemmas 2.2 and 2.4,

$$D_{\Gamma-A-U}(X)=D_{\Gamma-A}(X\cup U)=D(X\cup U)\setminus A.$$

Corollary 2.7. *If f, g are queues in W with disjoint ranges then* $\Delta_{\Gamma/f}(g)=\Delta(f*g)\setminus\Delta(f)$.

Proof. By Lemma 2.6,

$$\begin{aligned}\Delta_{\Gamma/f}(g)&=D_{\Gamma-\Delta(f)-\text{rge}\,f}(\text{rge } g)=D((\text{rge } f)\cup(\text{rge } g))\setminus\Delta(f)\\&=D(\text{rge }(f*g))\setminus\Delta(f)=\Delta(f*g)\setminus\Delta(f).\end{aligned}$$

Lemma 2.8. *If A is a finite subset of M and f is a queue in W then*

$$q_{\Gamma-A}(f)=q(f)+\|A\cap\Delta(f)\|. \tag{5}$$

Proof. By Lemma 2.2,

$$\begin{aligned}q_{\Gamma-A}(\square)&=-\|D_{\Gamma-A}(\emptyset)\|=-\|D(\emptyset)\setminus A\|\\&=-\|D(\emptyset)\|+\|A\cap D(\emptyset)\|=q(\square)+\|A\cap\Delta(\square)\|.\end{aligned}$$

Hence (5) is true when $f=\square$, i.e. when $\text{dom}\, f=0$.

To continue the proof by transfinite induction on $\text{dom}\, f$, suppose now that $\text{dom}\, f=\lambda>0$ and assume the inductive hypothesis that $q_{\Gamma-A}(g)=q(g)+\|A\cap\Delta(g)\|$ for every queue g in W with domain less than λ.

If λ is a successor ordinal $\kappa+1$, then the inductive hypothesis and Corollary 2.3 give

$$\begin{aligned}q_{\Gamma-A}(f)&=q_{\Gamma-A}(f_\kappa)+1-\|\Delta_{\Gamma-A}(f)\setminus\Delta_{\Gamma-A}(f_\kappa)\|\\&=q(f_\kappa)+\|A\cap\Delta(f_\kappa)\|+1\\&\quad-\|(\Delta(f)\setminus A)\setminus(\Delta(f_\kappa)\setminus A)\|\\&=q(f_\kappa)+\|A\cap\Delta(f_\kappa)\|+1-\|\Delta(f)\setminus\Delta(f_\kappa)\|\\&\quad+\|A\cap(\Delta(f)\setminus\Delta(f_\kappa))\|\\&=q(f_\kappa)+1-\|\Delta(f)\setminus\Delta(f_\kappa)\|+\|A\cap\Delta(f)\|\\&=q(f)+\|A\cap\Delta(f)\|.\end{aligned}$$

Now suppose that λ is a limit ordinal. Let ∇ denote $\bigcup_{\theta<\lambda}\Delta(f_\theta)$. Since A is finite, there exists $\varepsilon<\lambda$ such that $A\cap\nabla\subseteq\Delta(f_\varepsilon)$ and consequently $A\cap\nabla=A\cap\Delta(f_\theta)$ for $\varepsilon\leqslant\theta<\lambda$. Therefore

$$\begin{aligned}\liminf_{\theta\to\lambda} q(f_\theta)+\|A\cap\nabla\|&=\liminf_{\theta\to\lambda}\,(q(f_\theta)+\|A\cap\Delta(f_\theta)\|)\\&=\liminf_{\theta\to\lambda} q_{\Gamma-A}(f_\theta)\end{aligned}$$

by the inductive hypothesis. This observation and Corollary 2.3 give

$$\begin{aligned} q_{\Gamma-A}(f) &= \liminf_{\theta\to\lambda} q_{\Gamma-A}(f_\theta) - \|\Delta_{\Gamma-A}(f) \setminus \bigcup_{\theta<\lambda} \Delta_{\Gamma-A}(f_\theta)\| \\ &= \liminf_{\theta\to\lambda} q(f_\theta) + \|A\cap \nabla\| - \|(\Delta(f)\setminus A)\setminus \bigcup_{\theta<\lambda} (\Delta(f_\theta)\setminus A)\| \\ &= \liminf_{\theta\to\lambda} q(f_\theta) + \|A\cap \nabla\| - \|\Delta(f)\setminus \nabla\| + \|A\cap(\Delta(f)\setminus \nabla)\| \\ &= \liminf_{\theta\to\lambda} q(f_\theta) - \|\Delta(f)\setminus \nabla\| + \|A\cap \Delta(f)\| \\ &= q(f) + \|A\cap\Delta(f)\|. \end{aligned}$$

Lemma 2.9. *Let f, g be queues in W with disjoint ranges. Suppose that* dom $f=\delta$ *and* dom g *is a limit ordinal* λ. *Then*

$$\liminf_{\theta\to\delta+\lambda} q((f*g)_\theta) = \liminf_{\theta\to\lambda} q(f*g_\theta).$$

Proof. Let us write

$$\begin{aligned} q_\theta &= q((f*g)_\theta), & q'_\theta &= q(f*g_\theta), \\ Q_\phi &= \{q_\theta : \phi\leqslant\theta<\delta+\lambda\}, & Q'_\psi &= \{q'_\theta : \psi\leqslant\theta<\lambda\}, \\ i_\phi &= \inf Q_\phi, & i'_\psi &= \inf Q'_\psi, \\ l &= \liminf_{\theta\to\delta+\lambda} q_\theta, & l' &= \liminf_{\theta\to\lambda} q'_\theta. \end{aligned}$$

If $\theta<\lambda$ then $f*g_\theta=(f*g)_{\delta+\theta}$ and therefore $q'_\theta=q_{\delta+\theta}$. Therefore $Q'_\psi=Q_{\delta+\psi}$ and consequently

$$i'_\psi = i_{\delta+\psi} \quad \text{for every } \psi<\lambda. \tag{6}$$

Therefore

$$l' = \sup\{i'_\psi : \psi<\lambda\} = \sup\{i_\phi : \delta\leqslant\phi<\delta+\lambda\} \leqslant \sup\{i_\phi : \phi<\delta+\lambda\} = l.$$

For every ϕ such that $\delta\leqslant\phi<\delta+\lambda$, we have $\phi=\delta+\rho(\phi)$ for some ordinal $\rho(\phi)<\lambda$ and therefore, by (6),

$$i_\phi = i'_{\rho(\phi)} \leqslant \sup\{i'_\psi : \psi<\lambda\} = l'.$$

For every ordinal $\phi<\delta$, we have $Q_\delta\subseteq Q_\phi$ and therefore, using (6),

$$i_\phi \leqslant i_\delta = i'_0 \leqslant \sup\{i'_\psi : \psi<\lambda\} = l'.$$

It follows that $l'\geqslant \sup\{i_\phi : \phi<\delta+\lambda\}=l$. Since we have already proved that $l'\leqslant l$, we conclude that $l=l'$.

Lemma 2.10. *If* $u\in W$ *and f is a queue in* $W\setminus\{u\}$, *then*

$$q([u]*f) = 1 + q_{\Gamma-u}(f). \tag{7}$$

Proof. By the definition of $q([u])$ and Lemma 2.4,

$$q([u]*\square) = q([u]) = q(\square) + 1 - \|\Delta([u]) \setminus \Delta(\square)\|$$

$$= -\|D(\emptyset)\| + 1 - \|D(\{u\}) \setminus D(\emptyset)\| = 1 - \|D(\{u\})\|$$

$$= 1 - \|D_{\Gamma-u}(\emptyset)\| = 1 + q_{\Gamma-u}(\square).$$

Hence (7) is true when $\operatorname{dom} f = 0$.

To continue the proof by transfinite induction on $\operatorname{dom} f$, suppose now that $\operatorname{dom} f = \lambda > 0$ and assume the inductive hypothesis that $q([u]*g) = 1 + q_{\Gamma-u}(g)$ for every queue g in $W \setminus \{u\}$ with domain less than λ.

If λ is a successor ordinal $\kappa + 1$, then the inductive hypothesis and Corollary 2.5 give

$$q([u]*f) = q([u]*f_\kappa) + 1 - \|\Delta([u]*f) \setminus \Delta([u]*f_\kappa)\|$$

$$= 1 + q_{\Gamma-u}(f_\kappa) + 1 - \|\Delta_{\Gamma-u}(f) \setminus \Delta_{\Gamma-u}(f_\kappa)\|$$

$$= 1 + q_{\Gamma-u}(f).$$

Since $\Delta(([u]*f)_0) \subseteq \Delta(([u]*f)_1)$, it follows that

$$\bigcup_{\beta<1+\lambda} \Delta(([u]*f)_\beta) = \bigcup_{1\leq\beta<1+\lambda} \Delta(([u]*f)_\beta) = \bigcup_{\alpha<\lambda} \Delta([u]*f_\alpha). \tag{8}$$

If λ is a limit ordinal, then Lemma 2.9, (8), the inductive hypothesis and Corollary 2.5 give

$$q([u]*f) = \liminf_{\theta\to1+\lambda} q(([u]*f)_\theta) - \|\Delta([u]*f) \setminus \bigcup_{\beta<1+\lambda} \Delta(([u]*f)_\beta)\|$$

$$= \liminf_{\theta\to\lambda} q([u]*f_\theta) - \|\Delta([u]*f) \setminus \bigcup_{\alpha<\lambda} \Delta([u]*f_\alpha)\|$$

$$= \liminf_{\theta\to\lambda} (1 + q_{\Gamma-u}(f_\theta)) - \|\Delta_{\Gamma-u}(f) \setminus \bigcup_{\alpha<\lambda} \Delta_{\Gamma-u}(f_\alpha)\|$$

$$= 1 + \liminf_{\theta\to\lambda} q_{\Gamma-u}(f_\theta) - \|\Delta_{\Gamma-u}(f) \setminus \bigcup_{\alpha<\lambda} \Delta_{\Gamma-u}(f_\alpha)\|$$

$$= 1 + q_{\Gamma-u}(f).$$

(In fact, $1 + \lambda = \lambda$ when λ is a limit ordinal, but this fact is not needed in the above argument, although we use the fact that $1 + \lambda$ is a limit ordinal in applying the definition of $q([u]*f)$.)

Lemma 2.11. *If f, g are queues in W such that $q(f) = 0$ and $(\operatorname{rge} f) \cap (\operatorname{rge} g) = \emptyset$ then*

$$q_{\Gamma/f}(g) = q(f*g). \tag{9}$$

Proof. Using Lemma 2.6 we obtain

$$\begin{aligned} q_{\Gamma/f}(\Box) &= -\|D_{\Gamma/f}(\emptyset)\| = -\|D_{\Gamma-\Delta(f)-\mathrm{rge}\,f}(\emptyset)\| \\ &= -\|D(\emptyset \cup \mathrm{rge}\, f)\setminus \Delta(f)\| = 0 = q(f) = q(f*\Box). \end{aligned}$$

Hence (9) is true if dom $g=0$.

To continue the proof by transfinite induction on dom g, suppose now that dom $g=\lambda>0$ and assume the inductive hypothesis that $q_{\Gamma/f}(h)=q(f*h)$ for every queue h in $W\setminus(\mathrm{rge}\, f)$ with domain less than λ.

Let dom $f=\delta$. It is clear that

$$\Delta((f*g)_\theta)\subseteq \Delta((f*g)_\phi), \quad \text{if } \theta\leqslant\phi\leqslant\delta+\lambda. \tag{10}$$

In particular,

$$\Delta(f)\subseteq\Delta(f*g), \tag{11}$$

$$\Delta(f)\subseteq\Delta(f*g_\alpha), \quad \text{if } \alpha<\lambda. \tag{12}$$

Moreover, it follows from (10) that

$$\bigcup_{\beta<\delta+\lambda} \Delta((f*g)_\beta) = \bigcup_{\delta\leqslant\beta<\delta+\lambda} \Delta((f*g)_\beta) = \bigcup_{\alpha<\lambda} \Delta(f*g_\alpha). \tag{13}$$

If λ is a successor ordinal $\kappa+1$ then the inductive hypothesis, Corollary 2.7, (11) and (12) give

$$\begin{aligned} q_{\Gamma/f}(g) &= q_{\Gamma/f}(g_\kappa)+1-\|\Delta_{\Gamma/f}(g)\setminus\Delta_{\Gamma/f}(g_\kappa)\| \\ &= q(f*g_\kappa)+1-\|(\Delta(f*g)\setminus\Delta(f))\setminus(\Delta(f*g_\kappa)\setminus\Delta(f))\| \\ &= q(f*g_\kappa)+1-\|\Delta(f*g)\setminus\Delta(f*g_\kappa)\| = q(f*g). \end{aligned}$$

If λ is a limit ordinal, then the inductive hypothesis, Corollary 2.7, Lemma 2.9, (11), (12) and (13) give

$$\begin{aligned} q_{\Gamma/f}(g) &= \liminf_{\theta\to\lambda} q_{\Gamma/f}(g_\theta) - \|\Delta_{\Gamma/f}(g)\setminus \bigcup_{\alpha<\lambda}\Delta_{\Gamma/f}(g_\alpha)\| \\ &= \liminf_{\theta\to\lambda} q(f*g_\theta) - \|(\Delta(f*g)\setminus\Delta(f))\setminus \bigcup_{\alpha<\lambda}(\Delta(f*g_\alpha)\setminus\Delta(f))\| \\ &= \liminf_{\theta\to\delta+\lambda} q((f*g)_\theta) - \|\Delta(f*g)\setminus \bigcup_{\alpha<\lambda}\Delta(f*g_\alpha)\| \\ &= \liminf_{\theta\to\delta+\lambda} q((f*g)_\theta) - \|\Delta(f*g)\setminus \bigcup_{\beta<\delta+\lambda}\Delta((f*g)_\beta)\| \\ &= q(f*g). \end{aligned}$$

Corollary 2.12. *If Γ is good and f is a queue in W such that $q(f)=0$ then Γ/f is good.*

Proof. For every queue g in $W \setminus \operatorname{rge} f$, Lemma 2.11 and the goodness of Γ imply that $q_{\Gamma/f}(g) = q(f * g) \geqslant 0$.

Lemma 2.13. *If $V \subseteq W$ and $\Pi = \Gamma[D(V), V]$ and f is a queue in V then*

$$q_\Pi(f) = q(f). \tag{14}$$

Proof. Let $K_\Pi = K \cap (D(V) \times V)$. If $m \in D(V)$ then $K\langle m\rangle \subseteq V$ and therefore $K_\Pi\langle m\rangle = K\langle m\rangle$. Let X be a subset of V. If $m \in D(X)$ then $K\langle m\rangle \subseteq X \subseteq V$: therefore $m \in D(V)$ and $K_\Pi\langle m\rangle = K\langle m\rangle \subseteq X$ and so $m \in D_\Pi(X)$. If $m' \in D_\pi(X)$ then $K\langle m'\rangle = K_\Pi\langle m'\rangle \subseteq X$ and so $m' \in D(X)$. We conclude that, for every subset X of V,

$$D_\Pi(X) = D(X). \tag{15}$$

Consequently

$$q_\Pi(\square) = -\|D_\Pi(\emptyset)\| = -\|D(\emptyset)\| = q(\square),$$

and therefore (14) is true when $\operatorname{dom} f = 0$.

To continue the proof by transfinite induction on $\operatorname{dom} f$, suppose now that $\operatorname{dom} f = \lambda > 0$ and assume the inductive hypothesis that $q_\Pi(g) = q(g)$ for every queue g in V with domain less than λ. By (15), $\Delta_\Pi(f) = \Delta(f)$ and $\Delta_\Pi(f_\theta) = \Delta(f_\theta)$ for every $\theta < \lambda$. From this remark and the inductive hypothesis, we see that

(i) if λ is a successor ordinal $\kappa + 1$ then

$$\begin{aligned} q_\Pi(f) &= q_\Pi(f_\kappa) + 1 - \|\Delta_\Pi(f) \setminus \Delta_\Pi(f_\kappa)\| \\ &= q(f_\kappa) + 1 - \|\Delta(f) \setminus \Delta(f_\kappa)\| = q(f); \end{aligned}$$

(ii) if λ is a limit ordinal then

$$\begin{aligned} q_\Pi(f) &= \liminf_{\theta \to \lambda} q_\Pi(f_\theta) - \Big\|\Delta_\Pi(f) \setminus \bigcup_{\theta < \lambda} \Delta_\Pi(f_\theta)\Big\| \\ &= \liminf_{\theta \to \lambda} q(f_\theta) - \Big\|\Delta(f) \setminus \bigcup_{\theta < \lambda} \Delta(f_\theta)\Big\| = q(f). \end{aligned}$$

Corollary 2.14. *If Γ is good and g is a queue in W then $\Gamma[g]$ is good.*

Proof. Let f be any queue in $\operatorname{rge} g$. Then $q(f) \geqslant 0$ since Γ is good. Taking $V = \operatorname{rge} g$ in Lemma 2.13 gives $q_{\Gamma[g]}(f) = q(f) \geqslant 0$.

3. Selecting a wife for a

To prove the existence of an espousal in a good male-countable society Γ, we might begin by considering one man a and trying to find a woman $u \in K\langle a\rangle$ such that $\Gamma - a - u$ is good. Then a could be married to u and we could re-commence work in the good male-countable society $\Gamma - a - u$ by seeking a suitable wife for a second man. This process could be continued until all the men are married. This is approximately the method of proof which we shall use, but we shall need to replace the notion of a good society by an apparently more general notion of a "piecewise good" society in order to make the argument work. However, after Theorem 1.2 has been proved, it is fairly easy to show that a society is piecewise good if and only if it is good. This can be done by first deducing from Theorem 1.2 that a society $\Gamma=(M, W, K)$ is good if and only if $\Gamma[L, W]$ is espousable for every countable subset L of M—which is not surprising since the definition of goodness considers *countable* transfinite sequences only.

Lemma 3.1. *If Γ is good and male-countable, $a\in M$ and $|K\langle a\rangle|>\aleph_0$, then $\Gamma-a-u$ is good for some $u\in K\langle a\rangle$.*

Proof. Let $M^*=\{m\in M\colon |K\langle m\rangle|\leqslant\aleph_0\}$. Since $|M^*|\leqslant|M|\leqslant\aleph_0$ and $|K\langle m\rangle|\leqslant\aleph_0$ for every $m\in M^*$, it follows that $|K[M^*]|\leqslant\aleph_0<|K\langle a\rangle|$ and so we can select $u\in K\langle a\rangle\setminus K[M^*]$. Let f be a queue in $W\setminus\{u\}$. We shall show that

$$q_{\Gamma-u}(f)=q(f). \tag{16}$$

Let X be any countable subset of $W\setminus\{u\}$. Then $D(X)$ and $D(X\cup\{u\})$ are subsets of M^*. Moreover, since $u\notin K[M^*]$, it follows that an element of M^* belongs to $D(X\cup\{u\})$ if and only if it belongs to $D(X)$. Hence $D(X\cup\{u\})=D(X)$, from which it follows by Lemma 2.4 that

$$D_{\Gamma-u}(X)=D(X). \tag{17}$$

Taking $X=\emptyset$ in (17) gives

$$q_{\Gamma-u}(\square)=-\|D_{\Gamma-u}(\emptyset)\|=-\|D(\emptyset)\|=q(\square).$$

Hence (16) is true if $\operatorname{dom} f=0$.

To continue the proof of (16) by transfinite induction on $\operatorname{dom} f$, suppose now that $\operatorname{dom} f=\lambda>0$ and assume the inductive hypothesis that $q_{\Gamma-u}(g)=q(g)$ for every queue g in $W\setminus\{u\}$ with domain less than λ. The subsets $\operatorname{rge} f$, $\operatorname{rge} f_\alpha$ $(\alpha<\lambda)$ of $W\setminus\{u\}$ are countable by the definition of a queue, and so it follows from (17) that $\Delta_{\Gamma-u}(f)=\Delta(f)$ and $\Delta_{\Gamma-u}(f_\alpha)=\Delta(f_\alpha)$ for every $\alpha<\lambda$. From this remark and the inductive hypothesis, we see that

(i) if λ is a successor ordinal $\kappa+1$ then

$$\begin{aligned} q_{\Gamma-u}(f)&=q_{\Gamma-u}(f_\kappa)+1-\|\Delta_{\Gamma-u}(f)\setminus\Delta_{\Gamma-u}(f_\kappa)\| \\ &=q(f_\kappa)+1-\|\Delta(f)\setminus\Delta(f_\kappa)\|=q(f); \end{aligned}$$

(ii) if λ is a limit ordinal then

$$\begin{aligned} q_{\Gamma-u}(f)&=\liminf_{\theta\to\lambda} q_{\Gamma-u}(f_\theta)-\|\Delta_{\Gamma-u}(f)\setminus\bigcup_{\theta<\lambda}\Delta_{\Gamma-u}(f_\theta)\| \\ &=\liminf_{\theta\to\lambda} q(f_\theta)-\|\Delta(f)\setminus\bigcup_{\theta<\lambda}\Delta(f_\theta)\|=q(f). \end{aligned}$$

We have now proved (16) by transfinite induction.

If f is a queue in $W \setminus \{u\}$ then

$$q_{\Gamma-a-u}(f) = q_{\Gamma-u}(f) + \|\{a\} \cap \Delta_{\Gamma-u}(f)\| \geq q_{\Gamma-u}(f) = q(f) \geq 0$$

by Lemma 2.8, (16) and the hypothesis that Γ is good. Therefore $\Gamma - a - u$ is good.

Lemma 3.2. *If Γ is good, $a \in M$, $u \in W$ and $\Gamma - a - u$ is bad then there exists a queue g in W such that $u \in \operatorname{rge} g$, $a \notin \Delta(g)$ and $q(g) = 0$.*

Proof. Since $\Gamma - a - u$ is bad, there is a queue h in $W \setminus \{u\}$ such that $q_{\Gamma-a-u}(h) < 0$. By Lemmas 2.8 and 2.10,

$$\begin{aligned} q([u]*h) + \|\{a\} \cap \Delta([u]*h)\| &= q_{\Gamma-a}([u]*h) \\ &= 1 + q_{\Gamma-a-u}(h) \leq 0. \end{aligned}$$

From this and the fact that $q([u]*h) \geq 0$ since Γ is good, we conclude that $q([u]*h) = \|\{a\} \cap \Delta([u]*h)\| = 0$, so that $[u]*h$ is a queue g with the properties asserted by Lemma 3.2.

Lemma 3.3. *If Γ is good, f is a queue in W, $q(f) = 0$, $a \in M \setminus \Delta(f)$, $u \in W \setminus \operatorname{rge} f$ and $(\Gamma/f) - a - u$ is bad, then there exists a queue g in $W \setminus \operatorname{rge} f$ such that $u \in \operatorname{rge} g$, $a \notin \Delta(f*g)$ and $q(f*g) = 0$.*

Proof. By Corollary 2.12, Γ/f is good. Therefore, by Lemma 3.2 (with Γ/f replacing Γ) there exists a queue g in $W \setminus \operatorname{rge} f$ such that $u \in \operatorname{rge} g$, $a \notin \Delta_{\Gamma/f}(g)$ and $q_{\Gamma/f}(g) = 0$. Since $a \in M \setminus \Delta(f)$ and $a \notin \Delta_{\Gamma/f}(g)$, it follows by Corollary 2.7 that $a \notin \Delta(f*g)$. By Lemma 2.11, $q(f*g) = 0$.

Lemma 3.4. *If Γ is good and male-countable and $a \in M$ then there exist a queue f in W and a woman u such that $a \notin \Delta(f)$, $u \in K\langle a \rangle \setminus \operatorname{rge} f$ and $(\Gamma/f) - a - u$ is good.*

Proof. Suppose that

(†) *$(\Gamma/f) - a - u$ is bad for every pair f, u such that f is a queue in W and $a \notin \Delta(f)$ and $u \in K\langle a \rangle \setminus \operatorname{rge} f$.*

We observe that $-\|D(\emptyset)\| = q(\square) \geq 0$ since Γ is good. Therefore

$$\emptyset = D(\emptyset) = \Delta(\square). \tag{18}$$

Therefore $a \notin \Delta(\square)$ and $\Gamma/\square = \Gamma$ and so, taking $f = \square$ in (†), it follows that $\Gamma - a - u$ is bad for every $u \in K\langle a \rangle$. Therefore $|K\langle a \rangle| \leq \aleph_0$ by Lemma 3.1. Therefore there exists a queue h in $K\langle a \rangle$ such that $\operatorname{dom} h \leq \omega$, $\operatorname{rge} h = K\langle a \rangle$ [i.e. either the finite sequence $h(0), h(1), \ldots, h(|K\langle a \rangle| - 1)$ or the infinite sequence $h(0), h(1), \ldots$ is an enumeration of $K\langle a \rangle$].

By (18), $a \notin D(\emptyset)$. Therefore $K\langle a \rangle \neq \emptyset$ and so $h(0)$ exists. Let $i_1 = 0$. Since we have observed that $\Gamma - a - u$ is bad for every $u \in K\langle a\rangle$, it follows that $\Gamma - a - h(i_1)$ is bad. Therefore, by Lemma 3.2, there exists a queue f^1 in W such that $h(i_1) \in \operatorname{rge} f^1$, $a \notin \Delta(f^1)$ and $q(f^1) = 0$.

Since $a \notin \Delta(f^1)$, there exists an i for which $h(i) \notin \operatorname{rge} f^1$; let i_2 be the least such i.

By (†), $(\Gamma/f^1) - a - h(i_2)$ is bad. Therefore, by Lemma 3.3, there exists a queue f^2 in $W \setminus \operatorname{rge} f^1$ such that $h(i_2) \in \operatorname{rge} f^2$, $a \notin \Delta(f^1 * f^2)$ and $q(f^1 * f^2) = 0$.

Since $a \notin \Delta(f^1 * f^2)$, there exists an i for which $h(i) \notin \operatorname{rge}(f^1 * f^2)$; let i_3 be the least such i. By (†), $(\Gamma/(f^1 * f^2)) - a - h(i_3)$ is bad. Therefore, by Lemma 3.3, there exists a queue f^3 in $W \setminus \operatorname{rge}(f^1 * f^2)$ such that $h(i_3) \in \operatorname{rge} f^3$, $a \notin \Delta(f^1 * f^2 * f^3)$ and $q(f^1 * f^2 * f^3) = 0$.

Since $a \notin \Delta(f^1 * f^2 * f^3)$, there exists an i for which $h(i) \notin \operatorname{rge}(f^1 * f^2 * f^3)$; let i_4 be the least such i. By (†), $(\Gamma/(f^1 * f^2 * f^3)) - a - h(i_4)$ is bad. Therefore, by Lemma 3.3, there exists a queue f^4 in $W \setminus \operatorname{rge}(f^1 * f^2 * f^3)$ such that $h(i_4) \in \operatorname{rge} f^4$, $a \notin \Delta(f^1 * f^2 * f^3 * f^4)$ and $q(f^1 * f^2 * f^3 * f^4) = 0$.

Continuing this process, we define f^r, i_r for every positive integer r. Let F^0 denote $\square$, let F^r denote $f^1 * f^2 * \cdots * f^r$ for each positive integer r, and let $F = f^1 * f^2 * \cdots$ with the obvious interpretation, i.e. F is a queue with domain $\operatorname{dom} f^1 + \operatorname{dom} f^2 + \cdots$ such that $F(\operatorname{dom} F^{r-1} + \alpha) = f^r(\alpha)$ whenever r is a positive integer and $\alpha \in \operatorname{dom} f^r$.

Our procedure ensures that $h(i_r) \in \operatorname{rge} f^r$ for every positive integer r: therefore no f^r is $\square$ and so $\operatorname{dom} F$ is a limit ordinal. Our procedure also ensures that $q(F^r) = 0$ for every positive integer r and therefore

$$\liminf_{\alpha \to \operatorname{dom} F} q(F_\alpha) \leq 0. \tag{19}$$

Since $h(i_r) \in \operatorname{rge} f^r$ for each r, it follows that $i_1, i_2, \ldots$ are distinct. Since $h(i) \in \operatorname{rge} F^{r-1}$ whenever $i < i_r$, it follows that $h(0)$, $h(1), \ldots$ all belong to $\operatorname{rge} F$, i.e. $K\langle a\rangle \subseteq \operatorname{rge} F$ and consequently $a \in \Delta(F)$. On the other hand, since the element $h(i_r)$ of $K\langle a\rangle$ belongs to $\operatorname{rge} f^r$ for each positive integer r, there can be no $\theta < \operatorname{dom} F$ for which $K\langle a\rangle \subseteq \operatorname{rge} F_\theta$; and therefore $a \notin \bigcup_{\theta < \operatorname{dom} F} \Delta(F_\theta)$. Hence $a \in \Delta(F) \setminus \bigcup_{\theta < \operatorname{dom} F} \Delta(F_\theta)$ and so

$$\|\Delta(F) \setminus \bigcup_{\theta < \operatorname{dom} F} \Delta(F_\theta)\| > 0. \tag{20}$$

Since $\operatorname{dom} F$ is a limit ordinal, it follows from (19) and (20) that $q(F) < 0$, contradicting the goodness of Γ. Hence (†) leads to a contradiction and Lemma 3.4 is proved.

Lemma 3.5. *If Γ is good and male-countable and $a \in M$ then there exist $u \in K\langle a\rangle$ and disjoint sets M', M'', W', W'' such that*

$$M \setminus \{a\} = M' \cup M'', \qquad W \setminus \{u\} = W' \cup W''$$

and $\Gamma[M', W']$, $\Gamma[M'', W'']$ are both good.

Proof. There exist f, u as specified in Lemma 3.4. Let

$$M' = \Delta(f), \qquad M'' = M \setminus (\{a\} \cup \Delta(f)),$$
$$W' = \text{rge } f, \qquad W'' = W \setminus (\{u\} \cup \text{rge } f).$$

Then $\Gamma[M', W'] = \Gamma[f]$, which is good by Corollary 2.14, and $\Gamma[M'', W''] = (\Gamma/f) - a - u$, which is good since f, u are chosen in accordance with Lemma 3.4.

Definition 3.6. Γ is *piecewise good* if, for some positive integer r, there exist disjoint sets $M_1, W_1, M_2, W_2, \ldots, M_r, W_r$ such that

$$M_1 \cup M_2 \cup \cdots \cup M_r = M, \qquad W_1 \cup W_2 \cup \cdots \cup W_r = W$$

and $\Gamma[M_i, W_i]$ is good for $i = 1, \ldots, r$.

Lemma 3.7. *If Γ is piecewise good and male-countable and $a \in M$ then $\Gamma - a - u$ is piecewise good for some $u \in K\langle a \rangle$.*

Proof. Since Γ is piecewise good, there exist disjoint sets M_1, W_1, M_2, $W_2, \ldots, M_r, W_r$ such that

$$M_1 \cup \cdots \cup M_r = M, \qquad W_1 \cup \cdots \cup W_r = W$$

and the society $\Gamma_i = \Gamma[M_i, W_i]$ is good for $i = 1, \ldots, r$. For some k, $a \in M_k$ and, by Lemma 3.5 applied to Γ_k, there exist $u \in K\langle a \rangle \cap W_k$ and disjoint sets M', M'', W', W'' such that

$$M_k \setminus \{a\} = M' \cup M'', \qquad W_k \setminus \{u\} = W' \cup W''$$

and $\Gamma_k[M', W']$, $\Gamma_k[M'', W'']$ are both good. Let $\Gamma^* = \Gamma - a - u$. Then

$$\Gamma^*[M', W'] = \Gamma_k[M', W'], \qquad \Gamma^*[M'', W''] = \Gamma_k[M'', W''],$$
$$\Gamma^*[M_i, W_i] = \Gamma_i \quad (i \neq k)$$

and the goodness of these societies ensures that Γ^* is piecewise good.

4. Proof of Theorem 1.2

If Γ is espousable then it is good by Lemma 1.1.

To prove the converse for male-countable societies, suppose that Γ is good and male-countable. Let $M = \{a_i : i \in I\}$, where $I = \{1, 2, \ldots, |M|\}$ if M is finite, I is the set of all positive integers if M is infinite, and $a_i \neq a_j$ when $i \neq j$. Since Γ is good, it is piecewise good. Consequently, if $|M| \geqslant 1$, then $\Gamma - a_1 - u_1$ is by Lemma 3.7 piecewise good for some $u_1 \in K\langle a_1 \rangle$. Let $\Gamma - a_1 - u_1 = \Gamma_1 = (M_1, W_1, K_1)$. If $|M| \geqslant 2$, then $\Gamma_1 - a_2 - u_2$ is by Lemma 3.7 piecewise good for some $u_2 \in K_1\langle a_2 \rangle$. Let $\Gamma_1 - a_2 - u_2 = \Gamma_2 = (M_2, W_2, K_2)$. If $|M| \geqslant 3$, then $\Gamma_2 - a_3 - u_3$ is by Lemma 3.7 piecewise good for some $u_3 \in K_2\langle a_3 \rangle$. Let $\Gamma_2 - a_3 - u_3 = \Gamma_3 = (M_3, W_3, K_3)$. If $|M| \geqslant 4$, then $\Gamma_3 - a_4 - u_4$ is by Lemma 3.7 piecewise good for some $u_4 \in K_3\langle a_4 \rangle$, etc.

Continuing this argument, we obtain an espousal $\{(a_i, u_i): i \in I\}$ of Γ. Therefore Γ is espousable.

References

[1] R.M. Damerell and E.C.Milner, Necessary and sufficient conditions for transversals of countable set systems, J. Combinatorial Theory 17(A) (1974) 350–374.

[2] C.St.J.A. Nash-Williams, Which infinite set-systems have transversals?—a possible approach, in: Combinatorics, Proc. Conf. on Combinatorial Mathematics at Oxford (1972). Inst. of Mathematics and its Applications (1972) 237–253.

[3] C.St.J.A. Nash-Williams, Marriage in denumerable societies, J. Combinatorial Theory 19(A) (1975) 335–366.

Annals of Discrete Mathematics 3 (1978) 181–189.

SELECTIVE GRAPHS AND HYPERGRAPHS

Jaroslav NEŠETŘIL
Department of Mathematics, Charles University, Prague 8, Czechoslovakia

Vojtěch RÖDL
Department of Mathematics, Czech Technical University, Prague, Czechoslovakia

1. Introduction

One form of the Dirichlet's principle states the following.

For every positive integer n there exists a positive integer N such that for every set X with at least N elements the following holds: for every mapping $c: X \to \{1, 2\}$ (i.e. for every partition of X into two parts) there exists $Y \subseteq X, |Y| = n$, such that the mapping c restricted to the set Y is a constant mapping. (Of course, we may put $N = 2n - 1$.)

This theorem was generalized to the B-property of hypergraphs (see e.g. [3]), to the chromatic number of graphs and hypergraphs (a theorem typical for our purposes is the existence of highly chromatic graphs and hypergraphs which are locally sparse; this was started by [1, 2], and in full generality proved in [4, 6]), and to Ramsey theory (started by [12]).

All these concepts and theorems are dealing with partitions of subobjects of a certain "type" (such as vertices, edges) into a small number of parts. Shortly, the above concepts and theorems are related to *restricted partitions*.

In this paper we are interested in *unrestricted partitions*.

We try to develop results and theory analogous to the above one for restricted partitions. Some of the stones in this project are already known. These are Dirichlet's principle itself and the Erdös-Rado canonization Lemma which are stated below.

Dirichlet's principle. *For every positive integer n there a exists a positive integer N such that for every set X with at least N elements the following holds: for every mapping $c: X \to X$ (i.e. for every partition of a set X into any number of parts) there exists a subset $Y \subseteq X$, $|Y| = n$, such that the mapping c restricted to the set Y is either a constant or a 1-1 mapping. (Of course, it suffices to put $N = (n-1)^2 + 1$.)*

Erdös-Rado Canonization Lemma. *Put*

$$\binom{X}{k} = \{Y \subseteq X;\ |Y| = k\}.$$

For all positive integers n, k there exists a positive integer N such that for every set X with at least N elements the following holds: For every mapping

$$c:\binom{X}{k}\to\binom{X}{k}$$

there exists a subset $Y\subseteq X$, $|Y|=n$, *a (total) ordering* $\leqslant$ *of* Y, *and a set* $\omega\subseteq\{1,\ldots,k\}$ *such that*

$$c(\{m_1,\ldots,m_k\})=c(\{m'_1,\ldots,m'_k\}),\quad \{m_1,\ldots,m_k\}\in\binom{Y}{k},$$

$$\{m'_1,\ldots,m'_k\}\in\binom{Y}{k},\quad m_1<m_2<\cdots<m_k,\quad m'_1<m'_2<\cdots<m'_k,$$

if and only if $m_i=m'_i$ *for* $i\in\omega$.

We include these theorems into a more general framework. This can be done by means of the following definitions.

Definition 1.1. A hypergraph $(X,\mathcal{M})$ is a called *selective* if for every mapping $c:X\to X$ there exists an edge $M\in\mathcal{M}$ such that the mapping $c\,|_M$ is either a constant or a 1-1 mapping.

A constant with no stress on its actual value will be denoted by the symbol §. A 1-1 mapping will be denoted by 1-1. Using this convention we may rewrite the last part of Definition 1.1 as follows: "such that either $c\,|_M=\S$ or $c\,|_M=$ 1-1".

Definition 1.2. Let $(X,\mathcal{M})$ be a hypergraph. A hypergraph $(X',\mathcal{M}')$ is said to be *selective for* $(X,\mathcal{M})$ if for every mapping $c:X'\to X'$ there exists an embedding $f:(X,\mathcal{M})\to(X',\mathcal{M}')$ such that either $c\circ f=\S$ or $c\circ f=$ 1-1.

Here, a mapping $f:X\to X'$ is said to be an embedding if $f(M)\in\mathcal{M}'$ iff $M\in\mathcal{M}$ and f is 1-1.

The fact that $(X',\mathcal{M}')$ is selective for $(X,\mathcal{M})$ will be denoted by $(X,\mathcal{M})\to_{\text{sel}}(X',\mathcal{M}')$.

Definition 1.3. Let $\mathcal{K}$ be a class of hypergraphs. $\mathcal{K}$ is said to have the *selective property* if for every $B\in\mathcal{K}$ there exists a $C\in\mathcal{K}$ such that $B\to_{sel}C$.

Remark 1.4. In many respects the selective property of a hypergraph is analogous to the B-property of a hypergraph.

It is easy to see that $\chi(X,\mathcal{M})\geqslant k$ for every selective k-uniform hypergraph (X,m). On the other hand the hypergraph

$$\left(X,\binom{X}{k}\right)\text{ is }e\text{-selective iff }|X|\geqslant(k-1)^2+1.$$

From these two facts P.Erdös personal communication deduced that there are constants c_1, c_2 such that $c_2k \leqslant (s(k))^{1/k} \leqslant c_1k$ for all sufficiently large k. Here $s(k)$ is the minimal number of hyperedges of a k-uniform selective hypergraph. This is analogous to results related to B-property.

On the other hand a "large" chromatic number of a hypergraph does not imply selectivity. An example is provided by 3-uniform hypergraphs

$$\left(\binom{X}{2}, \left\{\binom{A}{2} : A \in \binom{X}{3}\right\}\right).$$

These hypergraphs fail to be selective for every set X.

In Section 3 we characterize nearly all classes of hypergraphs which have the selective property. This will be proved by means of the Existence Theorem of Section 2. In Section 4 we further generalize the above concepts so as to include Erdös-Rado Canonization Lemma. We prove also that the class of all finite graphs has the edge-selective property (see Definition 4.2 below).

2. Sparse and selective hypergraphs

Theorem 2.1. *For all positive integers k, r, $k \geqslant 2$, there exists a hypergraph $(X, \mathcal{M})$ with the following properties:*

(i) *$(X, \mathcal{M})$ is k-uniform;*
(ii) *$(X, \mathcal{M})$ is selective;*
(iii) *$(X, \mathcal{M})$ does not contain cycles of length $\leqslant r$.*

Proof. Let X be a set with n elements, $k \geqslant 2$. Let $c : X \to X$ be a colouring. A k-tuple $M \in \binom{X}{k}$ is said to be c-selective if either $c|_M = \S$ or $c|_M = $ 1-1.

We claim that there exists a positive constant $\alpha > 0$ which depends on k only such that the number of c-selective k-tuples is greater than $\alpha\binom{n}{k}$ for any colouring $c : X \to X$.

In order to prove this we distinguish two cases.

(i) $|c^{-1}(x)| < n/2k$ for every $x \in X$. Find a colouring $\bar{c} : X \to X$ such that c refines $\bar{c}$ (i.e. $c(x) = c(y)$ implies $\bar{c}(x) = \bar{c}(y)$) and such that $|\bar{c}^{-1}(x)| > n/k - n/2k$ whenever $\bar{c}^{-1}(x) \neq \emptyset$. Clearly every $\bar{c}$-selective edge is c-selective and the number of $\bar{c}$-selective edges is at least

$$\left(\frac{n}{k} - \frac{n}{2k}\right)^k > \alpha_1\binom{n}{k},$$

where α_1 does not depend on n.

(ii) $|c^{-1}(x)| \geqslant n/2k$ for an $x \in X$. Then the number of c-selective edges is at least

$$\binom{\left[\frac{n}{2k}\right]}{k} > \alpha_2\binom{n}{k}.$$

We put $\alpha = \min(\alpha_1, \alpha_2)$.

In the proof of Theorem we shall apply the method used by Erdös and Hajnal in [4].

Let $G=(X, \mathcal{M})$ be a hypergraph, $c: X \to X$ a colouring. Denote by G/c the set of all c-selective edges of G.

Let k, r be fixed. Let $\mathbb{G}_{n,k,p} = \mathbb{G}$ be a random subset of the set

$$\binom{\{1, 2, \ldots, n\}}{k} \quad (n \gg k,\ n \gg r),$$

where for each k-tuple M holds $\mathbf{P}[M \in \mathbb{G}] = p = n^{1-k+\delta}$, where $0 < \delta < r^{-1}$. Put

$$\gamma = \mathbf{P}[c: \{1, \ldots, n\} \to \{1, \ldots, n\} \text{ implies } |\mathbb{G}/c| \geq n]. \qquad (*)$$

Then

$$1 - \gamma \leq \sum_c (\mathbf{P}[\mathbb{G}/c < n]) < n^n \binom{\binom{n}{k}}{n} (1-p)^{\alpha\binom{n}{k}} = \mathrm{o}(1), \qquad (*)$$

where $\alpha > 0$ is a constant given by the above claim.

On the other hand if we denote by $c(G)$ the number of edges of a hypergraph G which are contained in cycles of length$t$, $2 \leq t \leq r$, then one can show that the expected value is

$$\mathbf{E}\,(c(\mathbb{G}_{n,k,p})) \sim \mathrm{o}(n) \qquad (**)$$

(see [4]).

It follows from $(*)$ and $(**)$ the existence of a k-uniform hypergraph which has for every partition of its vertices at least n selective edges and which has at most n edges in circular cycles of length $\leq r$. After deleting these edges we get a hypergraph with the required properties.

3. Selective classes of graphs and hypergraphs

We shall consider the following classes of hypergraphs (see [8, 10]).

Definition 3.1. Let $\mathfrak{A}$ be a set of k-uniform hypergraphs. Denote by Forb $(\mathfrak{A})$ the class of all k-uniform hypergraphs $(X, \mathcal{M})$ which do not contain any member of $\mathfrak{A}$ as an induced subhypergraph.

Explicitly: $(X, \mathcal{M}) \in$ Forb $(\mathfrak{A})$ iff $A \to (X, \mathcal{M})$ for no $A \in \mathfrak{A}$ ($A \to (X, \mathcal{M})$ means that there is an embedding A into $(X, \mathcal{M})$).

Theorem 3.2. *Let $k \geq 2$ be a positive integer. Let $\mathfrak{A}$ be a finite set of 2-connected k-uniform hypergraphs. Then the class* Forb $(\mathfrak{A})$ *has the selective property.*

A hypergraph $(X, \mathcal{M})$ is 2-connected if the hypergraph

$$(X \setminus \{x\}, \{M\colon\ x \notin M \in \mathcal{M}\})$$

is a connected hypergraph for every $x \in X$.

Proof. Put $r = \max\{|A|: A \in \mathfrak{A}\} + 2$. Let $(X, \mathcal{M}) \in \text{Forb}(\mathfrak{A})$. Put $|X| = K$. Let $(Y, \mathcal{N})$ be a k-uniform selective hypergraph without cycles of length $\leq r$ (use Theorem 2.1). Let $\mathcal{L}_N : X \to N$ be a fixed bijection for each $N \in \mathcal{N}$. Define a hypergraph $(Y, \mathcal{M}')$ by $M' \in \mathcal{M}'$ iff there exist $N \in \mathcal{N}$ and $M \in \mathcal{M}$ such that $\mathcal{L}_N(M) = M'$.

Claim 1. $(X, \mathcal{M}) \to_{sel} (Y, \mathcal{M}')$.

Let $c : Y \to Y$ be a mapping. By the selectivity of $(Y, \mathcal{N})$ there exists $N \in \mathcal{N}$ such that either $c|_N = \S$ or $c|_N = 1\text{-}1$. But $\mathcal{L}_N : X \to N$ is a desirable embedding $(x, \mathcal{M}) \to (Y, \mathcal{M}')$ as for $k \geq 2$ $(Y, \mathcal{N})$ does not contain 2-cycles: it is either $c \circ \mathcal{L}_N = \S$ or $c \circ \mathcal{L}_N = 1\text{-}1$.

Claim 2. $(Y, \mathcal{M}') \in \text{Forb}(\mathfrak{A})$.

Let there exist $A \in \mathfrak{A}$ and an embedding $f : A \to (Y, \mathcal{M}')$. As A is 2-connected and $(Y, \mathcal{N})$ does not contain a (non-trivial) 2-connected subhypergraph of size $|A|$, it follows that there exists $N \in \mathcal{N}$ such that $f(A) \subseteq N$. But $(Y, \mathcal{M}')$ restricted to the set N is a hypergraph isomorphic to $(X, \mathcal{M})$. This contradicts $(X, \mathcal{M}) \in \text{Forb}(\mathfrak{A})$.

Remark 3.3. Theorem 3.2 is, in several respects, the best possible. If $\mathfrak{A}$ fails to be finite than Forb $(\mathfrak{A})$ need not be selective. Consider $k = 2$, $\mathfrak{A} = \{C_{2k+1}: k \geq 1\}$. Then Forb $(\mathfrak{A})$ is the class of all bipartite graphs, which obviously fails to be a selective class. On the other side if $k = 2$, $\mathfrak{A} = \{P_n\}$ (P_n is the path of length n), then $G \in \text{Forb}(\mathfrak{A}) \Rightarrow \chi(G) \leq n$ and it is easy to see that Forb $(\mathfrak{A})$ fails to be a selective class for every $n \geq 1$.

Theorem 3.2. generalizes a previous result of the authors asserting that the class Gra (k) of all graphs which do not contain a complete graph with k vertices is a selective class (see [11]). This result was established by means of type representations of graphs (see also [9]).

4. A generalized selective property

For simplicity we relate concepts and results of this part to graphs only. In a full generality we hope to do this elsewhere.

Notation. For graphs G, H denote by $\binom{H}{G}$ the set of all (induced) subgraphs of H which are isomorphic to G.

Definition 4.1. Let F, G be graphs. We say that a graph H is *selective for G with respect to F* if for every mapping

$$c : \binom{H}{F} \to \binom{H}{F}$$

there exist a subgraph G' of H isomorphic to G (i.e. there exists $G' \in \binom{H}{G}$), an ordering $\leq$ of $V(G')$ and $\omega \subseteq \{1, 2, \ldots, n = |G|\}$ such that the following holds:

$$c(F_1) = c(F_2) \quad \text{iff} \quad V(F_i) = \{x_1^i, \ldots, x_n^i\},\ i = 1, 2, \quad \text{and} \quad x_j^1 = x_j^2 \quad \text{for } j \in \omega.$$

In this case we write shortly

$$c \Big|_{\binom{G'}{F}} = \text{can}.$$

The validity of the statement in Definition 4.1 will be denoted by $G \to_{\text{sel}}^{F} H$.

Remark. Clearly $G \to_{\text{sel}} H$ iff $G \to_{\text{sel}}^{K_1} H$.

Definition 4.2. A class $\mathcal{K}$ is said *to have the F-selective property* if $F \in \mathcal{K}$ and for every $G \in \mathcal{K}$ there exists $H \in \mathcal{K}$ such that $G \to_{\text{sel}}^{F} H$.

Theorem 3.2 describes many classes of graphs which have the K_1-selective property. For $F \neq K_1$ it is very hard to prove the F-selective property even for the simplest classes of graphs. In this paper we sketch a proof that the class Gra of all finite graphs has the K_2-selective property (that is the edge-selective property). In order to do this we have to introduce some more concepts.

Definition 4.3. Denote by Rel the class of all finite relations (X, R), $R \subseteq X \times X$, which do not contain a directed cycle.

A 1-1 mapping $f: X \to X'$ is called an *embedding of a relation* (X, R) *into a relation* (X', R') if

$$(f(x), f(y)) \in R' \Leftrightarrow (x, y) \in R.$$

(X, R) is said to be a *subrelation of* (X', R') if $X \subseteq X'$ and the inclusion is an embedding.

Denote by

$$\binom{(X', R')}{(X, R)}$$

the set of all subrelations of (X', R') which are isomorphic to (X, R).

We shall need the following special relations:

$$F_1 = (\{0, 1, 2\}, \{(0, 1), (1, 2), (0, 2)\});$$
$$F_2 = (\{0, 1, 2\}, \{(0,1), (1, 2)\});$$
$$F_3 = (\{0, 1, 2\}, \{(0, 1), (0, 2)\});$$
$$F_4 = (\{0, 1, 2\}, \{(0, 2), (1, 2)\}).$$

The following is a special form of a result proved in [10].

Theorem 4.4. *Let* $i \in \{1, 2, 3, 4\}$ *be fixed. Then for every* $(X, R) \in \mathrm{Rel}$ *there exists a relation* $(Y, S) \in \mathrm{Rel}$ *such that the following statement holds:*

For every mapping

$$c: \binom{(Y, S)}{F_i} \to \{0, 1\}$$

there exists

$$(X', R') \in \binom{(Y, S)}{(X, R)}$$

such that the mapping c restricted to the set $((X', R'), F_i)$ *is a constant mapping.*

The validity of the statement of Theorem 4.4 for particular (X, R) and (Y, S) will be denoted by $(X, R) \to_2^{F_i} (Y, S)$.

Using this theorem we prove the promised result.

Theorem 4.5. *The class* Gra *has edge-selective property.*

Proof. Let $G = (V, E)$ be a fixed graph, $V = \{v_1, \ldots, v_n\}$.

Define the relation (V, R) by

$$R = \{(v_i, v_j) : \{v_i, v_j\} \in E, i < j\}.$$

We may assume without loss of generality that the relation (V, R) has the following properties:

(i) there exists a vertex $a \in V$ such that $(a, v) \in R$ for all $a \neq v \in V$;

(ii) there exists a vertex $b \in V$ such that $(v, b) \in R$ for all $b \neq v \in V$;

(iii) (V, R) contains graphs G_1, G_2, G_3 depicted on Fig. 1 as induced subgraphs.

(i), (ii) and (iii) may be assumed as we may, eventually, enlarge G by new vertices and edges.

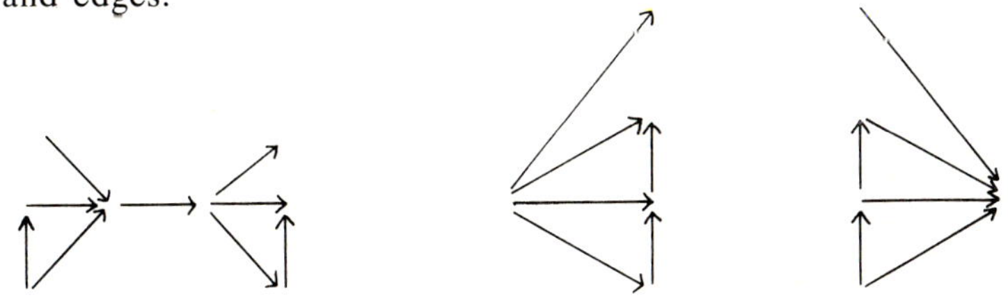

G_1 G_2 G_3

Fig. 1

Let (V_i, R_i), $i = 1, 2, 3, 4$, be relations satisfying

$$(V, R) \to_2^{F_1} (V_1, R_1) \to_2^{F_2} (V_2, R_2) \to_2^{F_3} (V_3, R_3) \to_2^{F_4} (V_4, R_4).$$

Put $(V_4, R_4) = (V', R')$ and define the graph (V', E') by $E' = \{\{x, y\} : (x, y) \in R'\}$.

We prove $(V, E) \to_{\mathrm{sel}}^{K_2} (V', E')$. Let $\leqslant$ be a fixed ordering of V' which satisfies $R' \subseteq \leqslant$.

This ordering $\leqslant$ exists since (V', R') is acyclic. Let $c: E' \to E'$ be a fixed mapping. Write the same letter c for the mapping $c: R' \to R'$ defined by $c((x, y)) = c(\{x, y\})$.

Each of the relations F_i, $i = 1, 2, 3, 4$, contains at most 3 arrows and for a partition $d: E(F_i) \to E(F_i)$, $i = 1, 2, 3, 4$, there are the following possibilities:

(1) d is a constant mapping;

(2) d is 1-1 mapping;

(3) $d((0, 1)) = d((0, 2))$ and $d((0, 1)) \neq d((1, 2))$;

(4) $d((0, 2)) = d((1, 2))$ and $d((0, 1)) \neq , 2))$.

Denote the partition which corresponds to a mapping d with property (1) by π_i, $i = 1, 2, 3, 4$. (The partitions π_3 and π_4 are related only to the arrows of F_1.)

Now define the mapping

$$c_i : \binom{(V', R')}{F_i} \to \{1, 2, 3, 4\}$$

as follows. Put

$$c_i(F') = k \quad \text{iff} \quad F' \in \binom{(V', R')}{F_i}$$

and the partition c restricted to the edge set of F' coincides with π_k.

Using the above definition of (V', R') we find a

$$(\bar{V}, \bar{R}) \in \binom{(V', R')}{(V, R)}$$

such that for every $i \in \{1, 2, 3, 4\}$ there exists a constant $\S_i$ with the property that the mapping c_i restricted to the set $((V, R), F_i)$ is the constant $\S_i$.

However the values $\S_1$, $\S_2$, $\S_3$ and $\S_4$ are not arbitrary. We distinguish four cases.

(a) $\S_2 = 1$.

Then, since the relations G_1 and G_2 which are subrelations of (V, R), we have $\S_1, = \S_2 = \S_3 = \S_4 = 1$. Consequently the mapping c restricted to the set $\bar{R}$ is a constant mapping.

(b) $\S_2 = 2$ and $\S_3 = 1$.

Then (because of the relation G_2) we have $\S_1 = 3$ and (because of the relation G_3) we have $\S_4 = 2$.

Finally, let $c((x, y)) = c((x', y'))$ and $x \neq x'$ for some (x, y), $(x', y') \in R$. It follows from $\S_3 = 1$ that

$$c((x, y)) = c((x, b)) = c((x', y')) = c((x', \bar{b})),$$

where $\bar{b}$ is the vertex of $(\bar{V}, \bar{R})$ which corresponds to the vertex b of (V, R). From this it follows that $\S_4 \neq 2$, $\S_1 \neq 3$, which is a contradiction.

Consequently $c((x, y)) = c((x', y'))$ iff $x = x'$ for all $(x, y) \in R$, $(x', y') \in R$.

(c) $\S_2 = 2$, $\S_3 = 2$ and $\S_4 = 1$.

Then $\S_1 = 4$ by the relations G_1 and G_3. As in the case (b) one can show that $c((x, y)) = c((x', y'))$ iff $y = y'$ for all $(x, y), (x', y') \in R$.

(d) $\S_2 = \S_3 = \S_4 = 2$.

It follows that $\S_1 = 2$ and it is easy to see that in this case c restricted to the set $\bar{R}$ is a 1-1 mapping.

As the above cases exhaust all possible cases, we have

$$(V, E) \to_{\text{sel}}^{K_2} (V', E').$$

Concluding remarks. (1) Using a similar argument we may prove that for every positive integer k the class Gra (k) of all graphs which do not contain a complete graph with k vertices, has the edge-selective property.

However for edge-selectivity we do not have a theorem similar to Theorem 3.2 for vertex-selectivity.

(2) We may prove that the class Gra has the F-selective property only if either $F = K_k$ or $F = D_k = (\{1, 2, \ldots, k\}, \emptyset)$.

This and other related results are going to appear elsewhere.

References

[1] B. Descartes, W.T. Tutte, A three colour problem, Eureka (March 1948).

[2] B. Descartes, W.T. Tutte, Solution to Advanced problem No. 4526, Amer. Math. Monthly, 61 (1954) 352.

[3] P. Erdös, On a Combinatorial problem II, Acta Math. Acad. Sci. Hung. 15 (1964) 445–447.

[4] P. Erdös, A. Hajnal, On chromatic Number of Graphs and Set Systems, Acta Math. Acad. Sci. Hungar. 17 (1966) 61–99.

[5] P. Erdös, R. Rado, Combinatorial Theorems on Classifications of Subsets of a Given Set, Proc. London Math. Soc. 3(2) (1952) 417–439.

[6] L. Lovász, On chromatic number of finite set-systems, Acta Math. Acad. Sci. Hung. 19 (1968) 59 67.

[7] J. Nešetřil, V. Rödl, Partitions of vertices, Comment. Math. Univ. Carolinae 17(1) (1976) 85–95.

[8] J. Nešetřil, V. Rödl, A structural generalization of the Ramsey theorem, Bull. Amer. Math. Soc. 83(1) (1977) 127–128.

[9] J. Nešetřil, V. Rödl, Ramsey property of graphs with forbidden complete subgraphs, J. Combinatorial Theory B, 20 (1976) 243–249.

[10] J. Nešetřil, V. Rödl, Partitions of relational and set systems, J. Combinatorial Theory A 22 (1977) 289–312.

[11] J. Nešetřil, V. Rödl, A selective theorem for graphs and hypergraphs, Proc. Paris Combinatorial Conf. 1976, to appear.

[12] F.P. Ramsey, On a problem of Formal Logic, Proc. London Math. Soc. 2nd Series 30 (1930) 264–286.

Annals of Discrete Mathematics 3 (1978) 191–194.

MONOCHROMATIC PATHS IN GRAPHS

Richard RADO

Department of Mathematics, University of Reading, Whiteknights, Reading RG6 2AX, Great Britain

P. Erdös kindly communicated to the author the following result.

Theorem 1. *Let Γ be a complete denumerable graph and suppose every edge of Γ receives one of the two colours c_0, c_1. Then there are two paths π_0, π_1, each finite or simply infinite, such that every vertex of Γ occurs exactly once in either π_0 or π_1 and every edge of π_i has the colour c_i for $i \in \{0, 1\}$.*

A close study of the proof sketched by Erdös reveals that Theorem 1 is a rather special case of a more general proposition, Theorem 2 below, in which the graph is directed and is only approximately complete and in which there may be more than two colours.

Italic capitals denote sets and $|A|$ is the cardinal of A. Put $L = \{1, 3, 5, \ldots\} \cup \{\omega\}$, where ω is the least infinite ordinal.

Theorem 2. *Let $A \subseteq V$ and $|A| \leq \aleph_0 \leq |V|$ and let Γ be a directed graph of the form*

$$\Gamma = (V, E),$$

where $E \subseteq A \times V$. Suppose that, for every $x \in A$,

$$|\{y \in V : (x, y) \notin E\}| < |V|, \tag{1}$$

and every edge of Γ receives a colour from the colour set I. Then there is a set $J \subseteq I$ and, for every $j \in J$, a $m_j \in L$ and $x_j(\nu) \in V$ for $0 \leq \nu < m_j$, such that

(i) *every $x \in A$ occurs among the $x_j(\nu)$,*

(ii) *$x_i(\mu) = x_j(\nu)$ implies $(i, \mu) = (j, \nu)$,*

(iii) *if $j \in J$ and $0 < \nu < m_j$ and ν is odd, then*

$$(x_j(\nu - 1), x_j(\nu)), (x_j(\nu + 1), x_j(\nu))$$

are two edges, each of colour j.

The condition (iii) means that the path belonging to a value $j \in J$ (monochromatic of colour j) has its edges alternately directed one way and the other:

$$\rightarrow \leftarrow \rightarrow \leftarrow \cdots$$

Theorem 1 is obtained by putting in Theorem 2 $A = V$; $|V| = \aleph_0$; $E = V \times V - \{(x, x) : x \in V\}$; $I = \{c_0, c_1\}$ and making the colour of every edge independent of its orientation.

Proof of Theorem 2. We begin by disposing of a trivial case. If $|I|\geq|A|$ then we can choose a set $J\subseteq I$ with $|J|=|A|$. We put $m_j=1$ for $j\in J$. We can write

$$A=\{x_j(0):j\in J\},$$

where $x_i(0)\neq x_j(0)$ for $i\neq j$. Then assertion (iii) holds vacuously because there is no odd number ν with $0<\nu<m_j$.

So we may suppose

$$|I|<|A|\leq\aleph_0\leq|V| \tag{2}$$

and write

$$A=\{a_\lambda:0\leq\lambda<p\} \tag{3}$$

for some $p\leq\omega$. For $x\in A$ put

$$f_i(x)=\{y\in V:(x,y)\text{ has colour } i\}\quad\text{for}\quad i\in I,$$

$$f(x)=\bigcup_{i\in I} f_i(x)=\{y\in V:(x,y)\in E\}.$$

We shall now define systems of paths which have some of the desired properties and which will be used to construct a system of paths as required in Theorem 2.

Denote by Ω the set of all triples of the form

$$(R,(m_\rho:\rho\in R),(x_\rho(\nu):\rho\in R;0\leq\nu<m_\rho)), \tag{4}$$

where: $R\subseteq I$; $m_\rho\in\{1,3,5,\ldots\}$ for $\rho\in R$; $x_\rho(\nu)\in V$ for $\rho\in R$; $0\leq\nu<m_\rho$; and we have the conditions: $x_\rho(\nu)=x_{\rho'}(\nu')$ implies $(\rho,\nu)=(\rho',\nu')$; if $\rho\in R$ and $0<\nu<m_\rho$; ν odd, then

$$x_\rho(\nu-1),\ x_\rho(\nu+1)\in A,$$

$$x\rho(\nu)\in f_\rho(x_\rho(\nu-1)\cap f_\rho(x_\rho(\nu+1));$$

and finally

$$|V\cap\bigcap_{\rho\in R} f_\rho(x_\rho(m_\rho-1))|=|V|. \tag{5}$$

We note that the paths constructed from the triple (4) already satisfy the conditions (ii) and (iii) of the theorem to be proved. The "factor" V on the left-hand side of (5) is added in order that the case $R=\emptyset$ should be covered. This enables us to assert that Ω is not empty, since the triple (ϕ,θ,θ) belongs to Ω, where θ denotes the empty family.

I now describe a recursive construction. Let the triple (4) belong to Ω. Moreover, let it be chosen in such a way that the set R is maximal by inclusion. If

$$A\subseteq\{x_\rho(\nu):\rho\in R;0\leq\nu<m_\rho\},$$

then the assertion holds with $J=R$. So we may suppose

$$A\not\subseteq\{x_\rho(\nu):\rho\in R;0\leq\nu<m_\rho\}.$$

Then there is a least number $\nu_0 < p$ such that $a_{\nu_0} \neq x_\rho(\nu)$ for all ρ, ν with $\rho \in R$ and $0 \leq \nu < m_\rho$. Put $a = a_{\nu_0}$.

Case 1. For every $\rho_0 \in I$ we have

$$|f_{\rho_0}(a) \cap \bigcap_{\rho \in R} f_\rho(x_\rho(m_\rho - 1))| < |V|. \tag{6}$$

Put

$$D = V \cap \bigcap_{\rho \in R} f_\rho(x_\rho(m_\rho - 1)).$$

Then $|V - f(a)| < |V|$ by (1); $|D| = |V|$ by (5); for every $\rho_0 \in I$, $|f_{\rho_0}(a) \cap D| < |V|$ by (6); from (2) we deduce that $|f(a) \cap D| < |V|$.

Now we obtain the contradiction

$$|V| = |D| = |D \cap f(a)| + |D - f(a)| \leq |D \cap f(a)| + |V - f(a)| < |V|.$$

Case 2. There is $\rho_0 \in I$ with

$$|f_{\rho_0}(a) \cap \bigcup_{\rho \in R} f_\rho(x_\rho(m_\rho - 1))| = |V|. \tag{7}$$

Case 2a. $\rho_0 \notin R$. Then put $m_{\rho_0} = 1$; $x_{\rho_0}(0) = a$; $R' = R \cup \{\rho_0\}$. It follows that the triple

$$(R', (m_\rho : \rho \in R'), (x_\rho(\nu) : \rho \in R'; 0 \leq \nu < m_\rho))$$

belongs to Ω, which contradicts the maximality of R.

Case 2b. $\rho_0 \in R$. By (7) we can choose an element $x_{\rho_0}(m_{\rho_0})$ of the set

$$f_{\rho_0}(a) \cap \bigcap_{\rho \in R} f_\rho(x_\rho(m_\rho - 1)) - \{x_\rho(\nu) : \rho \in R; 0 \leq \nu < m_\rho\}.$$

Put $x_{\rho_0}(m_{\rho_0} + 1) = a$; $m'_{\rho_0} = m_{\rho_0} + 2$; $m'_\rho = m_\rho$ for $\rho \in R - \{\rho_0\}$. Then the triple

$$(R, (m'_\rho : \rho \in R), (x_\rho(\nu) : \rho \in R; 0 \leq \nu < m'_\rho))$$

belongs to Ω.

It has now been shown that, if we are given any triple of Ω having a maximal set R and if the triple does not give rise to a system of paths with the required properties, then we can lengthen one of its paths by adding two more vertices. In doing this we shall have incorporated in our paths the first vertex of A in the enumeration (3) which has not yet occurred in our paths. A moment's consideration shows that if we iterate this procedure, starting with a triple of Ω whose R is maximal, we shall continue to lengthen some of our paths, always maintaining the truth of (ii) and (iii) of Theorem 2. We shall have satisfied the requirements of that theorem if we take as J our set R and as paths the limits of the paths which are being constructed by our procedure. □

It would be of interest to decide whether Theorem 1 can be extended to the case of monochromatic paths of length greater than ω. Let n be an ordinal number and let, for every ordinal $\nu < n$, x_ν be a vertex of the edge-coloured complete graph Γ. The family $(x_\nu : \nu < n)$ might be called a path in the colour c if $x_\mu \neq x_\nu$ for $\mu < \nu < n$ and if, for every ordinal ν with $\nu + 1 < n$, the edge $\{x_\nu, x_{\nu+1}\}$ has colour c and, for every limit ordinal $\nu < n$, the supremum of the set $\{\mu < \nu : \{x_\mu, x_\nu\}$ has colour $c\}$ has the value ν. The conditions are equivalent to saying that, whenever $\mu < \nu < n$, we have $x_\mu \neq x_\nu$ and $\{x_\lambda, x_\nu\}$ has colour c for some λ in $\mu \leq \lambda < \nu$.

Annals of Discrete Mathematics 3 (1978) 195–226.

ON THE PRINCIPAL EDGE TRIPARTITION OF A GRAPH*

P. ROSENSTIEHL

Ecole des Hautes Etudes en Sciences Sociales, 75270 Paris, Cédex 06, France

R.C. READ

Department of Combinatories and Optimization, University of Waterloo, Waterloo, Ontario N2L 3G1, Canada

Gallia est omnis divisa in partes tres (J. Caesar).

0. Introduction

In this paper we present a general study of a natural partition of the edges of a graph G into three classes — the principal tripartition of G — which is defined canonically from the cycle space of G.

For ease of visualization we shall employ the graph-theoretic concepts of cycles and cocycles, but unless otherwise stated the results hold also for binary matroids. We prove a number of theorems concerning this principal tripartition and indicate that the results are of more than merely academic interest by applying them to obtain a criterion for the planarity of a graph to solve a conjecture of Gauss concerning the sequence of crossing points of a closed curve. The tripartition can also be used to obtain many "Parity Theorems" — theorems concerning the parity of certain numbers associated with a graph, such as the number of spanning trees. A Parity Theory for graphs may be regarded as promising for networks engineering.

1. Definitions and notation

By a *graph* $G=(V,E)$ we shall mean a finite set E (the edges of G), each element of which is incident to two elements (not necessarily distinct) of another finite set V (the set of vertices). Thus we allow our graphs to have loops and multiple edges (see Berge [1]).

Let $\mathscr{E}=2^E$ be the free vector space over GF(2), with the elements e of E as basis. In other words, any subset A of E is represented by a vector whose e-component, A_e, is 1 if $e\in A$ and 0 otherwise. We shall use the same symbol to

* This research was supported by NATO Research Grant No. 637.

denote the set or the corresponding vector. Moreover, a lower-case letter denoting an edge in E will also denote the set consisting of that edge alone, as well as the corresponding vector. A set will be called '*even*' or '*odd*' according as it has even or odd cardinality.

Clearly $e+e=\mathbf{0}$ (the zero vector) for any $e\in E$ and $A+A=\mathbf{0}$ for any $A\in\mathscr{E}$. For A, $B\in\mathscr{E}$ we write

$$\langle A, B\rangle = \sum_{e\in E} A_e \cdot B_e$$

(the "scalar" product of A and B). If $\langle A, B\rangle = 0$, i.e., if A and B have an even number of common elements, we say that A and B are orthogonal.

If $\mathscr{D}$ is a subspace of $\mathscr{E}$, write

$$\mathscr{D}^\perp = \{B\in\mathscr{E} \mid \langle A, B\rangle = 0 \quad \text{for all} \quad A\in\mathscr{D}\}.$$

$\mathscr{D}^\perp$ is called the orthogonal subspace of $\mathscr{D}$. We have [6]

$$\dim \mathscr{D} + \dim \mathscr{D}^\perp = \operatorname{card} E$$

and

$$(\mathscr{D}^\perp)^\perp = \mathscr{D}.$$

Similar results to the above hold for the space $\mathscr{V}=2^V$ defined over the set V of vertices of G (or, for that matter, over any finite set).

The *boundary* ∂e of an edge e is defined as the sum of its incident vertices. The boundary mapping $\partial:\mathscr{E}\to\mathscr{V}$ is the linear mapping defined on the basis of $\mathscr{E}$ as above, i.e.,

$$\partial A = \sum_{e\in A} \partial e.$$

The *coboundary* δv of a vertex v is defined by the relation

$$\langle e, \delta v\rangle = \langle \partial e, v\rangle,$$

i.e., the edge $e\in\delta v$ if, and only if, $v\in\partial e$. The coboundary mapping $\mathscr{V}:\vartheta\to\mathscr{E}$ is the linear mapping defined on the basis of $\mathscr{V}$ as above, i.e.,

$$\delta S = \sum_{v\in S} \delta v.$$

It follows that for any $A\in\mathscr{E}$ and $S\in\mathscr{V}$

$$\langle A, \delta S\rangle = \langle \partial A, S\rangle. \tag{1.1}$$

The space $\mathscr{C}=\ker\partial$ will be called the *cycle space* of G, and the space $\operatorname{Im}\delta$ will be called the *cocycle space* of G. Since $\operatorname{Im}\delta$ is the orthogonal complement of $\ker\delta$ the cocycle space is denoted by $\mathscr{C}^\perp$. The elements of $\mathscr{C}$ are called *cycles* of G; those of $\mathscr{C}^\perp$ are called *cocycles*. There may be some sets of edges that are both cycles and cocycles. Such sets, belonging to the space $\mathscr{C}\cap\mathscr{C}^\perp$, are of particular importance and will be called *bicycles*. In Fig. 1.1. in heavy lines bicycles X are

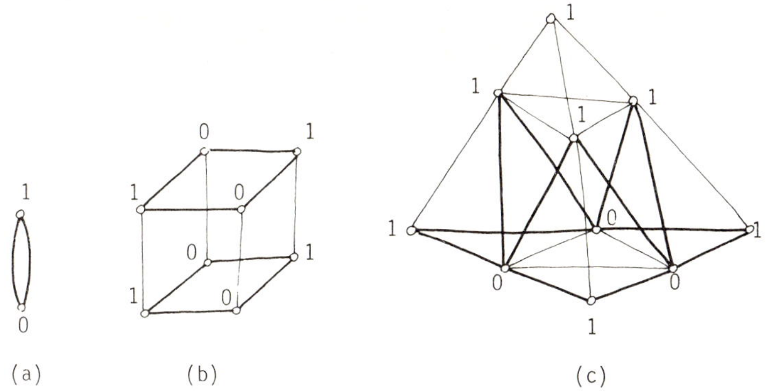

Fig. 1.1

displayed, which do not include other nonzero bicycles; vertices labels display S, such that $\delta S = X$. In Fig. 1.2. the sets $\{d, e\}$, $\{b, c, d\}$, $\{b, c, e, k, l\}$ are cycles, the sets $\{b, c\}$, $\{b, d, e, f\}$, $\{c, b, h, i, j, k, l\}$ are cocycles, and the set $\{b, c, f, g, h, i\}$ is a bicycle.

If $A \in \mathscr{C}$ and there is an edge e such that $A + e \in \mathscr{C}^{\perp}$, then A is called a *principal cycle* of G. The edge e and the principal cycle A will be said to be *associated*, each with the other. Similarly, if $B \in \mathscr{C}^{\perp}$ and there is an edge e such that $B + e \in \mathscr{C}$ then B is called *a principal cocycle* of G. In Fig. 1.2 the cycle $\{b, c, f, g, h, i, k, l\}$ is principal because the addition of edge j makes it into a cocycle; the cycle $\{b, c, d\}$ is also principal since the deletion of edge d makes it into a cocycle. The cocycles $\{b, c, f, g, h, i, j, k, l\}$ and $\{b, c\}$ are principal cocycles.

A *tree* in a graph is a minimal subset of E which meets every nonzero cocycle. A *cotree* is a minimal subset of E which meets every nonzero cycle. If Y is a tree then $Z = Y + E$ is a cotree. If $e \in Y$, denote by Y^e the unique cocycle which meets

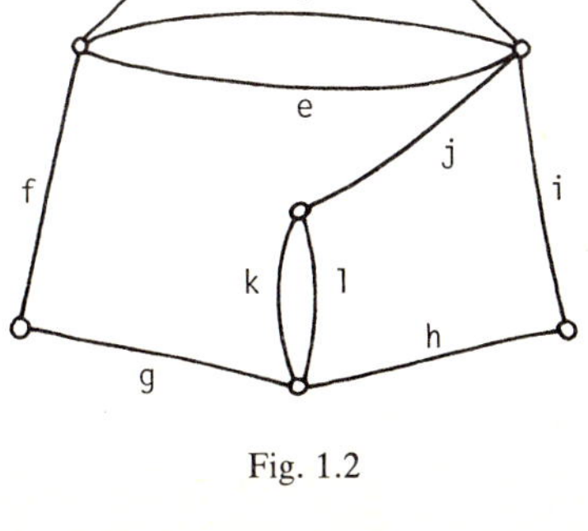

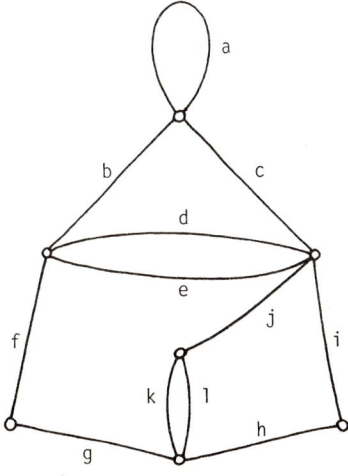

Fig. 1.2

Y in e alone; if $e \in Z$, denote by Z^e the unique cycle which meets Z in e alone. The family $\{Y^e\}_{e\in Y}$ is a basis – called a *fundamental basis* — for $\mathscr{C}^\perp$; similarly the family $\{Z^e\}_{e\in Z}$ is called a *fundamental basis* for $\mathscr{C}$.

A *dendroid* of $\mathscr{C} \cap \mathscr{C}^\perp$ is a minimal subset of E which meets every nonzero bicycle. A fundamental basis for $\mathscr{C} \cap \mathscr{C}^\perp$ is also defined as above.

2. The edge tripartition of a graph

By $\mathscr{A} + \mathscr{B}$ we mean the subspace of $\mathscr{E}$ which is the sum of the subspaces $\mathscr{A}$ and $\mathscr{B}$ of $\mathscr{E}$. Then it is known [6] that for any subspace $\mathscr{C}$ of $\mathscr{E}$ we have

$$(\mathscr{C} \cap \mathscr{C}^\perp)^\perp = \mathscr{C} + \mathscr{C}^\perp. \tag{2.1}$$

Let us now consider what status an edge e may have with respect to $\mathscr{C} + \mathscr{C}^\perp$.

If e belongs to a bicycle X then e is not orthogonal to X. Hence e is not a vector of $\mathscr{C} + \mathscr{C}^\perp$. If e does not belong to any bicycle, e is orthogonal to every bicycle, and hence e is a vector in $\mathscr{C} + \mathscr{C}^\perp$. In the latter case there exist a cycle $\gamma(e)$ and a cocycle $\omega(e)$ such that

$$e = \gamma(e) + \omega(e).$$

Clearly $\gamma(e)$ is a principal cycle and $\omega(e)$ is a principal cocycle.

Such an expression of e as the sum of a principal cycle and principal cocycle will be called a *decomposition* of e. In general it is not unique, for we also have

$$e = (\gamma(e) + X) + (\omega(e) + X), \tag{2.2}$$

where X is any bicycle. Moreover, it is easily verified that any decomposition of e is of the form (2.2) for some bicycle X.

From (2.2) we deduce that all principal cycles associated with e are of the form $\gamma(e) + X$, while all principal cocycles associated with e are of the form $\omega(e) + X$, where $X \in \mathscr{C} \cap \mathscr{C}^\perp$. Hence if e belongs to one of its associated principal cycles, it belongs to all of them, and similarly for its associated principal cocycles. Thus we have the following theorem.

Theorem 2.1. *For any edge e of G, exactly one of the following statements holds:*

(i) *e belongs to a cycle which becomes a cocycle when e is omitted from it,*

(ii) *e belongs to a cocycle which becomes a cycle when e is omitted from it,*

(iii) *e belongs to a bicycle.*

In this way we define the *principal tripartition* of G, denoted $\{P, Q, R\}$. We have

$$P = \{e \mid \exists \gamma \in \mathscr{C},\quad e \in \gamma \quad \text{and} \quad e + \gamma \in \mathscr{C}^\perp\},$$

$$Q = \{e \mid \exists \omega \in \mathscr{C}^\perp,\quad e \in \omega \quad \text{and} \quad e + \omega \in \mathscr{C}\},$$

$$R = \{e \mid \exists X \in \mathscr{C} \cap \mathscr{C}^\perp,\quad e \in X\}.$$

From Theorem 2.1 we have $P+Q+R=E$. Since $\mathscr{C}^{\perp}$ is defined from $\mathscr{C}$, Theorem 2.1 can be stated in terms of cycles only.

Theorem 2.2. *For any graph there exists a tripartition* $\{P, Q, R\}$ *of its edges such that*

(i) $e \in P$ *if there is a cycle, containing e, which is orthogonal to every cycle not containing e and is not orthogonal to any cycle containing e,*

(ii) $e \in Q$ *if there is a cycle, not containing e, which is orthogonal to every cycle not containing e, and is not orthogonal to any cycle containing e,*

(iii) $e \in R$ *if there is a cycle containing e which is orthogonal to every cycle.*

In Fig. 2.1 the principal tripartition of the edges of the graph is displayed. In this graph there is one bicycle $\{a, b, c, d, e, f, g, h, i, j\}$. Thus the edge p can be decomposed in two ways, viz.

$$p = \overbrace{a+c+e+f+g+i+k+l+n+q}^{\text{cycle}} \quad \overbrace{+a+c+e+f+g+i+k+l+n+p+q}^{\text{cocycle}},$$

and

$$p = b+d+h+j+k+l+n+q \qquad +b+d+h+j+k+l+n+p+q.$$

We give below some elementary properties of the principal tripartition. First, note that obviously every loop (edge which is a cycle) belongs to P, and every bridge (edge which is a cocycle) belongs to Q; more we have the following theorem.

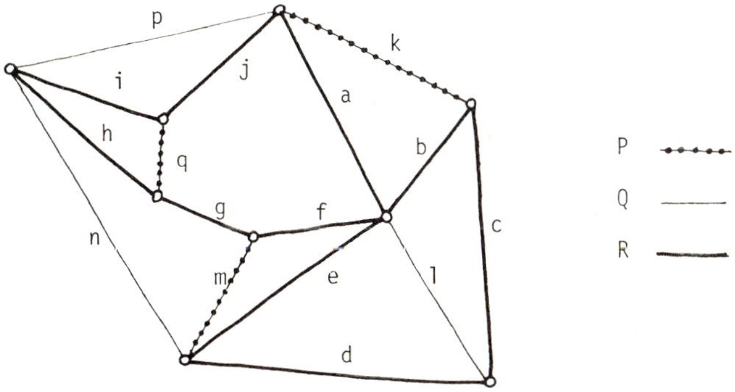

Fig. 2.1

Theorem 2.3. (i) *The zero cocycle is a principal cocycle associated with e if and only if e is a loop.*

(ii) *The zero cycle is a principal cycle associated with e if and only if e is a bridge.*

Theorem 2.4. (i) *If* $e \in P$ *every principal cycle associated with e is odd and every principal cocycle associated with e is even.*

(ii) *If* $e \in Q$ *the parities in* (i) *are reversed.*

(iii) *Every bicycle is even.*

Proof. This is a direct consequence of the orthogonality of cycles and cocycles.

Theorem 2.5. *Given* e, $f \in P+Q$, *and* $e=\gamma(e)+\omega(e)$ *and* $f=\gamma(f)+\omega(f)$ *two decompositions of* e *and* f:

(i) $e \in \gamma(f)$ *if, and only if* $f \in \gamma(e)$,

(ii) $e \in \omega(f)$ *if, and only if* $f \in \omega(e)$.

Proof. Consider

$$\langle\gamma(e), \gamma(f)\rangle = \langle e+\omega(e), \gamma(f)\rangle = \langle\gamma(e), f+\omega(f)\rangle.$$

Then, by orthogonality,

$$\langle e, \gamma(f)\rangle = \langle\gamma(e), f\rangle, \tag{2.3}$$

which is equivalent to (i). It follows immediately that

$$\langle e, \omega(f)\rangle = \langle\omega(e), f\rangle, \tag{2.4}$$

which is equivalent to (ii).

When $e \neq f$ then

$$\langle e, \gamma(f)\rangle = \langle e, \omega(f)\rangle = \langle\gamma(e), f\rangle = \langle\omega(e), f\rangle.$$

If these scalar products all have the value 1 we say that e and f are '*interlaced*'. The set of edges interlaced with e in G will be denoted by $\lambda(e)$. Thus, for $e \in P+Q$, we have

$$\lambda(e) = \gamma(e) \cap \omega(e) \cap (P+Q). \tag{2.5}$$

From (2.1) we also deduce that any set $A \in \mathscr{E}$ which is orthogonal to all bicycles can be decomposed, i.e. written as

$$A = \gamma(A) + \omega(A), \tag{2.6}$$

where $\gamma(A) \in \mathscr{C}$ and $\omega(A) \in \mathscr{C}^{\perp}$. $\gamma(A)$ is called a *decomposition cycle* associated with A, and $\omega(A)$ is called a *decomposition cocycle* associated with A.

Theorem 2.6. *Given two sets* A, $B \in (\mathscr{C} \cap \mathscr{C}^{\perp})^{\perp}$, *and any decomposition of them* $A=\gamma(A)+\omega(A)$, $B=\gamma(B)+\omega(B)$, *then*

$$\langle\gamma(A), B\rangle = \langle A, \gamma(B)\rangle \tag{2.7}$$

and

$$\langle\omega(A), B\rangle = \langle A, \omega(B)\rangle. \tag{2.8}$$

Proof. Consider

$$\langle \gamma(A),\ \gamma(B)\rangle = \langle \omega(A)+A,\ \gamma(B)\rangle = \langle \gamma(A),\ \omega(B)+B\rangle.$$

Then, by orthogonality

$$\langle A,\ \gamma(B)\rangle = \langle \gamma(A),\ B\rangle.$$

The same holds for $\omega(A)$, $\omega(B)$.

Since $\mathscr{C}^{\perp}$ is defined from $\mathscr{C}$, then the decomposition formula (2.6) can be stated in terms of cycles only, or in terms of cocycles only.

Theorem 2.7. (i) *Given $A \subset E$, a cycle $\gamma(A)$ is a decomposition cycle associated with A, if and only if $\gamma(A)$ is orthogonal to every cycle orthogonal to A, and not orthogonal to every cycle not orthogonal to A.*

(ii) *Given $A \subset E$, a cocycle $\omega(A)$ is a decomposition cocycle associated with A, if and only if $\omega(A)$ is orthogonal to every cocycle orthogonal to A, and not orthogonal to every cocycle not orthogonal to A.*

Proof. $\langle \gamma(A), \gamma\rangle = \langle A, \gamma\rangle$ is equivalent to $\langle \gamma(A)+A, \gamma\rangle = 0$. And $\langle \gamma(A)+A, \gamma\rangle = 0$ for every $\gamma \in \mathscr{C}$ means that $\gamma(A)+A$ is a cocycle, which also means that $\gamma(A)$ is a decomposition cycle associated with A. The proof of (ii) is similar.

Theorem 2.7 stated for $A = e$, or simply Theorem 2.2., gives the following theorem.

Theorem 2.8. (i) *A cycle $\gamma(e)$ is a principal cycle associated with e if and only if $\gamma(e)$ is orthogonal to every cycle not containing e and not orthogonal to every cycle containing e.*

(ii) *A cocycle $\omega(e)$ is a principal cocycle associated with e if and only if $\omega(e)$ is orthogonal to every cocycle not containing e and not orthogonal to every cocycle containing e.*

3. Graphs without bicycles

Let us consider graphs for which $R = \mathbf{0}$. The importance of this restriction will appear more clearly in Section 7 where it is shown that by a slight manipulation of its edges any graph can be converted into a graph without bicycles.

If $R = \mathbf{0}$, $\mathscr{E}$ is the direct sum of $\mathscr{C}$ and $\mathscr{C}^{\perp}$. The decomposition of any edge e,

$$e = \gamma(e) + \omega(e), \tag{3.1}$$

is unique, and we can speak of the principal cycle and cocycle of e. And for any

$A \in \mathscr{E}$, A has the unique decomposition,

$$A = \gamma(A) + \omega(A), \tag{3.2}$$

where

$$\gamma(A) = \sum_{e \in A} \gamma(e), \qquad \omega(A) = \sum_{e \in A} \omega(e). \tag{3.3}$$

This defines two linear projection mappings $\gamma: \mathscr{E} \to \mathscr{C}$ and $\omega: \mathscr{E} \to \mathscr{C}^\perp$. Clearly $\operatorname{Im} \gamma = \mathscr{C}$ and $\operatorname{Ker} \gamma = \mathscr{C}^\perp$, $\operatorname{Im} \omega = \mathscr{C}^\perp$, and $\operatorname{Ker} \omega = \mathscr{C}$, and also $\gamma^2 = \gamma$ and $\omega^2 = \omega$. In terms of principal cycles and cocycles we have the following theorem.

Theorem 3.1. *In a graph without bicycles*

(i) *$A \in \mathscr{E}$ is a cycle if and only if the principal cocycles of the edges of A have zero sum, i.e.,*

$$A \in \mathscr{C} \Leftrightarrow \sum_{e \in A} \omega(e) = \mathbf{0}.$$

(ii) *$B \in \mathscr{E}$ is a cocycle if and only if the principal cycles of the edges of B have zero sum, i.e.,*

$$B \in \mathscr{C}^\perp \Leftrightarrow \sum_{e \in B} \gamma(e) = \mathbf{0}.$$

Notice that this theorem can be used to derive principal cycles or cocycles from others. For example, the principal cycle for an edge belonging to a cocycle B can be found if the principal cycles for the other edges of B are known.

Theorem 3.2. *Let G be a graph without bicycles, and let Y be a tree of G and Z the corresponding cotree. The family of cycles $\{\gamma(e)\}_{e \in Z}$ is a basis for $\mathscr{C}$ and, similarly, $\{\omega(e)\}_{e \in Y}$ is a basis for $\mathscr{C}^\perp$.*

Proof. $\{\gamma(e)\}_{e \in E}$ generate $\mathscr{C}$ since $\operatorname{Im} \gamma = \mathscr{C}$. Let us consider, for $f \in Y$, the fundamental cocycle Y^f. Since $\operatorname{Ker} \gamma = \mathscr{C}^\perp$, we have $\gamma(Y^f) = \mathbf{0}$. Since $Y^f = f + Z \cap Y^f$, $\gamma(f)$ is expressible as a sum of $\gamma(e)$ for some edges $e \in Z$. Hence $\{\gamma(e)\}_{e \in Z}$ also generate $\mathscr{C}$. Moreover, $\{\gamma(e)\}_{e \in Z}$ is minimal since its cardinality is the dimension of $\mathscr{C}$.

Theorem 3.3. *If $R = \mathbf{0}$, then the number of edges interlaced with an edge e is always even, i.e.,*

$$0 \equiv \lambda(e) \pmod 2.$$

Proof. Since $P + Q = E$, Eq. (2.5) becomes $\lambda(e) = \gamma(e) \cap \omega(e)$. By orthogonality $\lambda(e)$ is even.

4. The generalized tree and cotree functions

Given a tree Y and its corresponding cotree Z in a graph G, consider the fundamental bases $\{Y^e\}_{e\in Y}$ of $\mathscr{C}^\perp$ and $\{Z^e\}_{e\in Z}$ of $\mathscr{C}$. For an edge $e \in E$ we define

$$\begin{aligned} y(e) &= Y^e && \text{if} \quad e \in Y, \\ y(e) &= \sum_{a\in Z^e+e} Y^a && \text{if} \quad e \in Z, \end{aligned} \tag{4.1}$$

and

$$\begin{aligned} z(e) &= Z^e && \text{if} \quad e \in Z, \\ z(e) &= \sum_{a\in Y^e+e} Z^a && \text{if} \quad e \in Y. \end{aligned} \tag{4.2}$$

By linearity we can extend these definitions from edges to sets of edges and obtain functions

$$y: \mathscr{E} \to \mathscr{C}^\perp, \qquad z: \mathscr{E} \to \mathscr{C}.$$

We call y the *generalized tree function* since it extends to edges not in Y the well-known isomorphism $y_{/Y}$ of 2^Y to $\mathscr{C}^\perp$. For a similar reason z is called the *generalized cotree function*, extending the isomorphism $z_{/Z}$ of 2^Z to $\mathscr{C}$. Clearly $\operatorname{Im} y = \mathscr{C}^\perp$ and $\operatorname{Im} z = \mathscr{C}$.

Theorem 4.1. (i) *$A \in \mathscr{E}$ is a cycle if and only if the cocycles $y(e)$ for $e \in A$ have zero sum, i.e.,*

$$A \in \mathscr{C} \Leftrightarrow \sum_{e\in A} y(e) = \mathbf{0}.$$

(ii) *$B \in \mathscr{E}$ is a cocycle if and only if the cycles $z(e)$ for $e \in B$ have zero sum, i.e.,*

$$B \in \mathscr{C}^\perp \Leftrightarrow \sum_{e\in B} z(e) = \mathbf{0}.$$

Proof. Left to the reader.

There exists a convenient relation between the generalized tree and cotree functions and the decomposition (2.6).

Theorem 4.2. *For any edge $e \in E$ of a graph without bicycles*

(i) *The cocycle $y(e)$ is a decomposition cocycle associated with $Y \cap y^2(e)$.*
(ii) *The cycle $z(e)$ is a decomposition cycle associated with $Z \cap z^2(e)$.*
(iii) *The union of the two families $\{Y \cap y^2(e)\}_{e\in Y}$ and $\{Z \cap z^2(e)\}_{e\in Z}$ is a basis of $\mathscr{E}$.*

Proof. By the definition of y, for any $B \in \mathscr{C}^\perp$ we have $B = y(Y \cap B)$. Hence, since $y^2(e) \in \mathscr{C}^\perp$, put $B = y^2(e)$:

$$y^2(e) = y(Y \cap y^2(e)),$$

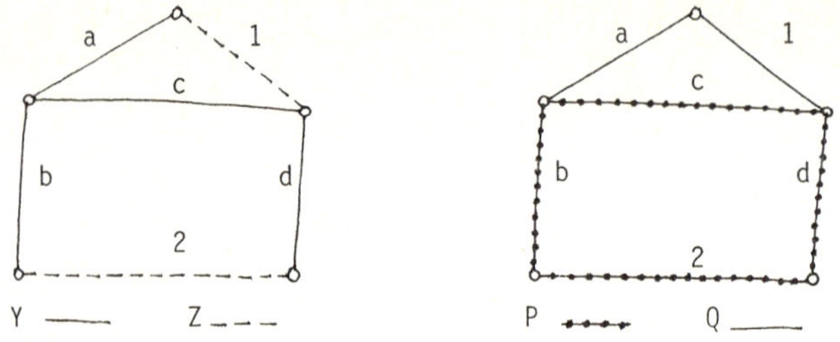

Fig. 4.1

which means that $y(e)$ and $Y \cap y^2(e)$ have the same image under y: since $\ker y = \mathscr{C}$, we have

$$Y \cap y^2(e) = A \dotplus y(e),$$

where $A \in \mathscr{C}$ (and $y(e) \in \mathscr{C}^\perp$): which is a decomposition (2.6) of $Y \cap y^2(e)$.

The proof of (ii) is similar, the proof of (iii) follows easily.

Theorem 4.3. *If G is a graph without bicycles,*

(i) *There exists an integer α such that $y^\alpha = \omega$.*
(ii) *There exists an integer β such that $z^\beta = \gamma$.*

Proof. Left to the reader.

Theorem 4.3 and Theorem 4.2 provide means for computing the projections γ and ω when $R = \mathbf{0}$.

For the graph of Fig. 4.1., take the tree $Y = \{a, b, c, d\}$ and the cotree $Z = \{1, 2\}$. Corresponding to Z the fundamental basis of cycles is displayed in the window of first column of Table 1. From it, the whole Table 1 is generated.

Table 1

$z(e)$	e	$y(e) = \omega(A)$	$\gamma(A)$	$\gamma(e)$	$\omega(e)$	PQR	e
$1ac$	a	$a1$	$ac1$	$bcd2$	$abcd2$	Q	a
$2bcd$	b	$b2$	$bcd2$	$abd12$	$ad12$	P	b
$12abd$	c	$c12$	$abd12$	$ac1$	$a1$	P	c
$2bcd$	d	$d2$	$bcd2$	$abd12$	$ab12$	P	d
$1ac$	1	$ac2$	$ac1$	$bcd2$	$bcd12$	Q	1
$2bcd$	2	$bcd12$	$bcd2$	$abd12$	$abd1$	P	2

5. Bipartition of the edges into a cycle and a cocycle

Let us consider the decomposition of E into a cycle and a cocycle.

Theorem 5.1.[1] *There exists a bipartition of the edges of a graph into a cycle and a cocycle, i.e.,*

$$E = (\gamma(E) + X) + (\omega(E) + X),$$

where $X \in \mathscr{C} \cap \mathscr{C}^{\perp}$.

Proof. Any bicycle X being even (Theorem 2.4), we have $\langle X, E\rangle = 0$ for any bicycle; thus E belongs to $(\mathscr{C} \cap \mathscr{C}^{\perp})^{\perp}$. Then, by Eq. (2.6), E has the decomposition given in the theorem.

It is usual to say that a graph (or a binary matroid) is *even* if E is a cycle, and *bipartite* if E is a cocycle. Then in an even graph there exists a decomposition with $\gamma(E) = E$ and $\omega(E) = \mathbf{0}$; and in a bipartite graph there exists a decomposition with $\gamma(E) = \mathbf{0}$ and $\omega(E) = E$.

Theorem 2.7 can be stated for $A = E$.

Theorem 5.2. (i) *A cycle* $\gamma(E)$ *is a decomposition cycle associated with E if and only if* $\gamma(E)$ *is not orthogonal to every odd cycle.*

(ii) *A cocycle* $\omega(E)$ *is a decomposition cocycle associated with E if and only if* $\omega(E)$ *is not orthogonal to every odd cocycle.*

And for graphs without bicycles we have the following.

Theorem 5.3. *Given a graph without bicycles,*

(i) *a nonzero cycle* $\gamma(E)$ *is a decomposition cycle associated with E if and only if* $\gamma(E)$ *is orthogonal to every even cycle,*

(ii) *a nonzero cocycle* $\omega(E)$ *is a decomposition cocycle associated with E if and only if* $\omega(E)$ *is orthogonal to every even cocycle.*

Theorem 5.4. *For any bipartition of a graph,* $E = \gamma(E) + \omega(E)$,

$$P \subset \gamma(E), \qquad Q \subset \omega(E).$$

In case $R = \mathbf{0}$,

$$P = \gamma(E), \qquad Q = \omega(E).$$

The proofs of Theorems 5.3 and 5.4 are immediate.

It follows from the theorem above that if R is known, and a bipartition is given, the principal tripartition is known. The knowledge of R for graphs comes simply

[1] The appearance of this Bipartition Theorem among graph theorists has been both recent and reticent. Pla proposed to us a proof in the context of flow theory, independently of the paper [2] by Chen. Later we learned from Lovász that Gallai had proved the theorem in 1965, and that Pósa had given a combinatorial algorithm for finding bipartitions. We shall describe Pósa's algorithm, with proof, in the next section.

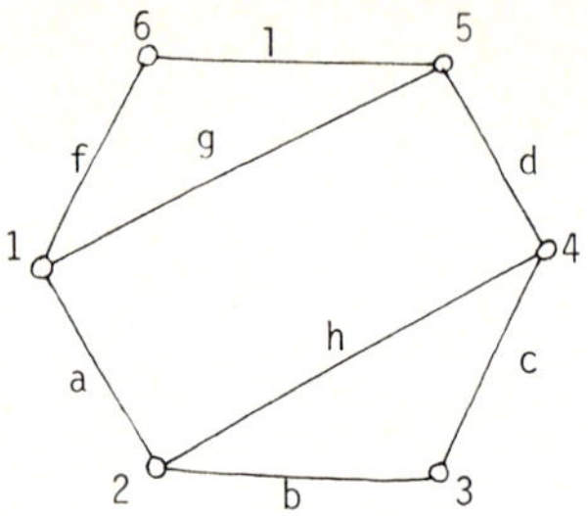

Fig. 5.1

from the fact that bicycles are coboundaries of the elements of Ker $\partial\delta = (\mathrm{Im}\ \partial\delta)^{\perp}$, and a generating system of Im $\partial\delta$ is obviously $(\partial\delta v)_{v\in V}$. For the graph of Fig. 5.1, while taking a fundamental basis for Im $\partial\delta$ (with dendroid $1+3+2+4$, in Table 2), bicycles appear given by δS, such that $\partial\delta S = \mathbf{0}$ ($S = 1+3+5$ or $2+4+6$; then $R = a+b+c+d+l+f$.). Also appears $\omega(E)$, given by δS such that $\partial\delta S = \partial E = 1+2+4+5$ ($S = 234$ or 156 or 23 or 1245; then $\omega(E) = a+d$). So the tripartition is:

$$P = g+h, \qquad Q = \mathbf{0}, \qquad R = a+b+c+d+l+f.$$

And also appears a principal cocycle associated with g, given by δS such that $\partial\delta S = \partial g = 1+5$ ($S = 2+4$; then $\omega(g) = a+b+c+d$).

Table 2

v	$\partial\delta v$	$\partial\delta S$	S
1	1256	1* 5[a]	24
2	1234	3* 5[a]	34
3	24	2* 6[a]	124
4	2345	4* 6[a]	1234
5	1456	—	135
6	15	—	246

[a] Stars indicate elements of the dendroid of Im $\partial\delta$.

The next section is devoted to an algorithm for computing combinatorially rather than algebraically the decomposition of E.

6. A combinatorial bipartition algorithm for graphs

Throughout this section we shall consider only graphs, rather than binary matroids in general, since we shall need to consider vertices and the boundary and coboundary mappings. We shall also assume that these graphs have no loops or multiple edges.

The problem of finding a bipartition of the edges of a graph G is that of finding a set F of vertices such that

$$E = \delta F + A, \tag{6.1}$$

where $A \in \mathscr{C}$, which is equivalent to finding $F \subset V$ such that

$$\partial E = \partial\delta F. \tag{6.2}$$

A set F of vertices satisfying (6.2) will be called a *foot* of G.

Now the edges in δF are those that join a vertex in F to a vertex in $V + F$, and the graph resulting from their removal (which has edge set A) is an even graph, since $A \in \mathscr{C}$. Hence a foot F of a graph G is a set of vertices such that the subgraphs induced by F and by $V + F$ are both even graphs.

The following theorem is a direct corollary of Theorem 5.1. The constructive proof of it given below is due to L. Pósa (personal communication).

Theorem 6.1. *Every graph has a foot.*

Proof. The proof is by mathematical induction on the number of vertices in the graph. The theorem is true for graphs with 1 edge and 2 vertices; suppose it true for graphs having $n-1$ vertices, and consider a graph G having n vertices.

If all the vertices of G have even degree then there is nothing to prove. If not, then let v be a vertex of odd degree, and let H be the set of vertices adjacent to v. We now modify G to obtain a new graph G', as follows. Delete the vertex v, and hence any edges incident with v, and replace the graph induced by H by its complement. Thus two vertices of H will be joined in G' if, and only if, they were not joined in G. Edges incident with at least one vertex not in $H \cup \{v\}$ are not affected.

Since G' has $n-1$ vertices it has a foot F'. The vertices of H are therefore divided into two classes, $H \cap F'$ and $H \cap (V' + F')$, where V' is the vertex set of G'. One of these has an odd number of vertices, the other an even number, since the cardinality of H is odd.

We now assert that the set (either F' or $V' + F'$) which contains an odd number of vertices of H is a foot F of G. For consider what happens when we reconstruct G from G'. In the subgraph induced by F only the vertices of $F \cap H$ are affected, each to the extent of reversing its adjacencies with the even number of other vertices of $F \cap H$. Hence the degrees of these vertices, and hence of all vertices in F remain even (see Fig. 6.1). In the subgraph induced by $V' + F$ the degrees of the vertices of $(V' + F) \cap H$ are changed in parity by the complementing operation; but the restoration of the vertex v to the set $V + F$ increases by 1 the degrees of these vertices, and moreover the degree of v in the subgraph induced by $V + F$ is even. Hence the subgraphs induced by F and $V + F$ are even, and F is a foot of G. This completes the proof of the theorem.

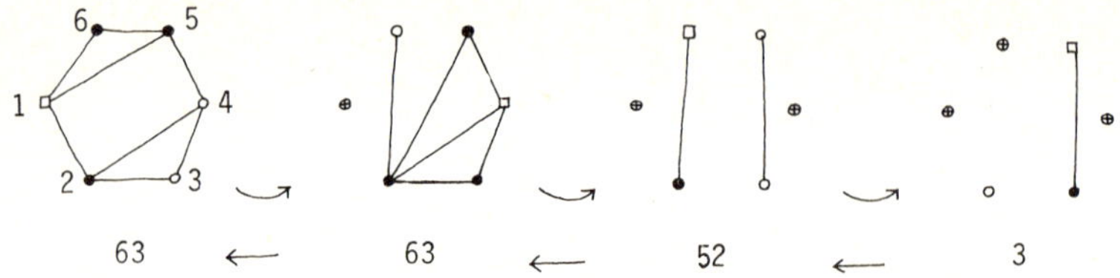

Fig. 6.1

It is clear that this theorem also gives rise to an algorithm (also due to Pósa) for determining a foot of a graph G on n vertices. We use the operation described above to construct a sequence of graphs of decreasing numbers of vertices until we obtain an even graph. This must happen in at most n steps — possibly with an empty graph. By retracing this sequence of graphs we can construct a foot of each from a foot of the preceding one and hence find a foot of G (see Fig. 6.1.).

The crucial step is that of taking the local complement, and this could require a time $O(n^2)$. Hence the whole algorithm can be completed in $O(n^3)$ time.

7. The effect of edge manipulation on the tripartition

We consider here four manipulations of an edge e (illustrated in Fig. 7.1) and introduce the following notation for them.

(a) $G{:}e$ denotes the graph obtained from a graph G by bisecting the edge e, i.e., replacing e by two edges, e_1 and e_2, in series.

(b) $G.e$ denotes the graph obtained from G by contracting the edge e, i.e., deleting e and identifying its ends.

(c) $G\&e$ denotes the graph obtained from G by doubling the edge e, i.e., replacing e by two edges, e_1 and e_2, in parallel.

(d) $G-e$ denotes the graph obtained from G by deleting the edge e.

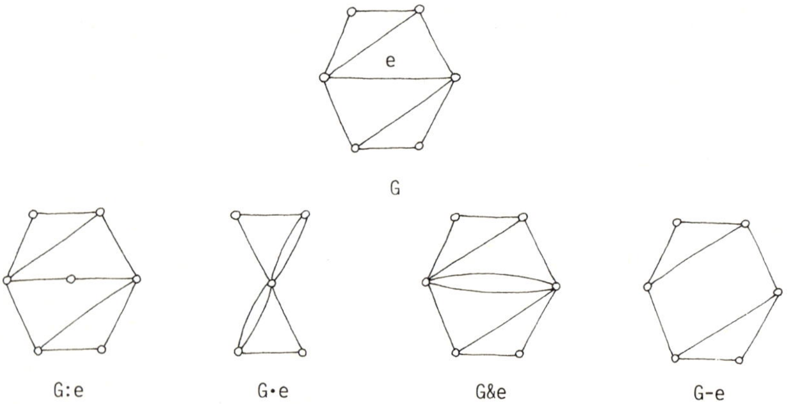

Fig. 7.1

Table 3

Manip	e in	e_1 & e_2 in	new P	new Q	new R	Δq
$G{:}e$	P	R	$P+e+P\cap\lambda(e)$	$Q+Q\cap\lambda(e)$	$R+e_1+e_2+\lambda(e)$	+1
$G.e$	P		$P+e+P\cap\lambda(e)$	$Q+Q\cap\lambda(e)$	$R+\lambda(e)$	+1[a]
$G\,\&\,e$	P	P	$P+e+e_1+e_2+\lambda(e)$	$Q+\lambda(e)$	R	0
$G-e$	P		$P+e+\lambda(e)$	$Q+\lambda(e)$	R	0
$G{:}e$	Q	Q	$P+\lambda(e)$	$Q+e+e_1+e_2+\lambda(e)$	R	0
$G.e$	Q		$P+\lambda(e)$	$Q+e+\lambda(e)$	R	0
$G\,\&\,e$	Q	R	$P+P\cap\lambda(e)$	$Q+e+Q\cap\lambda(e)$	$R+e_1+e_2+\lambda(e)$	+1
$G-e$	Q		$P+P\cap\lambda(e)$	$Q+e+Q\cap\lambda(e)$	$R+\lambda(e)$	+1[b]
$G{:}e$	R	P	$P+\mu^+(e)$	$Q+\mu^-(e)$	$R+\mu(e)$	−1
$G.e$	R		$P+\mu^+(e)$	$Q+\mu^-(e)$	$R+\mu(e)$	−1
$G\,\&\,e$	R	Q	$P+\mu^-(e)$	$Q+\mu^+(e)$	$R+\mu(e)$	−1
$G-e$	R		$P+\mu^-(e)$	$Q+\mu^+(e)$	$R+\mu(e)$	−1

[a] If e is not a loop.
[b] If e is not a bridge.

The effect of these four operations on the graph depends strongly on the status of the edge e relative to the tripartition. Hence there are 12 cases to consider. The results are displayed in Table 3; proofs are immediate.

The new symbols used in Table 3 are defined as follows. By Δq is meant the change in the dimension q of the bicycle space; by $\mu(e)$ is meant the set of edges isobicyclic to e, i.e. the set of $f\in R$ such that for any bicycle X, $e\in X\Leftrightarrow f\in X$; by $\mu^+(e)$ is meant the set of $f\in R$ such that in any decomposition of E, e and f occur in the same part (thus $\mu^+(e)$ is a subset of $\mu(e)$); by $\mu^-(e)$ is meant the set of edges f such that in any decomposition of E, e and f do not occur in the same part (thus $\mu^-(e)$ is also a subset of $\mu(e)$).

From Table 1 we see that for any one manipulation the change in the dimension of the bicycle space is -1, 0 or $+1$ depending on which of the classes P, Q, R (not necessarily in that order) contains the edge being manipulated. Hence we have:

Theorem 7.1. *Given any of the four manipulations and any edge e, the class (P, Q or R) to which e belongs can be determined from the change in the dimension of the bicycle space when the manipulation is performed.*

Manipulations produce relations between the principal cycles and the decomposition cycles associated with E, as follows.

Theorem 7.2. (i) *If* $e\in P$: $\gamma(e)=\gamma(E(G-e))+\gamma(E(G))$,
(ii) *If* $e\in Q$: $\omega(e)=\omega(E(G.e))+\omega(E(G))$.

Proof. Note that cycles of $G-e$ are cycles of G, and cocycles of $G.e$ are cocycles of G. Therefore Theorem 7.2. is a consequence of Theorems 2.8 and 5.2.

A manipulation can be performed on several edges, and is then called manipulation on a set of edges $A \subset E$ ($G:A$, $G.A$, $G \,\&\, A$, $G-A$).

Let us consider the problem of withdrawing all the bicycles from a graph.

Theorem 7.3. *Given a graph G with bicycles, and $A \subset E$; if A is a dendroid of the bicycle space of G, then $G:A$, $G\&A$, $G.A$ and $G-A$ are graphs without bicycles.*

Proof. The theorem is proved by simple inspection of Table 3 and the use of a fundamental basis of bicycles.

8. Bicycles and the parity of the tree number

In this section we shall make use of Table 3 of Section 7 to establish some assertions concerning the parity of the tree number $N(G)$—the number of spanning trees of the graph G. (In what follows the word 'spanning' will be understood).

The following theorem was first proved by W. K. Chen [2].

Theorem 8.1. *If G is connected, $N(G)$ is odd if and only if $R = \mathbf{0}$.*

Chen's proof of this theorem is algebraic, depending on the calculation of a determinant; here we give a combinatorial proof.

Proof. We know that if an edge e is not a bridge or a loop, then

$$N(G) = N(G.e) + N(G-e).$$

For any tree T of G either contains e, in which case it remains a tree when e is contracted, i.e., $T.e$ is a tree of $G.e$; or it does not, in which case it is also a tree of $G-e$. Moreover the number of trees is not changed when a bridge of a graph is contracted, or when a loop of a graph is deleted.

The proof of the theorem is made by mathematical induction on the number of edges. We first observe that the theorem is true for the connected graphs with two edges. Note also that the theorem is true for a connected graph where each edge is either bridge or loop.

Now suppose that the theorem is true for graphs having at most m edges, and consider any edge e which is not a bridge or a loop in a graph having $m+1$ edges.

If $e \in P$, then $G.e$ has bicycles and $N(G.e)$ is therefore even, by hypothesis. Hence $N(G)$ has the same parity as $N(G-e)$. But $G-e$ is connected, and G has the same bicycles as $G-e$, and hence the theorem is true for G.

If $e \in Q$, we obtain the same result by interchanging the operations of deletion and contraction.

If $e \in R$, $G.e$ and $G-e$ are connected and have same bicycle space dimension. Therefore $N(G.e)$ and $N(G-e)$ have the same parity; hence their sum is even.

Theorem 8.2. (a) *Let A be a non-empty subset of $E(G)$ not containing any cycle of G. The number of trees of G that contain A (or cotrees disjoint from A) is odd if and only if G.A has no bicycles.*

(b) *Let B be a nonempty subset of $E(G)$ not containing any cocycle of G. The number of trees of G that are disjoint from B (or cotrees containing B) is odd if and only if $G-B$ has no bicycles.*

Proof. In fact the trees containing A (or the cotrees disjoint from A) are in bijection with the trees (or cotrees) of $G.A$; the trees disjoint from B (or the cotrees containing B) are in bijection with the trees (or the cotrees) of $G-B$. Therefore Theorem 8.2. is a corollary of Theorem 8.1.

For other results on parity and principal tripartition, see de Fraysseix [3].

9. Bicycles and the Tutte[2] polynomial

We now show that the dimension q of the bicycle space of G is related to the Tutte polynomial (or dichromate, see [12]) of More, precisely:

Theorem 9.1. *If a graph has m edges and 2^q bicycles, then*

$$\chi(-1,-1)=(-1)^m(-2)^q.$$

Proof. For a graph in which every edge is either a bridge or a loop (say that there are α bridges and β loops, where $\alpha+\beta=m$) it is known that the Tutte polynomial reduces to $x^\alpha y^\beta$, and certainly $q=0$, so that the result of the theorem holds.

Suppose now that the theorem is true for graphs having at most λ edges which are neither bridge nor loop; let us consider a graph G with $\lambda+1$ edges which are neither bridge nor loop, and let e be one of them. Then it suffices to show that the recurrence relation for the Tutte polynomial is satisfied, that is to say that

$$(-1)^{m(G.e)}(-2)^{q(G.e)}+(-1)^{m(G-e)}(-2)^{q(G-e)}=(-1)^m(-2)^{q(G)}. \tag{9.1}$$

In $G.e$ and $G-e$ the number of edges which are neither bridge nor loop is smaller than or equal to λ. Then by recurrence hypothesis and from Table 3 we have the following values for the left-hand side of (9.1).

If $e\in P$: $(-1)^{m-1}(-2)^{q+1}+(-1)^{m-1}(-2)^q$,
if $e\in Q$: $(-1)^{m-1}(-2)^q+(-1)^{m-1}(-2)^{q+1}$, and
if $e\in R$: $(-1)^{m-1}(-2)^{q-1}+(-1)^{m-1}(-2)^{q-1}$.

[2] We dedicate a new value of the Tutte Polynomial to Professor W. T. Tutte on the occasion of his 60th birthday.

In each case this reduces to the right-hand side of (9.1) and the recurrence is therefore satisfied.

10. Characterization of planar graphs by the algebraic diagonal

A graph is planar in the sense of Whitney [13] if there exists a graph G^*, called the algebraic dual of G, and a bijection ϕ of $E(G)$ onto $E(G^*)$, called duality, such that

(i) every cocycle of a vertex of G has, as dual, a cycle of G^*, and

(ii) every elementary cycle (i.e. minimal under inclusion) of G^* is the dual of a cocycle of G.

By an *algebraic diagonal* of a connected graph G having no bicycles will be meant a walk S on G including each edge exactly twice, such that between two occurrences of the same edge e in the walk, the set of edges occurring exactly once makes, with e, the set $\gamma(e)$ if $e \in P$ and $\omega(e)$ if $e \in Q$.

Theorem 10.1. *A connected graph without bicycles is planar if and only if it has an algebraic diagonal.*

(The proof below has been very briefly described in [9]).

Proof. (a) Let G be a planar connected graph without bicycles. Then G and G^* admit dual plane realizations where the dual edges e and e^* cross each other and each is incident to two vertices (distinct or coincident) of one graph, corresponding to faces of the dual graph. We associate with these edges a set of four elements, denoted by Ie, Je, Te, Ue (represented by the arrows in Fig. 10.1) on which operates the Klein group (I, J, T, U) (I = identity, J = inverse, T = traverse and U = transverse). Each of the four elements is incident with a face and a vertex incident with e, such that, a being any one of them, a and Ta are incident with the same vertex, while a and Ja are incident with the same face. Thus with G and G^* represented in the plane, we associate a quadrialphabet $\mathscr{A}$ on which operates the Klein group (I, J, T, U). J and T interchange under duality.

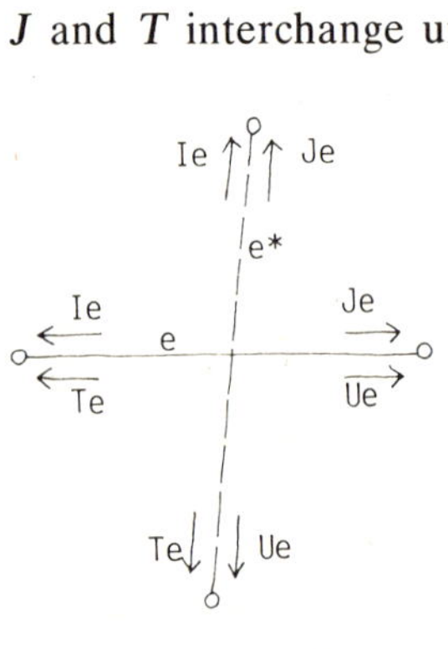

Fig. 10.1

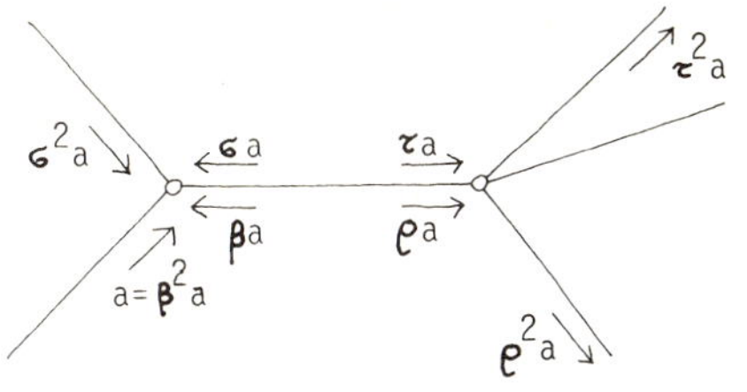

Fig. 10.2

A word in $\mathscr{A}$, $Z = c_0 c_1 \cdots c_i \cdots c_r$, such that every letter has its incident vertex in G (respectively G^*) in common with the inverse (respectively traverse) of the following letter is called a word of G; it describes a walk in G in one of two senses. The walk is closed if c_0 is considered as following c_r. Denote by $|Z|$ the set of edges encountered an odd number of times in Z. If Z is a closed walk in G then $|Z| \in \ker \partial$. For any permutation π on $\mathscr{A}$, if b is on the orbit of a, i.e. if $\pi^k a = b$ for some integer k, the interval from a (not included) to b (included), denoted by $(a, b]^\pi$ is defined as the shortest word $\pi a, \pi^2 a, \pi^3 a, \ldots, b$. $(a, a]^\pi$ denotes the orbit of π that contains a. Denote by $|a, b|^\pi$ the set of edges encountered an odd number of times in $(a, b]^\pi$.

It is known that for a plane representation of G there is associated with each vertex a cyclic permutation of the edges incident with that vertex, and hence [7] an involution β without fixed points on $\mathscr{A}$ which interchanges the letters having the same incidences (see Fig. 10.2). The permutation $\tau = U.\beta$ has the following property.

Property 10.2. *For every* $a \in \mathscr{A}$, $(U.\tau)^2 = I$ *and* $\tau a \neq Ua$.

It appears that '$a\ \tau a$' is a word of G and of G^*; in fact, a and $J.\tau a$ have the same incident vertex in G, since $J.\tau a = J.U.\beta a = T.\beta a$; moreover a and $T.\tau a$ have the same incident vertex in G^* (i.e. face of G) since $T.\tau a = T.U.\beta a = J.\beta a$. An orbit $(a, a]^\tau$ of τ therefore describes simultaneously a closed walk of G and of G^*. It follows that $|a, a|^\tau$ is the zero bicycle—the only bicycle in G, by hypothesis. Thus we have:

Property 10.3. *For every* $a \in \mathscr{A}$, $|a, a|^\tau = \mathbf{0}$.

It follows [8] that τ has exactly two orbits, transverses of each other, one being $\ldots abc \ldots$, the other $\ldots Uc\ Ub\ Ua \ldots$ (see Figs. 10.3 and 10.4) constituting a particular closed walk of G and G^* in which each edge occurs twice. This walk is called the Petrie polygon or *geometric diagonal* Δ of the plane representations of G and G^*. An argument similar to that used to obtain Property (10.3), taken together with Eq. (3.1), namely $e = \gamma(e) + \omega(e)$, shows that Δ has the following properties.

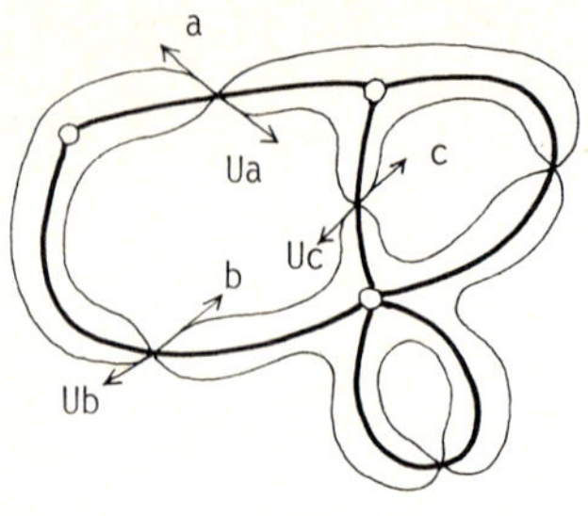

Fig. 10.3

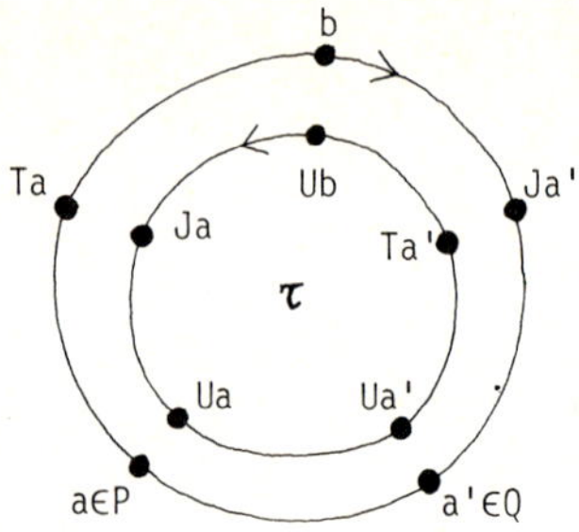

Fig. 10.4

Property 10.4. *For every* $a \in \mathscr{A}$

(i) *if* $|a| \in P$, *then* Ta *is on the orbit of* a *and* $|a, Ta|^\tau = \gamma(|a|)$,
(ii) *if* $|a| \in Q$, *then* Ja *is on the orbit of* a *and* $|a, Ja|^\tau = \omega(|a|)$.
Thus Δ *is an algebraic diagonal of* G.

For the proof of Property 10.4(i) note that if $|a| \in P$, then in G, $(a, Ta]^\tau$ is a closed walk. In G^* it is a walk plus the last edge $|Ta|$; Hence $(a, \tau^{-1}Ta]^\tau$ is a closed walk. Thus $|a, Ta|^\tau$ is a cycle of G and $|a, Ta|^\tau + |a|$ is a cocycle of G, whence $|a, Ta|^\tau = \gamma(|a|)$. The proof of (ii) is similar.

(b) Conversely we show that, given a connected graph G without bicycles, having an algebraic diagonal S, we can associate with it a graph $G^*(S)$ having the properties of an algebraic dual in the sense of Whitney. To do this we associate with $E(G)$ a quadrialphabet $\mathscr{A}$ on which operates a Klein group (I, J, T, U) with the same incidence conventions for each quadruplet and its edge as were given in Part (a) of the proof. The walk S, meeting each edge exactly twice, is closed. S induces a permutation τ (Fig. 10.4) on $\mathscr{A}$ consisting of two transverse orbits. One of these is constructed by describing S, starting at an initial edge and in an arbitrary sense, and by substituting for any edge e encountered for the first time the letter Ie (which fixes the vertex of incidence of Te and Ue), and by substituting for any edge e encountered for the second time the letter Te if the vertex of incidence of Te agrees, and the letter Je if not. Since, by construction, τ has the Properties (10.2), the permutation $\rho = T.\tau$ has the following property.

Property 10.5. *For every* $a \in \mathscr{A}$, $(J.\rho)^2 = I$ *and* $\rho a \neq Ja$.

Since '$a\,\tau a$' is a word of G, '$a\,\rho a$' is also a word of G; in fact $J.\rho a$ and $J.\tau a$ have the same incident vertex in G since $J.\rho a = J.T.\tau a = T.J.\tau a$. The orbits of ρ describe walks on G. It follows that the orbits of ρ containing a and Ja are distinct (inverses of each other) and constitute a particular closed walk of G, called a face of G relative to S. This face is said to be incident to the letters a and Ja. An arbitrary edge e is incident to two faces relative to S (distinct or coincident), said to be adjacent. Then $G^*(S)$ is, by definition, the graph whose vertices are the faces defined above and whose edges correspond to adjacent faces: an edge e^* of $G^*(S)$ corresponds to each edge e of G having the same incidence with the faces of G (that is to say with the vertices of G^*). The edges e and e^* are said to be duals.

It remains to verify that $G^*(S)$ satisfies the Whitney conditions (i) and (ii).

(i) Consider a vertex $v \in V(G)$. The permutation $\sigma = J.\tau$ has, by Property (10.2), the following properties:

$$\text{for every } a \in \mathcal{A},\ (T.\sigma)^2 = I \text{ and } \sigma a \neq Ta.$$

Since '$a\,\tau a$' is a word of $G^*(S)$ so is '$a\,\sigma a$'; in fact $T.\sigma a$ and $T.\tau a$ have the same incident vertex in $G^*(S)$ since $T.\sigma a = T.J.\tau a = J.T.\tau a$. It follows that the orbits of σ containing the letters a and Ta, incident with the vertex v, are distinct (each the transverse of the other), comprise only letters incident with v, and make up a closed walk of $G^*(S)$. The set of edges of G that are incident with v exactly once (a cocycle of G) therefore has, as dual, a cycle of $G^*(S)$.

(ii) An elementary cycle A^* of $G^*(S)$ corresponds, by definition, to a sequence of distinct faces of G, say $f_0, f_1, \ldots f_i, \ldots, f_n$, all distinct, where, if $n = 0$, there is an edge e_0 incident twice with f_0, and, if $n > 0$, there is an edge e_0 incident with f_n and f_0, and an edge $e_i (i = 1, 2, \ldots, n)$ incident with f_{i-1} and f_i (see Figs. 10.5 and 10.6). It is now a matter of showing that $A = (e_0, e_1, \ldots, e_n)$ is a cocycle of G.

Consider the following two words of G associated with A:

$$M(x) = (Ua_0, \rho^{-1}a_1]^\rho a_1 Ua_1 (Ua_1, \rho^{-1}a_2]^\rho a_2 Ua_2 \cdots$$
$$\cdots (Ua_{i-1}, \text{ (Fig. 10.5) or } x = Ja_0 \text{ (Fig. 10.6).}$$

and

$$N(x) = (Ua_0, \rho^{-1}a_1]^\rho (Ua_1, \rho^{-1}a_2]^\rho \cdots (Ua_{i-1}, \rho^{-1}a_i]^\rho \cdots (Ua_n, \rho^{-1}x]^\rho,$$

where a_0 is one of the two letters of e_0 incident with f_n, a_i, for $1 \leq i \leq n$, is the first letter of e_i encountered by the orbit of ρ starting at Ua_{i-1}, and x is the first letter of e_0 encountered by the orbit of p starting at Ua_n. $M(x)$ and $N(x)$ are not necessarily closed. Since x is, like a_0, a letter of e_0 incident with f_n we have two cases to consider: either $x = a_0$ (Fig. 10.5) or $x = Ja_0$ (Fig. 10.6).

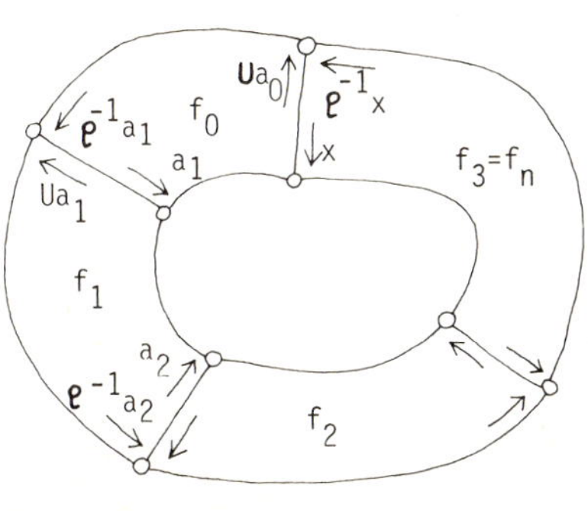

Fig. 10.5

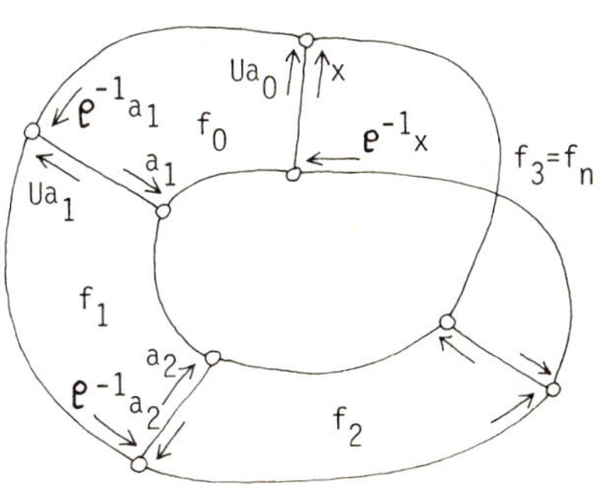

Fig. 10.6

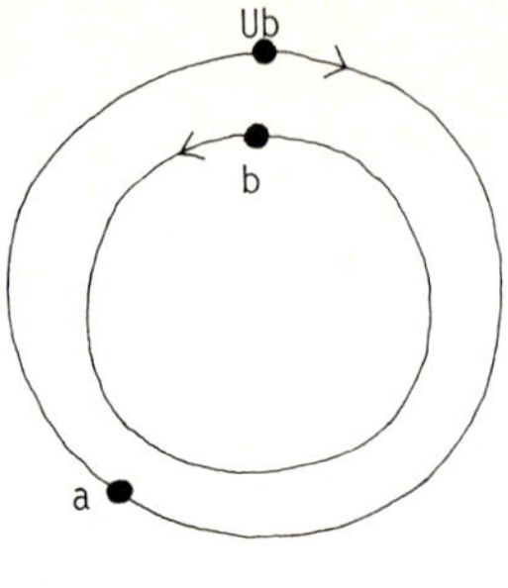

Fig. 10.7

At this point we digress slightly to extend our definition of an interval $(a, b]^\tau$ over τ. It was defined above as the portion of the orbit of τ from a to b, and this definition applies only when a and b are on the same orbit, i.e. when $b \in (a, a]^\tau$ (see Fig. 10.4). We now extend this definition to cover the case $b \notin (a, a]$ (see Fig. 10.7) as follows. If b is not on the same orbit as a then Ub is, i.e. $Ub \in (a, a]^\tau$, and we define $(a, b]^\tau$ to be $(a, Ub]^\tau b$.

Hence the interval from a to b on the diagonal includes the backtrack '$Ub\, b$' on the edge $|b|$, which is therefore traversed twice. This means that $|b| \notin |a, b|^\tau$.

In this new notation Properties (10.3) imply at once a 'Chasles formula', namely:

Property 10.6. *For* $a, b \in \mathcal{A}$, $|a, b|^\tau + |b, c|^\tau = |a, c|^\tau$.

Properties (10.4), which τ satisfies by hypothesis, generalize now to

Property 10.7. *For all* $a \in \mathcal{A}$, $|a, Ta|^\tau = \gamma(|a|)$, *and* $|a, Ja|^\tau = \omega(|a|)$.

For, if $|a| \notin P$ then $Ta \notin (a, a]^\tau$ and

$$|a, Ta|^\tau = |a, U.Ta|^\tau + |a| = \omega(|a|) + |a| = \gamma(|a|),$$

while if $|a| \notin Q$, then $Ja \notin (a, a]^\tau$ and

$$|a, Ja|^\tau = |a, U.Ja|^\tau + |a| = \gamma(|a|) + |a| = \omega(|a|).$$

We also have, as consequences, the following two useful results:

Property 10.8. *For every* $a \in \mathcal{A}$, $|a, Ua|^\tau = |a|$.

This property holds since $(a, Ua]^\tau = (a, a]^\tau\, Ua$, whence $|a, Ua|^\tau = 0 + |a|$.

Property 10.9. *For every* $a \in \mathcal{A}$, $|\rho^{-1}a, a|^\tau = \omega(|a|)$.

This property holds since $(\rho^{-1}a, a]^\tau = Ta(Ta, a]^\tau$, whence $|\rho^{-1}a, a|^\tau = |a| + \gamma(|a|)$ by virtue of Property (10.7).

We now resume the proof of the second part of the theorem.

In the case $x = a_0$, consider the two closed walks $K = M(a_0)Ua_0$ and $L = N(a_0)$, (see Fig. 10.5). Write $K = c_0 c_1 \cdots c_j \cdots c_r$, where $c_j \in \mathscr{A}$, and consider the expression

$$(c_0, c_1]^\tau (c_1, c_2]^\tau \cdots (c_{j-1}, c_j]^\tau \cdots (c_{r-1}, c_r]^\tau (c_r, c_0]^\tau,$$

which is a closed walk over the diagonal with occasional backtracking. The sum of the edges over the whole walk is zero, by virtue of Properties (10.6) and (10.3), i.e.

$$\begin{aligned} \mathbf{0} = |c_0, c_1|^\tau + |c_1, c_2|^\tau + \cdots + |c_{j-1}, c_j|^\tau \\ + \cdots + |c_{r-1}, c_r|^\tau + |c_r, c_0|^\tau. \end{aligned} \tag{10.1}$$

We distinguish two kinds of terms in (10.1). Corresponding to a factor $(Ua_{j-1}, \rho^{-1}a_j]^\rho$ in $N(a_0)$ we have terms with $c_{j-1} = \rho^{-1}c_j$ and thus $|c_{j-1}, c_j|^\tau = \omega(|c_j|)$, by virtue of Property (10.9). These terms therefore give on summation:

$$\sum_{c \in N(a_0)} \omega(|c|) = \sum_{c \in L} \omega(|c|) = \sum_{e \in |L|} \omega(e) = \omega(|L|).$$

Corresponding to the pair of terms $a_i \cup a_i$ in $M(x)$ we have, for $i = 0, 1, \ldots,$

$$|\rho^{-1}a_i, a_i|^\tau + |a_i, Ua_i|^\tau = \omega(|a_i|) + |a_i| = \gamma(|a_i|),$$

by virtue of Properties (10.1), (10.8) and Eq. (3.1). Summing for all these terms, we have

$$\sum_{i=0}^{n} \gamma(|a_i|) = \sum_{e \in A} \gamma(e) = \gamma(A).$$

Hence from (10.9) we have

$$\mathbf{0} = \omega(|L|) + \gamma(A). \tag{10.2}$$

L is a closed walk in G; $|L|$ is therefore a cycle of G, and belongs to the kernal of ω. Hence by (10.2) A belongs to the kernel of γ, i.e. it is a cocycle of G, which is what was to be proved.

In the case $x = Ja_0$, consider the two closed walks $K' = M(J_0)$ and $L' = N(Ja_0)Ua_0$ (see Fig. 10.6). Write $K' = c_0 c_1 \cdots c_j \cdots c_r$ and consider the closed walk

$$(c_0, c_1]^\tau (c_1, c_2]^\tau \cdots (c_{j-1}, c_j]^\tau \cdots (c_{r-1}, c_r]^\tau,$$

on which the sum of the edges is zero by Properties (10.6) and (10.3), i.e.

$$\begin{aligned} 0 = |c_0, c_1]^\tau + |c_1, c_2]^\tau + \cdots + |c_{j-1}, c_j|^\tau \\ + \cdots + |c_{r-1}, c_r|^\tau + |c_r, c_0|^\tau. \end{aligned} \tag{10.3}$$

Now we distinguish two kinds of terms in (10.3). There are the terms $|c_{j-1}, c_j|^\tau$ corresponding to the $c_j \in N(Ja_0)$, where, for $c_j \neq c_0$, $c_{j-1} = \rho^{-1}c_j$, whence

$$|c_{j-1}, c_j|^\tau = \omega(|c_j|),$$

by virtue of Property (10.9); while for $c_j = c_0$, $c_{j-1} = c_r = Ja_0 = T.\rho^{-1}c_0$, whence

$$|c_r, c_0|^\tau = |Ja_0, Ua_0|^\tau + |\rho^{-1}c_0, c_0|^\tau,$$

by virtue of Property (10.6), or again

$$|c_r, c_0|^\tau = \gamma(|a_o|) + \omega(|c_0|),$$

by virtue of Properties (10.7) and (10.9). On summation we get

$$\sum_{c\in N(Ja_0)} \omega(|c|) + \gamma(|a_0|) = \sum_{e\in|L'|} \omega(e) + e_0 = \omega(|L'|) + e_0,$$

by virtue of Eq. (3.1). The other terms, taken in pairs for $i = 1, 2, \ldots, n$ are equal to

$$|\rho^{-1}a_i, a_i|^\tau + |a_i, Ua_i|^\tau = \omega(|a_i|) + |a_i| = \gamma(|a_i|),$$

by virtue of Properties (10.9), (10.8) and Eq. (3.1), and in addition there is the term corresponding to $i = 0$, namely

$$|\rho^{-1}Ja_0, Ja_0|^\tau = \omega(|a_0|).$$

On summation we obtain

$$\sum_{i=1}^{n} \gamma(|a_i|) + \omega(|a_0|) = \sum_{i=0}^{n} \gamma(|a_i|) + |a_0| = \sum_{e\in A} \gamma(e) + e_0 = \gamma(A) + e_0$$

by virtue of Eqs. (3.1) and (3.3). The equation (10.3) becomes

$$\mathbf{0} = \omega(|L'|) + \gamma(A),$$

which is similar in form to (10.2) and implies the same conclusion. This completes the proof of Theorem 10.1.

Note. The above theorem applies only to graphs without bicycles, but it can be applied to graphs in general by making a slight modification to the graph in question as mentioned in Theorem 7.3.

11. Proof of the interlace conjecture of Gauss

Consider in the plane a closed curve C—a continuous image of a circle—which has a finite number of points of self intersection (see Fig. 11.1), each intersection being of two portions of the curve only. These points will be called *crossings*. Suppose that these crossings have been labelled with arbitrary distinct symbols, such as letters of the alphabet, as in Fig. 11.1, where the letters a, b, c, d and e have been used. Let the curve now be traversed exactly once, and the labels recorded in the order in which they are encountered during the traversal. Then we obtain a sequence of labels in which each symbol occurs exactly twice. Thus for

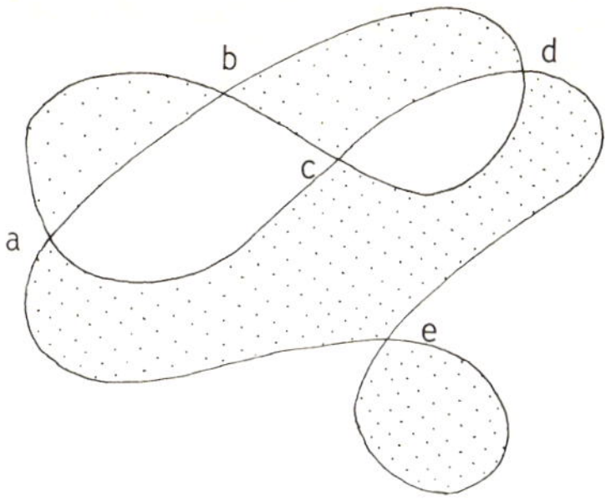

Fig. 11.1

the curve of Fig. 11.1, we obtain the sequence *abcdbaeedc*. The problem of Gauss, with which this section is concerned, is that of determining which sequences can arise in the above manner from some closed curve in the plane.

First we shall need a few definitions. In what follows the term *sequence* will be used to denote an ordered sequence of symbols in which each symbol occurs exactly twice. A *crossing sequence* is a sequence which can be obtained in the manner described above from some self-intersecting closed curve in the plane.

We shall denote by S^e the set of symbols that occur exactly once in S between the two occurrences of the symbol e. Clearly, if $f \in S^e$ then $e \in S^f$ (compare S^e to $\lambda(e)$ of Section 3.); we shall then say that e and f are "interlaced". Given a sequence S we define a graph, denoted by $I(S)$ and called the *interlace graph of S*, whose vertices correspond to the symbols in S and in which two vertices are adjacent if and only if the corresponding symbols are interlaced.

Gauss [5] observed that if a sequence is a crossing sequence then the following *interlace* property holds:

Property 11.1. *For every symbol e, S^e is even.*

This implies that the interlace graph of a crossing sequence is an even graph, i.e. every vertex has even degree. By exhibiting the sequence *abcdecdabe* Gauss showed that the necessary condition (11.1) is not sufficient, and propounded the problem of finding other interlace properties to be added to Property (11.1.) in order to form a necessary and sufficient condition for a sequence to be a crossing sequence. We call the conjecture that such interlace conditions exist *the interlace conjecture of Gauss.*

In 1936 Dehn gave a solution to the recognition of crossing sequences [4]. This solution was of an algorithmic nature, giving a method for determining whether or not a given sequence was a crossing sequence. This solution has been discussed by us elsewhere (see [7]), along with other approaches to this problem that rely, as does Dehn's, on topological transformations of the curve, and the corresponding transformations of the sequence S. Below we give an algebraic solution to the Gauss crossing problem, one that relies only on the properties of the sets S^e

defined above, in the spirit of the Gauss condition (11.1). Therefore the conjecture is proved that interlace properties can characterize crossing sequences.

The essentials of this solution are embodied in the following theorem.

Theorem 11.2. *A sequence S is a crossing sequence if and only if the interlace graph $I(S)$ has the following properties*:

(i) *for every e, S^e is even (this is Gauss's condition)*,

(ii) *for every e and f that are not adjacent in $I(S)$, $S^e \cap S^f$ is even*,

(iii) *the edges (e, f) for which $S^e \cap S^f$ is even form a cocycle.*

(The proof below has been very briefly described in [10]).

Proof. Let E denote the set of symbols occurring in S.

(a) We first show that properties (i), (ii) and (iii) are together equivalent to the following property.

Property 11.3. *There exists a bipartition (P, Q) of E, with $P \cap Q = \emptyset$ and $P \cup Q = E$, such that every element of the family $\{\gamma(e)\}_{e \in E}$ defined by*

$$\gamma(e) = S^e \cup \{e\} \quad \textit{if } e \in P,$$

and

$$\gamma(e) = S^e \quad \textit{if } e \in Q,$$

has an even intersection with every element of the family $\{\omega(e)\}_{e \in E}$ defined by

$$\omega(e) = S^e \quad \textit{if } e \in P,$$

and

$$\omega(e) = S^e \cup \{e\} \quad \textit{if } e \in Q.$$

On the one hand we show that Property (11.3) implies (i), (ii) and (iii) of Theorem 11.2. For every e, $\gamma(e) \cap \omega(e) = S^e$ is even, which gives (i). If $f \notin S^e$ then the two sets $S^e \cap (S^f \cup \{f\})$ and $S^e \cap S^f$ are equal. Hence since one of them is even, so is the second, and this gives (ii). Finally, if $f \in S^e$ then $S^e \cap (S^f \cup \{f\})$ and $S^e \cap S^f$ differ exactly in the element f. Hence e belongs to one class (P or Q) and f belongs to the other if and only if $S^e \cap S^f$ is even. This gives (iii).

On the other hand, (i), (ii) and (iii) of Theorem 11.2 together imply Property (11.3). The cocycle referred to in (iii) defines a partition of E into two classes P and Q. Consider the sets $\gamma(e)$ and $\omega(e)$ defined by P and Q in accordance with Property (11.3), and suppose that one intersection, say $\gamma(e) \cap \omega(f)$, is odd. Then $f \neq e$, by (i). Moreover, $f \in S^e$, since otherwise $\gamma(e) \cap \omega(f) = S^e \cap S^f$ would be even, by (ii). Hence $S^e \cap (S^f \cup \{f\})$ and $S^e \cap S^f$ have opposite parity. We now distinguish two cases. If e and f belong to the same class, then $S^e \cap S^f$ is odd, by (iii), and $\gamma(e) \cap \omega(f) = S^e \cap (S^f \cup \{f\})$ is even — which is not possible; but if e and f

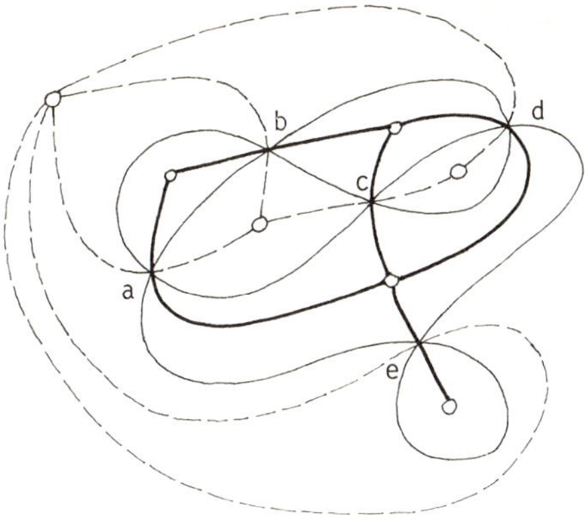

Fig. 11.2

belong to distinct classes, then $S^e \cap S^f$ is even, by (iii), and $\gamma(e) \cap \omega(f)$, which has the same parity as $S^e \cap S^f$, is also even — again a contradiction.

(b) We next show that a crossing sequence of a curve C satisfies Property (11.3). It is well-known that the curve C divides the plane into regions colourable in two colours, so that two regions incident with a common portion of C have different colours (see Fig. 11.1).

Consider the graph $G(S)$ whose vertices are the regions with colour different from that of the infinite region, and whose edge set E is defined as follows: $e \in E$ is incident with the two vertices of $G(C)$ which represent the regions which have the intersection point e in common. Then $G(C)$ is a plane graph which can be realized in the plane in a natural way by making the edge e pass through the point e of C (as in Fig. 11.2, heavy lines). Let $G^*(C)$ be the plane graph defined as for $G(C)$ but using the regions with the other colour (see Fig. 11.2, dotted lines). Then $G(C)$ and $G^*(C)$ are dual graphs, and C is seen to be the geometric diagonal of these two graphs, as defined in Section 10.

If the curve C is traversed once in the sense given by the sequence S, then the two directions in which a given edge of $G(C)$ is crossed by C can be described as being "the same" (as in Fig. 11.3a) or "opposite" (as in Fig. 11.3b). Let P be the set of edges that are crossed in the same direction by C — for example the edge 'd' in Fig. 11.2), and let Q be the set of edges that are crossed in opposite

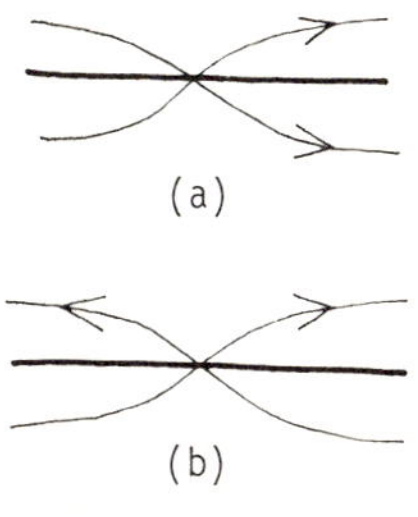

Fig. 11.3

directions by C—for example the edge 'b' in Fig. 11.2. Then the families $\{\gamma(e)\}_{e\in E}$ and $\{\omega(e)\}_{e\in E}$ defined by Property (11.3) are the families of cycles and cocycles of $G(C)$ respectively. This is clear for $\gamma(e)$, which, by definition, has zero boundary in $G(C)$; as for $\omega(e)$, since it has, in the same way, zero boundary in $G^*(C)$, it is a cycle in $G^*(C)$, and hence is a cocycle in $G(C)$.

Hence Property (11.3) is satisfied by S by virtue of the orthogonality of cycles and cocycles in $G(C)$.

(c) Finally we show, conversely, that if a sequence S satisfies Property (11.3) then it is a crossing sequence.

Let P and Q be defined for S as in Property (11.3), giving rise to the mutually orthogonal sets $\gamma(e)$ and $\omega(e)$. Associate with S the graph $G(S)$ induced by the sequence S on the set E considered as a set of edges, S being a walk on the graph $G(S)$ such that an edge e is described twice in the same direction if $e\in P$, and once in each direction if $e\in Q$. To see that this is always possible, consider the process of drawing such a graph $G(S)$ edge by edge, adding the edges in the order in which the symbols occur in S. If an edge x is about to be drawn, and the next edge e has already been drawn, then the membership of e (in P or in Q) determines at which end vertex of e the edge x must end. Moreover, if the walk so far drawn ends at a vertex A, and the next edge f to be included in the walk is already drawn but is not incident with A, then one end of f (and again the membership of f indicates which one) must be identified with A. It follows from this that the vertex set V of $G(S)$ is also determined.

We now show that the sets $\gamma(e)$ and $\omega(e)$ are respectively the cycles and cocycles of $G(S)$.

First, $\gamma(e)\in\ker\partial$, since the boundary of $\gamma(e)$ is that of a closed walk in $G(S)$, by the definition of $\gamma(e)$ in Property (11.3).

Second, $\omega(e)\in(\ker\partial)^{\perp}$, since $\omega(e)$ is, by virtue of Property (11.3), orthogonal to each element of the family $\{\gamma(e)\}_{e\in E}$, which includes a base for $\ker\partial$, as we shall now show.

To this end, describe the sequence S from the beginning, and note the incidences of the edges that appear in S. Their $2|E|$ extremities reduce, by successive identifications as indicated above, to $|V|$ vertices of $G(S)$. With the exception of the first edge, every edge occurring for the first time in S gives rise to one such identification (with one end of the preceding edge)—$|E|-1$ identifications in all. An edge occurring for the second time may or may not imply an identification; let $e_1, e_2, \ldots, e_r$ be those that do. Hence the total number of identifications $2|E|-|V|$ is $|E|-1+r$, whence $r=|E|-|V|+1=\dim\ker\partial$. The second occurrence of an edge e_i produces the cycle $\gamma(e_i)$ and a vertex identification, and it follows that the cycle $\gamma(e_i)$ has a nonzero boundary in the graph induced by the portion of S up to, but not including, e_i, and hence $\gamma(e_i)$ is independent of the $\gamma(e_j)$ for $j<i$. Thus $\{\gamma(e_i)\}_{0\leqslant i\leqslant r}$ is a base for $\ker\partial$ by virtue of this independence and the dimension; and it is certainly included in $\{\gamma(e)\}_{e\in E}$.

Finally we show that $G(S)$ is a planar graph. From Property (11.3) we have, for every e,

$$\gamma(e)+\omega(e)=e,$$

where $\gamma(e)$ is a cycle of $G(S)$, and $\omega(e)$ a cocycle of $G(S)$, as proved above. Thus in virtue of Theorem 2.1, $G(S)$ is a graph without bicycles whose principal cycles are the $\gamma(e)$ and whose principal cocycles are the $\omega(e)$. Moreover

$$\{e \mid e \in \gamma(e)\} = P, \qquad \{e \mid e \in \omega(e)\} = Q.$$

Hence S is an algebraic diagonal for $G(S)$ as defined in Section 10. Hence, by Theorem 10.1, $G(S)$ is planar. It admits a plane representation in which the geometric diagonal crosses the edges in the order given by the sequence S. Hence S is a crossing sequence.

Theorem 11.2 above does more than just give a necessary and sufficient condition for a sequence to be a crossing sequence; it gives also a practical method for constructing a curve C of which the sequence S is the crossing sequence. This can be done in three stages.

Stage 1. Find classes P and Q. From Theorem {11.2(iii) we see that if e and f are interlaced, then they will belong in the same class, if and only if $S^e \cap S^f$ is odd. Choose any partition (P, Q) consistent with these requirements. (If $I(S)$ is not connected then each of its components will be partitioned, and there will be more than one way of putting the parts together to form P and Q). Thus for the sequence *abcdbaeedc* of Fig. 11.1 we may take $P=\{c, d, e\}$ and $Q=\{a, b\}$.

Stage 2. Construct $G(S)$. In S, underline the second occurrence of elements of Q, to indicate which edges are to be traversed once in each direction. Place a prime (′) against every other element of S, to indicate that an edge following a primed edge is the first one encountered when one turns in an agreed sense at their common vertex (any, for example, that it is the 'first on the left' following the primed edge), and that an edge following an unprimed edge will be the 'first on the right'. From the sequence above we now have $a'bc'd\underline{b}'\underline{a}c'ed'c$.

If $G(S)$ exists it is determined uniquely by this sequence (see Fig. 11.4) and S is

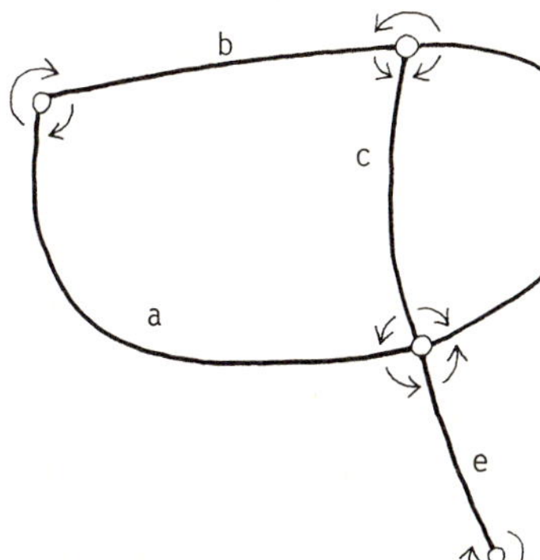

Fig. 11.4

a walk described by the 'first on the left–first on the right' algorithm of Shank [11].

Stage 3. Construct C. The geometric diagonal C can now be traced by going along each edge, first along one side as far as the midpoint and then crossing to the other side to complete the traversal of the edge. Since every edge is traversed twice, this procedure effectively replaces each edge by two portions of C which cross at the midpoint, these portions being joined up near the vertices of $G(S)$ in the manner shown in Fig. 11.5.

In this way a curve C which gives rise to the sequence S is constructed.

The analogous problem for several closed curves in the plane that intersect themselves and each other can be solved by reducing it to the problem for a single curve. We now have a family $C=\{C_i\}$, $i=1,2,\ldots,k$, of curves, supposed oriented, and for each curve C_i there is a sequence S_i of points of intersection in the cyclic order in which they are encountered on C_i, these points being represented by symbols from a set E. Each symbol in E occurs exactly twice in the collection $S=\{S_i\}$.

As before, these curves define regions of the plane, colourable in two colours, and we denote by $G(C)$ the graph defined by the regions of one colour. This graph has, as diagonals, the family C of curves. Denote by $|S_i|$ the set of symbols occurring exactly once in S_i. We may assume, without loss of generality, that the family of sets $\{|S_i|\}$ is connected, since otherwise the problem will break up into smaller independent problems for which the assumption is true.

Now any $k-1$ sets $|S_i|$ form a base for the bicycle space of $G(C)$. Choose a dendroid of this space by taking a minimal transversal of the $|S_i|$. For example, choose an element a_1 of $|S_1|$; it will belong to some other set, say $|S_2|$. If $k=2$, we are finished. If not, then $|S_1|+|S_2|\neq \mathbf{0}$, and we choose $a_2\in|S_1|+|S_2|$. Now a_2 belongs to some other $|S_i|$, say $|S_3|$. We now choose a_3 in $|S_1|+|S_2|+|S_3|$, and so on.

If, in $G(C)$, we bisect the edge corresponding to the intersection a_1, replacing it by two edges, a_1' and a_1'' in series, the the two curves C_1 and C_2 are 'merged' to

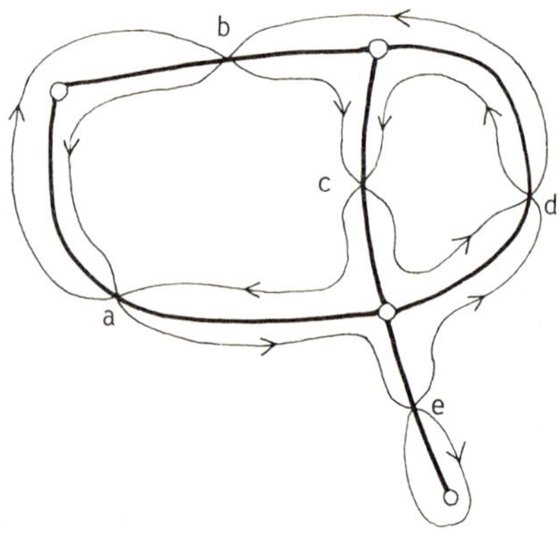

Fig. 11.5

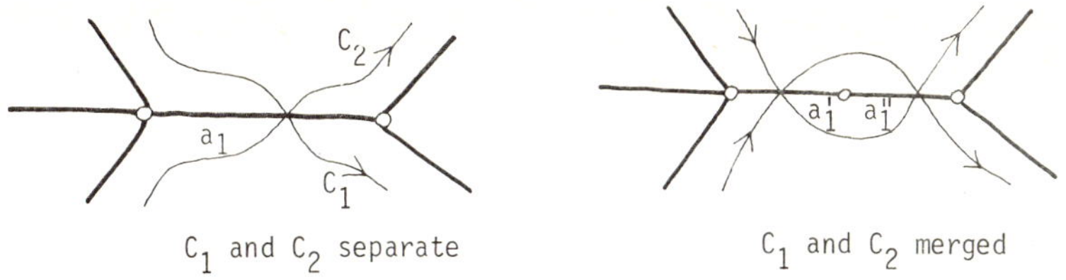

Fig. 11.6

become a single curve, and the element pair $a_1'a_1''$ is traversed twice in this order (see Fig. 11.6). Bisection of a_2 will now merge this curve with C_3, and so on. In this way, by successive mergings, we eventually obtain a single curve.

Conversely, given a family of k sequences $\{S_i\}$ over a set E, such that each element of E occurs exactly twice, we can determine whether it is representable as the set of points of intersection of k closed curves in the plane. We do so by reducing the problem to that of a single curve by successive mergings of one curve with another, as described above. The corresponding operation on the sequences is as follows. If a_1 is the chosen element of $|S_1|+|S_2|$, we replace its occurrence in S_2 by the sequence $a_1'S_1a_1''$, and drop the sequence S_1 from the family. This process is repeated with the chosen elements a_2, a_3, etc. until a single sequence S^* is obtained. From this we construct the graph $G(S^*)$, and, on suppressing the midpoints of the chosen edges, we obtain the graph $G(S)$ whose set of diagonals is the required family of curves.

This process of reducing a set of curves to a single curve is similar, though not identical, to the procedure, based on Dehn's algorithm, described in [7] for the same problem.

Acknowledgement

We thank our colleagues D. Bresson, J.C. Fournier, G.Th. Guilbaud and H. Shank, who all encouraged the supremacy of bicycles.

References

[1] C. Berge, Graphes et Hypergraphes (Dunood, Paris, 1973).
[2] W.K. Chen, On vector spaces associated with a graph, SIAM J. Appl. Math. 20 (3) (May 1971).
[3] H.de Fraysseix, Thèse de doctorat de troisième cycle, Ecole des Hautes Etudes en Sciences Sociales, Paris (1977).
[4] M. Dehn, Uber Kombinatorische Topologie, Acta Math. 67 (1936) 123–168.
[5] C.F. Gauss, Werke (Teubner, Leipzig, 1900) 272, 282–286.
[6] S. Lang, Algebra (Addison Wesley, Reading, MA, 1965).
[7] R.C. Read and P. Rosenstiehl, On the Gauss crossing problem, Proc. Fifth Hungarian Comb. Colloq. (to appear).
[8] P. Rosenstiehl, Bicycles et diagonales des graphes planaires, Cahiers du Cero 17 (Brussels 1975) 365–383.

[9] P. Rosenstiehl, Caractérisation des graphes planaires par une diagonale algébrique, Comp. Rend. Acad. Sci. Paris Sér. A. 283 (1976) 417–419.
[10] P. Rosenstiehl, Solution algébrique du problème de Gauss sur la permutation des points d'intersection d'une ou plusieurs courbes fermées du plan, Comp. Rend. Acad. Sci. Paris Sér. A. 283 (1976) 551–553.
[11] H. Shank, The theory of left-right paths, Combinatorial Mathematics III, Lecture notes in Mathematics 452 (Springer Berlin, 1975) 42–54.
[12] W.T. Tutte, On dichromatic polynomials, J. Combinatorial Theory 2 (1967) 301–320.
[13] H. Whitney, Non-separable and planar graphs Trans. Amer. Math. Soc. 34 (1932) 339–362.

Annals of Discrete Mathematics 3 (1978) 227–245.

PERCOLATION PROBABILITIES ON THE SQUARE LATTICE

P.D. SEYMOUR and D.J.A. WELSH

Merton College, Oxford, England

1. Introduction

This paper deals mainly with bond percolation on the square lattice. This model is a special but perhaps the most interesting case of the general theory of percolation introduced by Broadbent and Hammersley [4] in 1957. In Section 2 we review briefly the general percolation model; for further details see Frisch and Hammersley [13], Shante and Kirkpatrick [24], Essam [9] or Welsh [29].

In Section 3 we introduce the FKG inequality of Fortuin, Kasteleyn and Ginibre [12]. In Section 4 we introduce the problem of percolation through an $n \times n$ sponge (loosely speaking, when is it possible to move from one side to another of a randomly dammed chessboard?). We examine two of the possible critical probabilities $\boldsymbol{p}_{\mathrm{T}}$, $\boldsymbol{p}_{\mathrm{H}}$ defined in [29] and use the theory developed for the sponge problem to prove the result

$$\boldsymbol{p}_{\mathrm{T}} + \boldsymbol{p}_{\mathrm{H}} = 1.$$

Since Harris [18] has proved $\boldsymbol{p}_{\mathrm{H}} \geq \frac{1}{2}$ and since intuitively one expects the numbers to be equal this suggests that all the critical probabilities for bond percolation on the square lattice have the common value $\frac{1}{2}$.

2. The percolation model

If G is a graph, finite or infinite, we let $V = V(G)$ be its set of vertices and $E = E(G)$ its set of edges. The little graph terminology we use is standard (see for example Berge [2] or Bondy and Murty [3]).

By the *percolation model* on G we mean the assignment of *open* or *closed* to each edge of G with probabilities $\boldsymbol{p}$ and $\boldsymbol{q} = 1 - \boldsymbol{p}$ respectively, the assignments to be independent for each edge. If an edge is open we picture it as allowing fluid to pass along it; if closed it does not allow fluid to move along it. Thus if A is any subset of edges of the finite graph G, the probability that A is exactly the set of open edges is

$$\pi(A) = \boldsymbol{p}^{|A|} \boldsymbol{q}^{|E \setminus A|}.$$

If Ω denotes the set of all possible assignments, we identify a typical member ω of Ω with the subset of edges which are open in ω. We shall be dealing throughout with a graph G in which $E(G)$ is at most countable and the random variables are on the space Ω. There is never any problem with the measurability or lack of it for the random variables which we shall be discussing and hence we shall usually write X for $X(\omega)$ and so on. For details of similar such arguments see for example [17].

If G is a graph and A, B are subsets of $V(G)$ and U is a subgraph of G,

$$\{A \overset{U}{\rightsquigarrow} B\}$$

denotes the fact that there is a path lying entirely in U which connects some vertex x in A to some vertex y in B. Occasionally we abuse notation and U is not a subgraph of G but just a set of vertices. In such cases we interpret the expression as $\{A \overset{\hat{U}}{\rightsquigarrow} B\}$ where $\hat{U}$ is the graph induced by U.

If Ω is the probability space of the percolation model on G the *event* $\{A \rightsquigarrow B\}$ is the event of Ω that there is some path of open edges linking a vertex of A to a vertex of B.

Throughout $\mathscr{L}$ will denote the square lattice, that is the set of points (x, y) of the plane having integer coordinates x and y and having edges joining each point (x, y) to its nearest neighbours $(x+1, y)$, $(x-1, y)$, $(x, y-1)$, $(x, y+1)$.

As usual in this theory it is convenient to regard $\mathscr{L}$ as the "limit" of a sequence of finite graphs. One suitable sequence is $(\mathscr{L}_n\colon 0 \leq n < \infty)$ where $\mathscr{L}_n$ is the restriction of $\mathscr{L}$ to the set of vertices $\{(x, y)\colon -n \leq x \leq n, -n \leq y \leq n\}$. $\mathscr{L}$ itself is self-dual; that is, if we consider a new infinite graph $\mathscr{L}^*$ whose vertices are the points $(x+\frac{1}{2}, y+\frac{1}{2})$ where x, y run through the integers, and whose edges are again those lines joining nearest neighbours, then $\mathscr{L}^*$ has the following properties.

(a) It is isomorphic to $\mathscr{L}$.

(b) There is an obvious geometric duality between $\mathscr{L}$ and $\mathscr{L}^*$ inasmuch as they can be drawn as geometric duals in the plane, see for example [3].

Almost exclusively in this paper we shall restrict ourselves to percolation on $\mathscr{L}$, or some sequence of subgraphs of $\mathscr{L}$ which approach $\mathscr{L}$.

Suppose we now regard the origin O as a source of fluid. We say that a point v of $\mathscr{L}$ is *wet* by fluid from the origin if there is a path consisting of open edges from O to v, and otherwise v is *dry*.

Let us now fix $\boldsymbol{p}$, $0 \leq \boldsymbol{p} \leq 1$. We let $\boldsymbol{P}_n(\boldsymbol{p})$ be the probability that at least n points are wet by fluid from the origin. Clearly

$$\boldsymbol{P}_n(\boldsymbol{p}) \geq \boldsymbol{P}_{n+1}(\boldsymbol{p})$$

so that

$$\boldsymbol{P}(\boldsymbol{p}) = \lim_{n\to\infty} \boldsymbol{P}_n(\boldsymbol{p})$$

exists, and satisfies

$$0 \leqslant \boldsymbol{P}(\boldsymbol{p}) \leqslant 1.$$

However, though each $\boldsymbol{P}_n(\boldsymbol{p})$ is a polynomial in $\boldsymbol{p}$ and can be calculated, it still is not known for example whether or not $\boldsymbol{P}(\boldsymbol{p})$ is continuous in $\boldsymbol{p}$. Broadbent and Hammersley [4] show that there exists a *critical probability* $\boldsymbol{p}_{\mathrm{H}}$ defined by

$$\boldsymbol{p}_{\mathrm{H}} = \inf \boldsymbol{p} \colon \boldsymbol{P}(\boldsymbol{p}) > 0.$$

Harris [18] proved that

$$\boldsymbol{p}_{\mathrm{H}} \geqslant \frac{1}{2} \tag{1}$$

and Hammersley [16] that

$$\boldsymbol{p}_{\mathrm{H}} \leqslant 0.646790. \tag{2}$$

As pointed out in [29] there are several other "critical probabilities" in the literature, and the relationships among them are obscure to say the least. First consider $V(\boldsymbol{p})$, the expected number of points wet by the source at the origin—that is,

$$V(\boldsymbol{p}) = \sum_{n \geqslant 1} \boldsymbol{P}_n(\boldsymbol{p}).$$

We define $\boldsymbol{p}_{\mathrm{T}}$ by

$$\boldsymbol{p}_{\mathrm{T}} = \inf \boldsymbol{p} \colon V(\boldsymbol{p}) = \infty.$$

Since $V(\boldsymbol{p})$ is infinite if $\boldsymbol{P}(\boldsymbol{p}) > 0$, we have immediately that

$$\boldsymbol{p}_{\mathrm{T}} \leqslant \boldsymbol{p}_{\mathrm{H}}. \tag{3}$$

One of our results below will be that

$$\boldsymbol{p}_{\mathrm{T}} \leqslant \frac{1}{2}. \tag{4}$$

This has an easy proof, but also follows from our main theorem:

Theorem 2.1. *In percolation on the square lattice the critical probabilities* $\boldsymbol{p}_{\mathrm{T}}$, $\boldsymbol{p}_{\mathrm{H}}$ *satisfy* $\boldsymbol{p}_{\mathrm{T}} + \boldsymbol{p}_{\mathrm{H}} = 1$.

Our proof of this is quite long and is given in Section 5. Thus on an intuitive level at least there is strong evidence to support the following conjecture.

Conjecture 2.2. $p_T = p_H = \frac{1}{2}$.

We should emphasize that for several years there has been a folklore belief that the above conjecture was proved by Sykes and Essam [25] in 1964. Sykes and Essam in fact show that under certain (as yet unproven) assumptions a *third* quantity p_E associated with percolation on the square lattice is equal to $\frac{1}{2}$. Although various attempts have been made (see for example Grimmett [14]) to prove that the assumptions demanded by Sykes and Essam are correct it is a much more difficult (in fact, as far as we can see, hopelessly intractable) problem to relate p_E with p_H or p_T. Even the very definition of p_E is shrouded with mystery.

3. The FKG inequality

In 1971 Fortuin, Kasteleyn and Ginibre [12] proved a remarkable inequality showing that non-decreasing functions on a finite distributive lattice are positively correlated by all positive measures which have a certain convexity property. This inequality was originally applied to Ising ferromagnets in an arbitrary magnetic field, but as pointed out in [12] it is also closely related to a lemma used by Harris [18] in proving Theorem 2.1. In [23] we showed that the inequality has diverse applications in combinatorial theory, and Kempermann [20] has given some new applications in probability theory. In this section we shall use it to obtain some new results in percolation, first passage percolation, and random graph theory. It is also used repeatedly in the proof of our main result in Section 5.

Two random variables X and Y are *covariant* if $\mathscr{E}(XY) \geq (\mathscr{E}X)(\mathscr{E}Y)$. Two events A, B are *covariant* if their respective indicator functions are covariant. Clearly (if $\boldsymbol{P}(B) \neq 0$) A, B are covariant if and only if

$$\boldsymbol{P}(A \mid B) \geq \boldsymbol{P}(A).$$

A set $\{X_1, \ldots, X_k\}$ of random variables is *covariant* if for any subset $I \subseteq \{1, \ldots, k\}$, $\mathscr{E}(\prod_{i \in I} X_i) \geq \prod_{i \in I} \mathscr{E}(X_i)$.

Let D be a distributive lattice, where obviously we are using "lattice" in its algebraic sense. A function $f: D \to R$ is called *increasing* if $f(x) \leq f(y)$ for any pair of elements x, y of D such that $x \leq y$. A function f is *decreasing* if $-f$ is increasing.

When D is finite and $\mu: D \to R^+$, the *μ-average* of a function $f: D \to R$ is given by

$$\langle f \rangle = \langle f \rangle_\mu = \Big(\sum_{x \in D} f(x)\mu(x) \Big) \Big/ \sum_{x \in D} \mu(x).$$

The original version of the FKG inequality proved in [12] is as follows.

Theorem 3.1 (The FKG inequality). *Let D be a finite distributive lattice and let $\mu : D \to R^+$ satisfy*

$$\mu(x)\mu(y) \leq \mu(x \wedge y)\mu(x \vee y) \quad (x, y \in D). \tag{5}$$

Then if f, g are both increasing or both decreasing functions, then

$$\langle fg \rangle \geq \langle f \rangle \langle g \rangle. \tag{6}$$

An obvious corollary of this is that if f and g are functions on D which are monotone but in the opposite sense, then

$$\langle fg \rangle \leq \langle f \rangle \langle g \rangle.$$

Before proceeding to give some applications of Theorem 3.1 we prove a lemma. The proof is elementary, but we give it because we use the result several times later.

Lemma 3.2. *If A_1, A_2 are covariant events in Ω with $\boldsymbol{P}(A_1) = \boldsymbol{P}(A_2)$ then*

$$\boldsymbol{P}(A_1) \geq 1 - [1 - \boldsymbol{P}(A_1 \cup A_2)]^{1/2}$$

Proof.

$$\begin{aligned}\boldsymbol{P}(A_1 \cup A_2) &= \boldsymbol{P}(A_1) + \boldsymbol{P}(A_2 \mid \Omega \setminus A_1)\boldsymbol{P}(\Omega \setminus A_1)\\ &\leq \boldsymbol{P}(A_1) + \boldsymbol{P}(A_2)[1 - \boldsymbol{P}(A_1)],\end{aligned}$$

since A_1, A_2 are covariant. Hence since $\boldsymbol{P}(A_1) = \boldsymbol{P}(A_2)$ we have

$$1 - \boldsymbol{P}(A_1 \cup A_2) \geq [1 - \boldsymbol{P}(A_1)]^2,$$

which completes the proof.

Example 3.3 (Random graphs). For each positive integer n let D_n be the lattice of subsets of E_n, the set of edges of the complete graph K_n. Now let μ be defined as

$$\mu A = \boldsymbol{p}^{|A|}\boldsymbol{q}^{|E \setminus A|}.$$

Consider the following events about the random graphs ω on n vertices in which each edge of K_n exists or does not exist with probabilities $\boldsymbol{p}$, $1 - \boldsymbol{p}$;

A: ω is planar,

B: ω is hamiltonian,

C: ω is 4-colourable.

It is clear that whereas A and C have decreasing indicator functions, B has an increasing indicator function. Hence the FKG inequality gives such statements as

$$\begin{aligned}&\boldsymbol{P}[\text{random graph } \omega \text{ is hamiltonian} \mid \omega \text{ is planar}]\\ &\quad \leq \boldsymbol{P}[\text{random graph } \omega \text{ is hamiltonian}].\end{aligned} \tag{7}$$

$$\boldsymbol{P}[\text{random graph } \omega \text{ is 4 colourable} \mid \omega \text{ is hamiltonian}]$$
$$\leq \boldsymbol{P}[\text{random graph } \omega \text{ is 4 colourable}]. \tag{8}$$

Although intuitively appealing, such results do not seem easy to prove directly and serve to indicate the power of the FKG inequality.

Now the reader will notice that in the FKG inequality as stated in Theorem 3.1 the lattice D is restricted to being finite. Various infinite extensions of the inequality and of a stronger result of Holley [19] have been made recently by Batty [1], Cartier [5], Edwards [7], Kempermann [20] and Preston [22]. However, as far as the main theorems of this paper are concerned the only infinite extension we need is the following covariance inequality first proved by Fortuin [10].

Theorem 3.4. *Let G be a countable graph and let* $\boldsymbol{P}$ *be the probability measure induced by a percolation model on G. Let f and g be increasing functions on the partially ordered probability space associated with this model. Then if $\mathscr{E}$ is the expectation operator associated with* $\boldsymbol{P}$,

$$\mathscr{E}(fg) \geq \mathscr{E}(f)\mathscr{E}(g)$$

whenever the expectations exist.

Immediately from this we see that the results obtained in Example 3.3 above hold when G is a countably infinite graph.

We close this section by sketching a proof of an extension of Harris' correlation result to first passage percolation theory as defined by Hammersley and Welsh [17]. One interest of this extension is that Theorem 3.5 below was the original "physical result" which motivated Batty's infinite extension [1] of the FKG inequality.

Let G be a (finite or countably infinite) graph directed or undirected, with vertex set V and edge set E. Suppose that to each edge e_i of G we assign a random variable u_i drawn, independently for each edge, from a distribution $F(x)$. We call u_i the *time coordinate* of e_i.

The set Ω of E-tuples ω, defined by $\omega(e_i) = u_i$, $e_i \in E$, is called the *phase space* and can be ordered by

$$\omega \leq \omega' \Leftrightarrow \omega(e_i) \leq \omega'(e_i) \quad \forall e_1 \in E.$$

If x, y are any two vertices of G we write $t_{xy}(\omega)$ to denote the *first passage* (*shortest*) *time* between x and y over paths of G, when it is in state ω. More precisely

$$t_{xy}(\omega) = \inf t(P, \omega),$$

where $t(P, \omega)$ is the sum of the time coordinates of the edges making up the path P, and the infimum is over all paths P joining x and y.

Now for any points x_1, x_2, y_1, y_2 of $V(G)$ it is obvious that $t_{x_1x_2}(\omega)$ and $t_{y_1y_2}(\omega)$

are monotone on Ω, in the sense that

$$\omega \leq \omega' \Rightarrow t_{x_1x_2}(\omega) \leq t_{x_1x_2}(\omega').$$

Thus we can apply the infinite version of the FKG inequality implicit in the work of Batty [1] and Edwards [7] to get the result that the pair of random variables $t_{x_1x_2}$ and $t_{y_1y_2}$ are covariant. More generally, if A, B are two subsets of V and

$$t_{AB}(\omega) = \inf_{\substack{x \in A \\ y \in B}} t_{xy}(\omega)$$

represents the first passage time between A and B when G is in state ω we have the following general result:

Theorem 3.5. *For any sets A, B, C, D of vertices of the countable graph G the first passage times t_{AB} and t_{CD} are covariant random variables.*

4. The sponge problem

In this section we consider a new variant of the percolation problem. It is of some interest in its own right; indeed we studied it purely for its own sake before realising that it was a useful tool in giving insight into the relationship between $\boldsymbol{p}_T$ and $\boldsymbol{p}_H$. Most of the results of this section will be used in proving our main result, Theorem 2.1. The vertex or atom percolation version of this problem has also been studied numerically by Kurkijarvi and Padmore [21]. However, they assume as physically obvious certain results which we have found impossible to prove rigorously.

The $m \times n$ *sponge* consists of the subgraph $T(m, n)$ of $\mathcal{L}$ induced on the mn points

$$\{(x, y): 1 \leq x \leq n, 1 \leq y \leq m\}.$$

Each of the m points $(1, y)$, $1 \leq y \leq m$, is regarded as an infinite source of fluid which may percolate through those edges of the sponge which are open. The probability that any edge is open is $\boldsymbol{p}$, independently for each edge.

We let $S_{\boldsymbol{p}}(m, n) = S(m, n)$ denote the probability that some of the points (n, k), $1 \leq k \leq m$, become wet by fluid.

Trivial inequalities are

$$S(m, n+1) \leq S(m, n), \tag{9}$$

$$S(m, n) \leq S(m+1, n). \tag{10}$$

A basic, but extremely useful, result is the following.

Theorem 4.1. *For all* $\mathbf{p}$, $0 \leq \mathbf{p} \leq 1$, *and all positive integers* $m \geq 1$, $n \geq 2$,

$$S_{\mathbf{p}}(m, n) + S_{\mathbf{q}}(n-1, m+1) = 1,$$

where $\mathbf{q} = 1 - \mathbf{p}$.

Proof. Construct a new graph $G(m, n)$ from the $m \times n$ sponge $T(m, n)$ as follows. Identify all the vertices $(1, y)$, $1 \leq y \leq m$, in a new vertex x_1. (Remove all edges which become loops.) Similarly identify all vertices (n, y), $1 \leq y \leq m$, in a vertex x_2. Add a new edge e joining x_1 and x_2. The graph $G(m, n)$ is planar, and its planar dual G^* is isomorphic to $G(n-1, m+1)$. Now consider any assignment ω of open and closed values to the edges of $T(m, n)$. There is a path of open edges from one of the vertices $(1, y)$, $1 \leq y \leq m$, to one of (n, y), $1 \leq y \leq m$, if and only if there is a cycle in $G(m, n)$ consisting of e and otherwise edges which are open in ω. But, by the elementary max-flow min-cut theorem, either there is such a cycle in $G(m, n)$, or there is a cycle in G^* consisting of e and otherwise edges closed in ω (and not both). But since G^* is isomorphic to $G(n-1, m+1)$, and an edge of $T(m, n)$ is closed with probability $\mathbf{q}$, the result follows.

Hence if we define

$$S_n(\mathbf{p}) = S_{\mathbf{p}}(n, n+1),$$

we have for all positive integers n,

$$S_n(\mathbf{p}) + S_n(1-\mathbf{p}) = 1 \quad (0 \leq \mathbf{p} \leq 1).$$

In particular

$$S_n(\tfrac{1}{2}) = \tfrac{1}{2} \quad (1 \leq n < \infty). \tag{11}$$

It is also clear that $S_n(\mathbf{p})$ is a monotonic increasing function of $\mathbf{p}$, satisfying for all n,

$$S_n(0) = 0, \qquad S_n(1) = 1.$$

However we have *not* been able to prove:

Conjecture 4.2. *For all* $\mathbf{p}$, $0 \leq \mathbf{p} \leq 1$, $\lim S_n(\mathbf{p})$ *exists. (We have shown that, even if the limit always exists, it is not continuous.)*

Conjecture 4.3. *For* $\mathbf{p} < \frac{1}{2}$ *(respectively* $> \frac{1}{2}$*),* $S_n(\mathbf{p})$ *is a monotone decreasing (respectively increasing) function of* n.

We now relate $S_n(\mathbf{p})$ with $\mathbf{P}_n(\mathbf{p})$.

Theorem 4.4. *For any positive integer* n *and* $0 \leq \mathbf{p} \leq 1$

$$S_n(\mathbf{p}) \leq 1 - (1 - \mathbf{P}_{n+1}(\mathbf{p}))^n.$$

Proof. Consider the $n \times (n+1)$ sponge and let

$$X = \{(x, y): x = 1, 1 \leq y \leq n\},$$
$$Y = \{(x, y): x = n+1, 1 \leq y \leq n\}.$$

Then

$$1 - S_n(\boldsymbol{p}) = \boldsymbol{P}(X \overset{T_n}{\nrightarrow} Y)$$
$$\geq \boldsymbol{P}\Big(\bigcap_{i=1}^{n} A_i\Big),$$

where $A_i = \{(1, i) \overset{\mathscr{L}}{\nrightarrow} Y\}$, $1 \leq i \leq n$.

But by the FKG inequality the A_i are covariant events, each having probability $\geq 1 - \boldsymbol{P}_{n+1}(\boldsymbol{p})$. Hence

$$1 - S_n(\boldsymbol{p}) \geq (1 - \boldsymbol{P}_{n+1}(\boldsymbol{p}))^n$$

and the result follows.

Suppose now we define the *critical sponge probability* $\boldsymbol{p}_s$ by

$$\boldsymbol{p}_s = \inf \boldsymbol{p}: \limsup_{n \to \infty} S_n(\boldsymbol{p}) > 0.$$

Then we know from (11) that

$$\boldsymbol{p}_s \leq \tfrac{1}{2}. \tag{12}$$

It will follow from the proof of Theorem 2.1 that

$$0.353210 \leq \boldsymbol{p}_T \leq \boldsymbol{p}_s. \tag{13}$$

One final result which we need before proving the main theorem is the following: For any n,

$$S(n, n) \leq 8S(n-1, n-1). \tag{14}$$

To see this consider the $n \times n$ sponge. If there is a path across it, then there must be a path across one of the four $(n-1) \times (n-1)$ sponges inside it or there must be a path from the top to the bottom of one of these sponges. Considering the union of these events gives (14).

5. Proof of Theorem 2.1

We shall prove Theorem 2.1 by the series of Lemmas 5.1–5.6 below. However, it is probably instructive to show the broad outline here.

First in Lemma 5.1, which is relatively straightforward, we show that

$$\boldsymbol{p}_T + \boldsymbol{p}_H \leq 1. \tag{15}$$

In Lemmas 5.2–5.4 we prove various inequalities about the $S(m, n)$ which enable us to show in Lemma 5.5 that if $\boldsymbol{p} < 1 - \boldsymbol{p}_{\mathrm{H}}$ not only does the sequence $S_n(\boldsymbol{p})$ converge, but

$$\lim_{n\to\infty} S_n(\boldsymbol{p}) = 0.$$

In Lemma 5.6, we show that if $\boldsymbol{p} > \boldsymbol{p}_{\mathrm{T}}$,

$$\limsup_{n\to\infty} S_n(\boldsymbol{p}) \geq \delta > 0.$$

Thus we have $1 - \boldsymbol{p}_{\mathrm{H}} \leq \boldsymbol{p}_{\mathrm{T}}$ which with (15) proves our final result that

$$\boldsymbol{p}_{\mathrm{T}} + \boldsymbol{p}_{\mathrm{H}} = 1.$$

Lemma 5.1. $\boldsymbol{p}_{\mathrm{T}} + \boldsymbol{p}_{\mathrm{H}} \leq 1$.

Proof. Let L be the set of points $\{(i, 0)\colon i \geq 0\}$ of $\mathscr{L}$ and for $1 \leq n < \infty$ let L_n be the set of points $\{(-i, 0)\colon i \geq n\}$. We choose a fixed $\boldsymbol{p} < \boldsymbol{p}_{\mathrm{T}}$; and then

$$\sum_{i=1}^{\infty} \boldsymbol{P}_i(\boldsymbol{p}) < \infty.$$

Choose N so that $\sum_{i \geq N} \boldsymbol{P}_i(\boldsymbol{p}) < 1$. Now

$$\boldsymbol{P}\{(-i, 0) \overset{\mathscr{L}}{\rightsquigarrow} L\} = \boldsymbol{P}\{(0, 0) \overset{\mathscr{L}}{\rightsquigarrow} L_i\} \leq \boldsymbol{P}_i(\boldsymbol{p}).$$

Hence

$$\sum_{i \geq N} \boldsymbol{P}((-i, 0) \overset{\mathscr{L}}{\rightsquigarrow} L\} < 1.$$

Hence

$$\boldsymbol{P}\{L_N \overset{\mathscr{L}}{\rightsquigarrow} L\} < 1.$$

Now let B_i $(1 \leq i < \infty)$ be the points $(-i + \frac{1}{2}, \frac{1}{2})$ $(0 \leq i < \infty)$ which are the vertices of the dual lattice $\mathscr{L}^*$. For each assignment ω of open and closed to the edges of $\mathscr{L}$ we will consider $\mathscr{L}^*$ in state ω^*, where if e is closed in $\mathscr{L}$ under ω then the corresponding edge e^* of $\mathscr{L}^*$ is closed in $\mathscr{L}^*$.

Let $B^* = B^*(\omega)$ be the set of points of the dual lattice which are joined by a path of closed edges of $\mathscr{L}^*$ to one of $B_1, \ldots, B_N$. Suppose that we assume that with probability one B^* is finite. Then if B^* is finite let P^* be those edges of $\mathscr{L}^*$ joining vertices of B^* to vertices of $\mathscr{L}^* \backslash B^*$. Then every edge in P^* must be open in $\mathscr{L}^*$.

Now since B^* is finite P^* must cut L_N and must also cut L. Hence by elementary graph theory arguments there is an open path in $\mathscr{L}$ connecting L_N with L. But we have chosen N so that the event $L_N \rightsquigarrow L$ has probability strictly less than 1. Hence the assumption that B^* is finite with probability one is false and we must have

$$\boldsymbol{P}(|B^*|=\infty)>0.$$

But if for $1 \leqslant i \leqslant N$ we let

$$A_i = \{\omega: B_i \text{ is connected in } \mathscr{L}^* \text{ by a closed path to an infinite number of points of } \mathscr{L}^*\},$$

then

$$\boldsymbol{P}(|B^*|=\infty) \leqslant \sum_{i=1}^{N} \boldsymbol{P}(A_i).$$

Since $\boldsymbol{P}(A_i)=\boldsymbol{P}(\boldsymbol{q})$, we must have

$$N\boldsymbol{P}(\boldsymbol{q})>0,$$

which implies

$$\boldsymbol{q} \geqslant \boldsymbol{p}_{\mathrm{H}}$$

so that $\boldsymbol{p}_{\mathrm{T}}+\boldsymbol{p}_{\mathrm{H}} \leqslant 1$ as required.

Lemma 5.2. *If* $S(2n, 2n)=\tau$, *then* $S(2n, 4n)) \geqslant \tau(1-(1-\tau)^{1/2})^8$.

Proof. Consider the following regions of the square lattice.

$$R=\{(x, y): 1 \leqslant x \leqslant 4n, 1 \leqslant y \leqslant 2n\},$$
$$X=\{(x, y): x=1, 1 \leqslant y \leqslant 2n\},$$
$$Z=\{(x, y): x=2n, 1 \leqslant y \leqslant 2n\},$$
$$W=\{(x, y): x=n+1, 1 \leqslant y \leqslant 2n\},$$
$$W_1=\{(x, y): x=n+1, 1 \leqslant y \leqslant n\},$$
$$W_2=\{(x, y): x=n+1, n+1 \leqslant y \leqslant 2n\},$$
$$U_1=\{(x, y): n+1 \leqslant x \leqslant 3n, y=1\},$$
$$U_2=\{(x, y): n+1 \leqslant x \leqslant 3n, y=2n\},$$
$$S_1=\{(x, y): 1 \leqslant x \leqslant 2n, 1 \leqslant y \leqslant 2n\},$$
$$S=\{(x, y): n+1 \leqslant x \leqslant 3n, 1 \leqslant y \leqslant 2n\}.$$

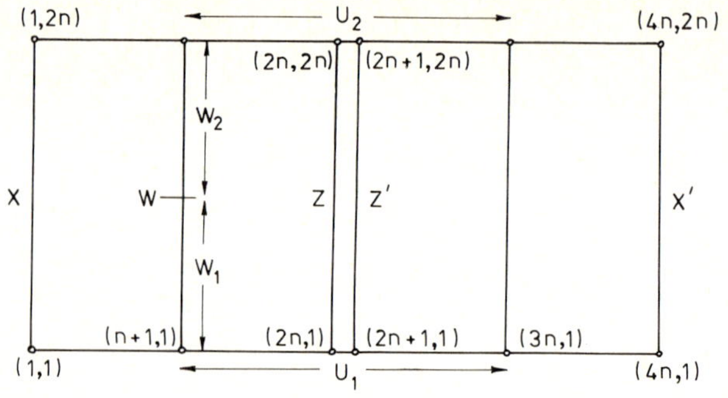

Fig. 1.

We illustrate the situation in Fig. 1.

Now for any subset of vertices A of $\mathscr{L}$ let A' be defined by

$$A' = \{(4n+1-x,\ y): (x, y) \in A\}.$$

so that for example

$$W' = \{(x, y): x = 3n,\ 1 \leq y \leq 2n\},$$

$$X' = \{(x, y): x = 4n,\ 1 \leq y \leq 2n\},$$

$$S_1' = \{(x, y): 2n+1 \leq x \leq 4n,\ 1 \leq y \leq 2n\}.$$

Consider now the events A_1, A_2, A_3 of Ω defined by

$$A_1 = \{\omega: W \overset{S}{\rightsquigarrow} W'\}$$

$A_2 = \{\omega$: there is an open path from X to Z in S_1 which meets an open path from U_1 to U_2 in $S\}$,

$A_3 = \{\omega$: there is an open path from X' to Z' in S_1' which meets an open path from U_1 to U_2 in $S\}$.

Then since A_1, A_2, A_3 are monotone in the same sense they are covariant and since also

$$A_1 \cap A_2 \cap A_3 \subseteq \{X \overset{R}{\sim} X'\},$$

we have

$$S(2n, 4n) \geq \boldsymbol{P}(A_1 \cap A_2 \cap A_2)$$
$$\geq \boldsymbol{P}(A_1)(\boldsymbol{P}(A_2))^2$$
$$= S(2n, 2n)(\boldsymbol{P}(A_2))^2.$$

We now consider $\boldsymbol{P}(A_2)$. We wish to show

$$\boldsymbol{P}(A_2) \geq (1-(1-\tau)^{1/2})^4.$$

Let $(P_i: 1 \leq i \leq k)$ be the collection of paths in S_1 which join X to Z and which have the additional property that their last point Q_i of intersection with W is a point of W_1. For $1 \leq i \leq k$ let F_i be the section of P_i from Q_i to Z. Then each F_i is a path from W_i to Z.

Let X_i be the event that there is an open path in S from F_i to U_2 which uses only one vertex of F_i and no vertex of F_i'. Let X_i' be the event that there is an open path in S from F_i' to U_2 which uses only one vertex of F_i' and no vertex of F_i.

Now the set of points $F_i \cup F_i'$ separates U_1 from U_2 in S. Hence if there is an open path in S from U_1 to U_2 then either X_i or X_i' occurs. Hence

$$\begin{aligned} \boldsymbol{P}(X_i \cup X_i') &\geq \boldsymbol{P}(U_1 \overset{S}{\rightsquigarrow} U_2) \\ &= S(2n, 2n) = \tau. \end{aligned}$$

But X_i, X_i' are covariant, and by symmetry have equal probabilities; hence by Lemma 3.2,

$$\boldsymbol{P}(X_i) = \boldsymbol{P}(X_i') \geq 1 - \surd(1-\tau).$$

Let us now fix i and consider the three events,

$$\begin{aligned} B_1 &= B_1^{(i)} = \{\omega\colon \text{path } P_i \text{ is open}\}, \\ B_2 &= B_2^{(i)} = \{\omega\colon \text{for each } j \neq i \text{ such that } P_j \text{ lies in the} \\ &\qquad\qquad \text{region bounded by } P_i \text{ and } y = 1,\ P_j \text{ is not open}\}, \\ B_3 &= B_3^{(i)} = X_i. \end{aligned}$$

We assert

$$\boldsymbol{P}(B_1 \cap B_2 \cap B_3) \geq (1 - \surd(1-\tau))\boldsymbol{P}(B_1 \cap B_2).$$

For

$$\boldsymbol{P}(B_1 \cap B_2 \cap B_3) = \boldsymbol{P}(B_2 \cap B_3 \mid B_1)\boldsymbol{P}(B_1),$$

and if B_1 occurs, then the occurrence of B_2 depends only on the state of the edges of $\mathscr{L}$ strictly below P_i in S_1, and the occurrence of B_3 depends only on the state of edges strictly above $F_i \cup F_i'$ in S. Since these two sets of edges are disjoint

$$\boldsymbol{P}(B_2 \cap B_3 \mid B_1) = \boldsymbol{P}(B_2 \mid B_1)\boldsymbol{P}(B_3 \mid B_1).$$

Hence

$$\begin{aligned} \boldsymbol{P}(B_1 \cap B_2 \cap B_3) &= \boldsymbol{P}(B_2 \mid B_1)\boldsymbol{P}(B_3 \mid B_1)\boldsymbol{P}(B_1) \\ &= \boldsymbol{P}(B_2 \cap B_1)\boldsymbol{P}(B_3 \mid B_1). \end{aligned}$$

But since B_1 and B_3 are covariant

$$\boldsymbol{P}(B_3 \mid B_1) \geq \boldsymbol{P}(B_3) = \boldsymbol{P}(X_i) \geq 1 - \surd(1-\tau).$$

Hence

$$\boldsymbol{P}(B_1 \cap B_2 \cap B_3) \geq (1 - \surd(1-\tau))\boldsymbol{P}(B_1 \cap B_2).$$

Now note that

$$\boldsymbol{P}((B_1^{(1)}\cap B_2^{(1)})\cup(B_1^{(2)}\cap B_2^{(2)})\cup\cdots\cup(B_1^{(k)}\cap B_2^{(k)}))$$
$$=\sum_1^k \boldsymbol{P}(B_1^{(i)}\cap B_2^{(i)}).$$

But

$$\boldsymbol{P}((B_1^{(1)}\cap B_2^{(1)})\cup\cdots\cup(B_1^{(k)}\cap B_2^{(k)}))=\boldsymbol{P}(B_1^{(1)}\cup\cdots\cup B_1^{(k)}),$$

and

$$\boldsymbol{P}\Big(\bigcup_i (B_1^{(i)}\cap B_3^{(i)})\Big)\geq\sum_{i=1}^k \boldsymbol{P}(B_1^{(i)}\cap B_2^{(i)}\cap B_3^{(i)})$$
$$\geq(1-\sqrt{(1-\tau)})\sum_{i=1}^k \boldsymbol{P}(B_1^{(i)}\cap B_2^{(i)})$$
$$=(1-\sqrt{(1-\tau)})\boldsymbol{P}(\text{at least one } P_i \text{ is open}).$$

Now consider the event C that at least one of the P_i is open. Let $\{\hat{P}_i: 1\leq i\leq k\}$ be the collection of paths in S_1 which join X to Z and which have the property that their last point of intersection with W is a point of W_2.

The event $\hat{C}$ that at least one of the $\hat{P}_i$ is open is covariant with C and by symmetry

$$\boldsymbol{P}(\hat{C})=\boldsymbol{P}(C).$$

Also $\boldsymbol{P}(\hat{C}\cup C)=S(2n,2n)=\tau$ so that by Lemma 5.1,

$$\boldsymbol{P}(C)\geq 1-\sqrt{(1-\tau)},$$

and

$$\boldsymbol{P}(B_1^{(i)}\cap B_3^{(i)} \text{ for some } i)\geq(1-\sqrt{(1-\tau)})^2 \tag{16}$$

Let E_1 be the event that there is a point $w\in W_1$ such that

$$w\overset{S_1}{\rightsquigarrow}X,\qquad w\overset{S}{\rightsquigarrow}U_2,\qquad w\overset{S}{\rightsquigarrow}Z.$$

Let E_2 be the event that there is a point $v\in W_2$ such that

$$v\overset{S_1}{\rightsquigarrow}X,\qquad v\overset{S}{\rightsquigarrow}U_1,\qquad v\overset{S}{\rightsquigarrow}Z.$$

Now

$$E_1\cap E_2\subseteq A_2$$

and hence the proof of Lemma 5.2 is complete if we show that

$$\boldsymbol{P}(E_1\cap E_2)\geq(1-\sqrt{(1-\tau)})^4.$$

But E_1, E_2 are covariant and by symmetry

$$\boldsymbol{P}(E_1)=\boldsymbol{P}(E_2).$$

Hence it is enough to show that

$$\boldsymbol{P}(E_1)\geqslant(1-\surd(1-\tau))^2.$$

But (by drawing a picture) E_1 occurs if, for some i, P_i is open and F_i is joined to U_2 by an open path in S. That is

$$\boldsymbol{P}(E_1)\geqslant\boldsymbol{P}\left(\bigcup_i(B_1^{(i)}\cap B_3^{(i)})\right).$$

Thus with (16) we have the required result.

Lemma 5.3. $S(2n,6n)\geqslant[S(2n,2n)]^3\ (1-\surd(1-S(2n,2n)))^{16}$

Proof. Consider the following regions of $\mathscr{L}$ (see Fig. 2):

$U=\{(x,y)\colon y=2n,\ 2n+1\leqslant x\leqslant 4n\}$,
$V=\{(x,y)\colon y=1,\ 2n+1\leqslant x\leqslant 4n\}$,
$S=\{(x,y)\colon 2n+1\leqslant x\leqslant 4n,\ 1\leqslant y\leqslant 2n\}$,
$R=\{(x,y)\colon 1\leqslant x\leqslant 6n,\ 1\leqslant y\leqslant 2n\}$,
$X=\{x,y)\colon x=1,\ 1\leqslant y\leqslant 2n\}$,
$Z=\{(x,y)\colon x=2n,\ 1\leqslant y\leqslant 2n\}$,
$W=\{(x,y)\colon x=4n,\ 1\leqslant y\leqslant 2n\}$,
$Y=\{(x,y)\colon x=6n,\ 1\leqslant y\leqslant 2n\}$.

Let

$$A=\{\omega\colon X\overset{R}{\rightsquigarrow}W\},$$

$$B=\{\omega\colon Z\overset{R}{\rightsquigarrow}Y\},$$

$$C=\{\omega\colon U\overset{S}{\rightsquigarrow}V\}.$$

Then

$$A\cap B\cap C\subseteq\{\omega\colon X\overset{R}{\rightsquigarrow}Y\},$$

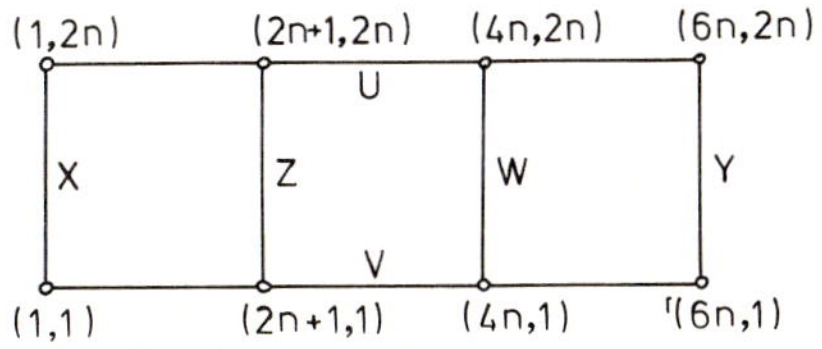

Fig. 2.

and since A, B, C are monotone in the same sense and hence covariant we have

$$\boldsymbol{P}(X \overset{R}{\rightsquigarrow} Y) \geq \boldsymbol{P}(A)\boldsymbol{P}(B)\boldsymbol{P}(C)$$

$$= [S(2n, 4n)]^2 S(2n, 2n).$$

which with Lemma 5.2 proves the result.

Let $R(n)$ be the annulus of the square lattice $\mathscr{L}$ shown in Fig. 2 bounded by the squares C_n, D_n, where C_n consists of the lines

$$y = -3n+1, \quad x = 3n, \quad y = 3n, \quad x = -3n+1$$

and D_n consists of the lines

$$y = -n, \quad x = n+1, \quad y = n+1, \quad x = -n.$$

Lemma 5.4. *The probability that there is an open cycle around the annulus $R(n)$, that is a cycle of open edges encircling the square D_n and encircled by C_n, is at least*

$$S(2n, 2n)^{12}(1 - \sqrt{(1 - S(2n, 2n))})^{64}.$$

Proof. Let A, B, C, D be the regions of $R(n)$ defined as follows (see Fig. 3):

$$A = \{(x, y): -3n+1 \leq x \leq -n, -3n+1 \leq y \leq 3n\},$$
$$B = \{(x, y): -3n+1 \leq x \leq 3n, -3n+1 \leq y \leq -n\},$$
$$C = \{(x, y): n+1 \leq x \leq 3n, -3n+1 \leq y \leq 3n\},$$
$$D = \{(x, y): -3n+1 \leq x \leq 3n, n+1 \leq y \leq 3n\}.$$

Let

$$X = \{(x, y): -3n+1 \leq x \leq -n, y = -3n+1\},$$
$$X' = \{(x, y): -3n+1 \leq x \leq -n, y = 3n\},$$
$$Y = \{(x, y): x = -3n+1, n+1 \leq y \leq 3n\},$$
$$Y' = \{(x, y): x = 3n, n+1 \leq y \leq 3n\},$$
$$U = \{(x, y): x = -3n+1, -3n+1 \leq y \leq -n\},$$
$$U' = \{(x, y): x = 3n, -3n+1 \leq y \leq -n\},$$
$$W = \{(x, y): n+1 \leq x \leq 3n, y = -3n+1\},$$
$$W' = \{(x, y): n+1 \leq x \leq 3n, y = 3n\}.$$

Then if F_n is the event that there is an open cycle around $R(n)$ we have

$$F_n \supseteq \{X \overset{A}{\rightsquigarrow} X'\} \cap \{U \overset{B}{\rightsquigarrow} U'\} \cap \{W \overset{C}{\rightsquigarrow} W'\} \cap \{Y \overset{D}{\rightsquigarrow} Y'\}.$$

Now the events on the right hand side are monotone in the same sense and thus covariant and each has probability $S(2n, 6n)$, so that

$$\boldsymbol{P}(F_n) \geq (S(2n, 6n))^4$$

which with Lemma 5.3 proves Lemma 5.4.

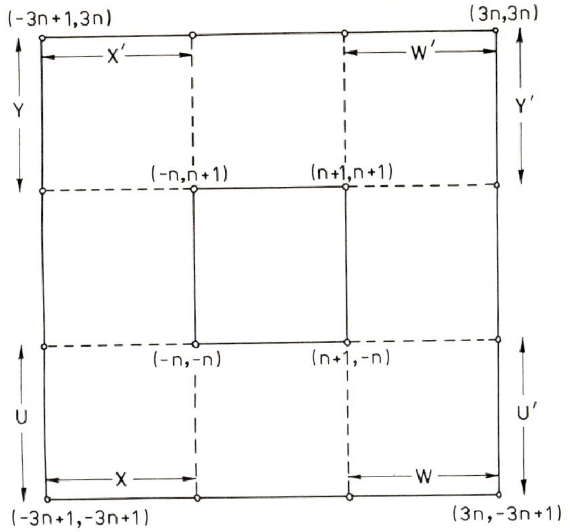

Lemma 5.5. *If* $\boldsymbol{p}<1-\boldsymbol{p}_{\mathrm{H}}$, *then* $\lim_{n\to\infty} S(n, n)$ *exists and is zero, in other words*

$$\boldsymbol{p}_{\mathrm{s}} \geqslant 1-\boldsymbol{p}_{\mathrm{H}}.$$

Proof. If $\boldsymbol{p}<1-\boldsymbol{p}_{\mathrm{H}}$, then $\boldsymbol{q}>\boldsymbol{p}_{\mathrm{H}}$ so that there is a positive probability of an infinite closed path from the origin in the dual lattice $\mathscr{L}^*$. Suppose that for some $\epsilon>0$, $S(n, n)>8\epsilon$ for infinitely many n. Choose $n_1, n_2, \ldots$ so that $R(n_1)$, $R(n_2), \ldots$, are disjoint annuli and $S(2n_i, 2n_i)>\epsilon$ for each i. This is possible by (14).

Now by Lemma 5.4, the probability that there is an open path around $R(n_i)$ is at least

$$\epsilon^{12}(1-\sqrt{(1\quad\epsilon)})^{64}$$

for each i. Since the disjointness of the $R(n_i)$ makes these events independent, the Borel–Cantelli lemmas imply that with probability one there can be no closed infinite path from the origin in $\mathscr{L}^*$. Thus we have a contradiction.

Lemma 5.6. *If* $\epsilon>0$ *and* $\boldsymbol{p}\geqslant \boldsymbol{p}_{\mathrm{T}}$, *then for infinitely many values of* n,

$$(1-S(2n, 2n))^{12}(1-\sqrt{S(2n, 2n)})^{64} \leqslant \tfrac{8}{9}+\epsilon.$$

In other words if $\boldsymbol{p}\geqslant \boldsymbol{p}_{\mathrm{T}}$, *then* $\limsup_{n\to\infty} S(2n, 2n) \geqslant \delta$, *where* δ *is a little bit bigger than* 5×10^{-6}.

Proof. Suppose the lemma is false, and choose N so that for all $n\geqslant 3^N$ the inequality fails. Now by Theorem 4.1 if $\boldsymbol{q}=1-\boldsymbol{p}$,

$$S_{\boldsymbol{q}}(2n, 2n) \geqslant S_{\boldsymbol{q}}(2n-1, 2n+1) = 1-S_{\boldsymbol{p}}(2n, 2n).$$

Hence by Lemma 5.4 if $3^t \geq 3^N$ then the probability that there is a closed cycle around the annulus $R(3^t)$ is at least $\frac{8}{9}+\epsilon$.

Hence by the duality theory of graphs the probability of the event D_t that in $\mathscr{L}^*$ there is an open path from the origin through $R(3^t)$ is not more than $\frac{1}{9}-\epsilon$.

If the number of points in $\mathscr{L}^*$ which are wet by a source at the origin is N_p^*, then we have

$$\mathscr{E}(N_p^*) \leq 4\times 3^{2N} + \sum_{t\geq N} |R(3^t)| \prod_{N\leq k\leq t-1} \boldsymbol{P}(D_k)$$

$$\leq 4\times 3^{2N} + \sum_{t\geq N} 4\times 3^{2N}(\tfrac{1}{9}-\epsilon)^{n-N}$$

$$<\infty$$

which contradicts $\boldsymbol{p} \geq \boldsymbol{p}_{\mathrm{T}}$.

As our final corollary note that from Lemma 5.6 we know that

$$\boldsymbol{p} > \boldsymbol{p}_{\mathrm{T}} \Rightarrow \limsup_{n\to\infty} S_n(\boldsymbol{p}) \geq \delta > 0,$$

whereas if $\boldsymbol{p} < \boldsymbol{p}_{\mathrm{T}}$ then

$$\lim_{n\to\infty} S_n(\boldsymbol{p}) = 0.$$

Hence we have shown that even if Conjecture 4.2 is true and $\lim_{n\to\infty} S_n(\boldsymbol{p})$ exists and equals $S(\boldsymbol{p})$ say, then $S(\boldsymbol{p})$ must be a discontinuous function of $\boldsymbol{p}$.

Note also that our proof gives the result (13) and we can sum up the situation with the set of inequalities

$$0.353210 \leq \boldsymbol{p}_{\mathrm{T}} \leq \boldsymbol{p}_{\mathrm{S}} \leq \tfrac{1}{2} \leq \boldsymbol{p}_{\mathrm{H}} \leq 0.646790 \tag{17}$$

which with Lemma 5.5 imply that for $\boldsymbol{p} > 0.646790$,

$$\lim_{n\to\infty} S_n(\boldsymbol{p}) = 1. \tag{18}$$

Acknowledgement

We would like to thank R.T. Smythe for pointing out an error in an earlier version of the paper, and G.R. Grimmett for some helpful comments.

References

[1] C.J.K. Batty, An extension of an inequality of R. Holley, Quart. J. Math. Oxford. Ser, 27 (1976) 457–461.

[2] C. Berge, Graphs and Hypergraphs (North-Holland, Amsterdam, 1973).

[3] J.A. Bondy and U.S.R. Murty, Graph Theory with Applications (Macmillan, New York, 1976).

[4] S.R. Broadbent and J.M. Hammersley, Percolation processes I. Crystals and mazes, Proc. Cambridge Philos. Soc. 53 (1957) 626–641.

[5] P. Cartier, Inégalités de correlation en mécanique statistique, Seminaire Bourbaki 25 (1973) 401–423.

[6] C. Domb, Some statistical problems connected with crystal lattices, J. Roy. Statist. Soc. 26 (1964) 367–397.

[7] D.A. Edwards, On the Holley–Preston inequalities, Proc. Edinburgh. Math. Soc., to appear.

[8] J. W. Essam, Graph theory and statistical physics, Discrete Math. 1 (1971) 83–112.

[9] J.W. Essam, Percolation and cluster size, in: C. Domb and M.S. Green, eds., Phase Transition and Critical Phenomena 2 (1972) 197–270.

[10] C.M. Fortuin, On the random cluster model. II. The percolation model, Physica 58 (1972) 393–418.

[11] C.M. Fortuin and P.W. Kasteleyn, On the random cluster model I. Introduction and relation to other models, Physica 57 (1972) 536–564.

[12] C.M. Fortuin, P.W. Kasteleyn and J. Ginibre, Correlation inequalities on some partially ordered sets, Comm. Math. Phys. 22 (1971) 89–103.

[13] H.L. Frisch and J.M. Hammersley, Percolation processes and related topics, SIAM J. 11 (1963) 894–918.

[14] G.R. Grimmett, On the number of clusters in the percolation model, J. London Math. Soc. 2 (1976) 346–350.

[15] J.M. Hammersley, Percolation processes: lower bounds for the critical probability, Ann. Math. Statist. 28 (1957) 790–795.

[16] J.M. Hammersley, Bornes superieures de la probabilité critique dans un processus de filtration, Proc. 87th Int. Colloq. CNRS (Paris, 1957) 17–37.

[17] J.M. Hammersley and D.J.A. Welsh, First passage percolation, subadditive processes, stochastic networks and generalised renewal theory. in: Bernoulli (1723) Bayes (1763) Laplace (1813) Anniversary Volume, University of California Berkeley (1963) (Springer, Berlin, 1965) 61–110.

[18] T.E. Harris, A lower bound for the critical probability in a certain percolation process, Proc. Cambridge Philos. Soc. 56 (1960) 13–20.

[19] R. Holley, Remarks on the FKG inequalities. Comm. Math. Phys. 36 (1974) 227–231.

[20] J.H.B. Kemperman, On the FKG-inequality for measures on a partially ordered space, (to appear).

[21] J. Kurkijarvi and T.C. Padmore, Percolation in two-dimensional lattices, J. Phys. A 8 (1975) 683–696.

[22] C.J. Preston, A generalization of the FKG inequalities, Comm. Math. Phys. 36 (1974) 233–241.

[23] P.D. Seymour and D.J.A. Welsh, Combinatorial applications of an inequality from statistical mechanics, Math. Proc. Cambridge Philos. Soc. 77 (1975) 485–495.

[24] V.K.S. Shante and S. Kirkpatrick, An introduction to percolation theory, Adv. Phys. 20 (1971) 325–356.

[25] M.F. Sykes and J.W. Essam, Exact critical percolation probabilities for the site and bond problems in two dimensions, J. Math. Phys. 5 (1964) 1117–1132.

[26] M.F. Sykes, D.S. Gaunt and M. Glen, Percolation processes in two dimensions II–IV, J. Phys. A. Math. Gen. 9 (1976) 97–103, 715–724, 725–730.

[27] H.N.V. Temperley and E.H. Lieb, Relations between the "percolation" and "colouring" problem and other graph-theoretical problems associated with regular planar lattices: some exact results for the "percolation" problem, Proc. Roy. Soc. London Ser. A 322 (1971) 251–280.

[28] W.T. Tutte, A ring in graph theory, Proc. Cambridge Philos. Soc. 43 (1947) 26–40.

[29] D.J.A. Welsh, Percolation and related topics, Science Progress 64 (1977) 67–85.

Annals of Discrete Mathematics 3 (1978) 247–257.

ON TUTTE'S DICHROMATE POLYNOMIAL

Cedric A.B. SMITH
Galton Laboratory, University College London, London NW1 2HE, England

A simplified method of defining the Tutte dichromate polynomial is described, together with direct proofs of its principal properties. It is related to the fragmentation of a matroid. The method includes graphs and sets of vectors as special cases.

1. Introduction

The object of this paper is to give a short, unified and simplified presentation of the definition and properties of Tutte's dichromate polynomial. Comparatively few completely new results are obtained.

Tutte [10] introduced this polynomial in graph theory. (It is now commonly known simply as the "Tutte polynomial", but Tutte's own choice of name reminds us on the one hand that it is a polynomial in 2 variables related to the chromatic polynomial, and on the other hand that Tutte graduated in chemistry before he graduated to mathematics. It is also clear from an earlier paper of Tutte's [9] that he was partly led to the discovery of the polynomial through the theory of the solution of electrical networks by means of trees (Kirchhoff [5], Brooks et al. [1]), where the recurrence relations for the complexity (= number of spanning trees) are similar to those for the dichromate.)

It is surprising that Tutte, who is a pioneer in the study of matroids, should not have noticed that the dichromate immediately generalizes to them in the obvious way. An extension to sets of vectors was given by Smith [6] and the extension to arbitrary matroids by Crapo [3]. In addition, Smith [7] also showed that the electrical network theory generalizes to regular matroids. However, Tutte's and Crapo's approaches are complicated, involving a theory of "internal" and "external" activity, See also Heron [4]. A rather simpler approach is given by Heron [4] and Welsh [11], but this is still fairly complicated. We show here that the existence and standard properties of the dichromate polynomial are straightforward consequences of standard matroid theorems.

2. Definitions

We briefly recall the standard definitions and properties of matroids, in the form which will be most convenient to us.

Throughout the paper the following symbolism is used consistently (and is to be understood in this way, even where, for brevity, this is not explicitly stated). E stands for a set of *elements* e_j, M for a (typical) matroid over E. S is a subset of E. We write S^{k-j} for $\{e_k\} \cup S \setminus \{e_j\}$, $B_b(\subseteq E)$ is a (typical) *base* of M(=maximal independent set). R_j, C_j stand respectively for the operations of *removing* and *contracting* the element e_j.

The simplest definition of a matroid is that its independent sets obey the axioms:

Axiom 2.1. *A subset of an independent set is independent.*

Axiom 2.2. *The maximal independent subsets of any given* $S(\subseteq E)$ *have the same number of elements, called rank S.*

An element e_h of M is a *loop* if it belongs to no independent set, and an *isthmus* or *coloop* if it belongs to every base (maximal independent set). Loops and coloops are *degenerate* elements.

The following standard properties of bases are immediate consequences of Axioms 2.1 and 2.2.

Property 2.3. *If nondegenerate* e_j *belongs to* B_b, *then there exists* $e_h \notin B_b$ *such that* $B_b^{h-j} = e_h \cup B_b \setminus e_j$ *is a base.*

(N.B. conventionally this is written $\{e_h\} \cup B_b \setminus \{e_j\}$, but the notation used here seems unambiguous.)

Property 2.4. *If nondegenerate* $e_h \notin B_b$, *there exists* $e_j \in B_b$, *such that* B_b^{h-j} *is a base.*

The matroids R_jM(="*M with* e_j *removed*") and C_jM (="*M with* e_j *contracted*") are defined over the set $E \setminus e_j$. The matroid R_jM is defined only when e_j is not an isthmus; its bases are identical with the bases of M which do not include e_j. The matroid C_jM is defined only when e_j is not a loop. Its bases are obtained from the bases of M which do include e_j by deleting e_j.

The *Tutte dichromate polynomial* $TM = TM(x, y)$ of a matroid M is a polynomial in 2 variables, x, y, satisfying the relations:

2.5. *If* e_j *is nondegenerate,*

$$TM = TR_jM + TC_jM$$

2.6. *If all elements* e_j *are degenerate, I of them being isthmuses (coloops) and L being loops,*

$$TM(x, y) = x^I y^L.$$

If, for the moment, we take for granted that 2.5 and 2.6 define a unique polynomial, then the following well-known properties follow straightforwardly by induction:

Property 2.7. *T in invariant. That is, with the obvious definition of isomorphism,*

$$M \simeq M' \Rightarrow TM = TM'.$$

Property 2.8. *All coefficients in* $TM(x, y)$ *are nonnegative.*

Property 2.9. $TM(1, 1)$ = *the complexity of* M = *the number of bases.*

Property 2.10. *If* M^{D} = *the dual of* M, (*i.e. the bases of* M^{D} *are the complements of the bases of* M),

$$TM^{\mathrm{D}}(y, x) = TM(x, y).$$

Notice that all this discussion applies equally well to sets of vectors (considered as "representable matroids") and to connected graphs (with "element" = edge, "base" = spanning tree). For the relation to the chromatic polynomial of a graph, see, for example, Tutte [10], Heron [4], or Welsh [11].

3. Existence and uniqueness

We now demonstrate inductively.

Theorem 3.1. *There exists a unique polynomial satisfying* 2.5–7 (*and hence also* 2.3–10).

Proof. This is trivially true for $|E| \leqslant 1$. Hence set $|E| = n \geqslant 2$, and assume true for all smaller matroids. If all elements of M are degenerate, the existence and uniqueness follow from 2.6. Otherwise we have to prove

Lemma 3.2. *If* e_j, e_h *are nondegenerate elements,*

$$TR_jM + TC_jM = TR_hM + TC_hM.$$

Proof. Note first that if R_hR_jM is defined, it is the matroid on the set $E' = E \setminus (e_j \cup e_k)$, whose bases are identical with the bases of M which do not include either e_h or e_j.

Hence $R_hR_jM = R_jR_hM$, provided that both are defined. (Actually, it is sufficient that either should be defined, but we do not need this stronger formulation.) Similarly, provided that both are defined, $C_hR_jM = R_jC_hM$ = the Matroid on E' whose bases are those subsets S which become bases of M after adding e_h. And similarly $C_hC_jM = C_jC_hM$ when both are defined. It follows that, provided that all terms are defined, we have a "one-line" proof of Lemma 3.2

$$\begin{aligned} TR_jM + TC_jM &= (TR_hR_jM + TC_hR_jM) + (TR_hC_jM + TC_hC_jM) \\ &= (TR_jR_hM + TC_jR_hM) + (TR_jC_hM + TC_jC_hM) \\ &= TR_hM + TC_hM. \end{aligned} \tag{3.1}$$

The proof is now completed by:

Lemma 3.3. *If any of the terms in eq.* (3.1) *are not defined, then interchange of* e_j *and* e_h *(i.e. interchanging suffixes h, j) is an isomorphism of M.*

(For then it immediately follows from Property 2.7 that $R_jM \simeq R_hM$, $C_jM \simeq C_hM$, *whence Lemma 3.2 follows.)*

Proof. When, for example, is R_jR_hM not defined? When e_j is an isthmus (coloop) in R_hM. That is, when every base B_b of M, which does not include e_h must include e_j; that is, every base of M includes at least one of e_h, e_j. If, as above, B_b inclues e_j but not e_h, then by Property 2.3 there exists e_k such that $B_c = B_b^{k-j} = e_k \cup B_b \setminus e_j$ is a base; and we must have $k = h$, otherwise B_c would contain neither e_j nor e_h. Conversely, if B_c is a base containing e_h but not e_j, then $B_b = B_c^{j-h}$ contains e_j but not e_h. Hence an interchange of e_h and e_j interchanges the bases B_b, B_c. The remaining bases B_d are those containing both e_h and e_j, which are left invariant under the interchange. Hence the interchange of e_h and e_j maps bases onto bases, i.e., is an isomorphism.

A similar argument, but using Property 2.4, shows that the interchange is an isomorphism when C_hC_jM is undefined, i.e., when no base contains both e_j, e_h. And (remembering that by hypothesis e_h, e_j are nondegenerate in M), it is easy to see that Properties 2.3 and 2.4 imply that R_jC_hM, C_hR_jM, etc., always exist. This completes the proof of Lemma 3.3, hence also of Lemma 3.2, hence also of existence and uniqueness (Theorem 3.1).

4. Fragmentation of sets of matroids

The argument developed above proves some slightly stronger results than Theorem 3.1. We begin with some definitions.

Let Z be a set of matroids, $\{M_m\}$. (The argument which follows will remain valid for multisets, i.e., sets with repetitions allowed.) We define the *dichromate of* Z to be the sum of the dichromates of its members,

$$TZ = \sum TM_m.$$

If e_j is a nondegenerate element in some M_m in Z, we define the *fracture of Z* (*through* e_j) to be the replacement of M_m in Z by the pair of matroids R_jM_m, C_jM_m (i.e. formally, replacing Z by

$$Z' = Z \cup \{R_jM_m\} \cup \{C_jM_m\} \setminus \{M_m\}.$$

a new set with one more matroid member than Z.) By (2.5), this fracture leaves the dichromate unchanged: $TZ' = TZ$. If there is any nondegenerate element in Z', we choose one such, say e_h, and fracture Z' through e_h, to give Z''. We continue in this way until all remaining elements are degenerate; the set $Z''^{\cdots\prime}$ of matroids then remaining is a *fragmentation* FZ of Z, and its elements are the *fragments* of Z; each fragment contains no nondegenerate elements, and therefore has a dichromate of the form (2.6).

Given Z, we can in general perform the fractures leading to a fragmentation in various orders; instead of beginning with a fracture through Z_j; we could fracture through some other nondegenerate edge Z_h, and so on. But the argument leading to Theorem 3.1 is readily modified to give

4.1. *The fragments of Z are uniquely determined by Z to within an isomorphism, i.e., they do not depend in structure on the particular sequence of fractures used.*

Also we have

4.2. *The dichromate TZ of Z is the sum of the dichromates of all the fragments of Z.*

(Note: 4.2 can be used as an alternative definition of the dichromate, and 4.1 immediately shows that it gives a unique value.)

In particular, when the set Z consists of a single matroid M, we have:

4.3. *The numbers of fragments of M (in one fragmentation)$= TM(1,1) =$ the complexity of $M =$ the number of bases of M.*

This suggests that there may be a one-one correspondence between the fragments of M and the bases of M. To verify this, we proceed as follows. We assume that the order in which the fractures of M are done is given. We note that the process of repeatedly fracturing M can be represented by a rooted directed tree, t_{FM}, as follows. The root node is the original matroid M.

If, first M is fractured through e_j, we introduce two new nodes, R_jM and C_jM, and two directed edges from M to R_jM and M to C_jM respectively. If, subsequently, R_jM is fractured through e_h, we introduce two new nodes, R_hR_jM and C_hR_jM, and two new edges from R_jM to R_hR_jM and from R_jM to C_hRjM respectively. We do similarly if C_jM is fractured through e_k.

When this construction has been carried as far as it will go, each tip of the directed graph ($=$node with no outgoing edge) is a fragment of M, and there is a unique directed path from M to this fragment, representing the sequence of operations of removal and contraction of elements by which this fragment is obtained from M. Hence the number of tips of t_{FM} = the number of fragments of M.

First, suppose that a base B_b is given (and hence the complementary cobase $\bar{B}_b = E \setminus B_b$); we require to construct a corresponding fragment f_b. We begin with the complete matroid M = the root of digraph. If the first edge through which we fracture, in the fragmentation process is e_j, we move from M to R_jM or C_jM according as e_j belongs to $\bar{B}_b$ or to B_b respectively.

If, say, we have gone to R_jM, and e_h is the next edge through which we fracture, we go to R_hR_jM or C_hR_jM according as e_h belongs to $\bar{B}_b$ or B_b. We continue in this way until the process ends at a fragment f_b. Now the removal R_u of the element e_u deletes the element e_u from any cobase to which it belongs, but leaves bases not including e_u unaltered, and a similar remark applies to contraction C_u, with the words "base" and "cobase" interchanged. Hence any element of the base B_b of the original matroid M will either be deleted (by contraction) at some stage of the sequence of operations leading to f_b, or will be a member of a base of f_b, i.e., an isthmus in f_b. Similarly every element of the cobase $\bar{B}_b$ will either be removed or will survive as a loop in f_b.

Conversely, if we are given a fragment f, we can divide the elements of M into 4 classes, K_{fC}, K_{fR}, K_{fI}, K_{fL}, namely those elements which are respectively contracted or removed in the sequence of operations leading from M to f, and those which are respectively isthmuses and loops in f. The converse argument (starting from f and replacing elements in turn to go back to M) then shows that $B = K_{fC} \cup K_{fI}$ is a base of M, and $\bar{B} = K_{fR} \cup K_{fL}$ a cobase. Thus these constructions give a bijection of bases onto fragments.

All these arguments extend straightforwardly to "matroids with multiple elements" (*integral polymatroids*, see Welsh [11]).

Suppose that, in an ordinary matroid M, there is a set η of m elements, $e_1, e_2, \ldots, e_m$, such that any permutation of these elements is an isomorphism of the matroid. Then we can (if we wish) call η a *multiple element* of multiplicity m. If all elements of M are removed, other than η, then η will have a certain rank ρ_η, its *autorank*, and a certain corank, κ_η, its *autocorank*.

Clearly

$$\rho_\eta + \kappa_\eta = m. \tag{4.1}$$

A multiple element with $\kappa_\eta = 0$ is a *multiple isthmus*, with $\rho_\eta = 0$, a *multiple loop*, otherwise it is *nondegenerate*. (A set of *parallel* edges in a graph, often known as a *multiple edge*, has autorank 1). If η_λ is a multiple element, we can naturally define $R_\lambda M$, $C_\lambda M$ to mean R_jM, C_jM, respectively. where e_j is some ordinary element belonging to the set η_λ. That is, the removal operation R_λ replaces η_λ by a new multiple element η'_λ whose autorank is 1 less than the autorank of η_λ, but whose autocorank is unaltered. The multiplicity is therefore reduced by 1. (If η'_λ is the null set, it can be omitted from the matroid). Similar remarks (with "autorank" and "autocorank" interchanged) apply to C_λ.

Using these ideas, we can extend the theory developed above to matroids with multiple edges in the obvious way: for example, all elements of fragments of M

will be multiple isthmuses and multiple loops, and the dichromate of M can be defined as the sum of the dichromates of its fragments.

The theory of fragmentation can also be used to give another definition of the *bichromate polynomial* (Smith [7]), and its relation to patroid nets (Smith [7, 8]) although the treatment is not so simple as for the dichromate, but corresponds more closely to Heron's [4] and Welsh's [11] definition of a dichromate.

In bichromate theory, each element e_j is assigned two numbers, its *proconductance* c_j and its *proresistance* r_j. Now we have seen that every fragment f divides the elements of M into 4 classes, K_{fC}, K_{fR}, K_{fI}, K_{fL}. With this fragment we can associate the product

$$\varphi(f)=\prod_{e_a\in K_{fC}}(c_a)\prod_{e_b\in K_{fR}}(r_b)\prod_{e_i\in K_{fI}}(c_i+r_iX)\prod_{e_l\in K_{fL}}(c_lY+r_l). \tag{4.2}$$

The bichromate is then the sum of the $\varphi(f)$ taken over all fragments f; it is a polynomial in the variables X, Y. To show that it is independent of the particular fragmentation, we can observe that this definition is equivalent to Smith's [7] form. For any subset $S\subseteq E$, write $\bar{S}=E\setminus S$, and let $\nu=$ the nullity of S in $M=|S|-\text{rank } S$, and similarly $\mu=$ the nullity of $\bar{S}$ in the dual of M.

Then the bichromate of M is

$$\text{bic } M(X,Y)=\sum_{S\subseteq E}\left[Y^{\nu}\prod_{e_i\in S}(c_i), X^{\mu}\prod_{e_i\in S}(r_j)\right]. \tag{4.3}$$

From these forms it is easy to prove the main properties of the bichromate, namely that if e_j is nondegenerate

$$\text{bic } M=\text{bic } R_jM+\text{bic } C_jM_i \tag{4.4}$$

that if M is separable into M_1 and M_2,

$$\text{bic } M=\text{bic } M_1\cdot\text{bic } M_2; \tag{4.5}$$

that if M is a one-element matroid consisting of a single isthmus e_i,

$$\text{bic } M=c_i+r_iX; \tag{4.6}$$

(and similarly c_lY+r_l for a single loop); and that if we set all $c_j=1=r_j$, the polynomial $\text{bic } M(1+x,1+y)$ is the dichromate. We will not go into the formal proofs of these results here: for one method of proof, see Smith [7].

However, by taking e_j to be the *first* element to be removed or contracted in the fragmentation process, the expression (4.2) gives an immediate proof of (4.4); and clearly (4.5) and (4.6) are also immediate consequences of (4.2).

These relations can also be extended in the obvious way to matroids with multiple elements, thus if $e_1, e_2, \ldots, e_m$ all have the same proconductance c and the same proresistance r, and the matroid is invariant under any permutation of these edges, they may together be considered as a multiple element with proresistance r and proconductance c. For example, a matroid consisting of one triple element will have bichromate

$$(c+rX)^3 = c^3+3c^2rX+3cr^2X^2+r^3X^3,$$

if this is a triple isthmus, but

$$c^3Y+3c^2r+3cr^2X+r^3X^2,$$

if it has autorank 2 and autocorank 1, and so on.

5. Separation

If E is partitioned into (at least two disjoint) subsets E_η in such a way that, for each E_η, every base B_b meets E_η in the same number of elements, $\rho_\eta = |E_\eta \cap B_b|$, then the matroid M is *separable*, and is in fact *separated by the partition* $\{E_\eta\}$ *of* E. The following standard properties of separation are simple consequences of this definition and the matroid axioms:

Property 5.1. $\sum \rho_\eta = \text{rank } M$.

Property 5.2. *There is a matroid M_η over E_η, having rank ρ_η, whose bases are the intersections of the bases of M with E_η. we call M_η the component of M in E_η.*

Property 5.3. *Conversely, if we choose, for each M_η, an arbitary base B_η, the union of the B_η is a base of M.*

Property 5.4. *If $|E| \geqslant 2$, every degenerate edge is a component of M.*

Property 5.5. *If M is separated by the $\{E_\eta\}$, and e_j is in the component M_λ, then R_jM and C_jM are matroids also separated by the $\{E_\eta\}$ and having the same components M_η as M, except that M_λ is replaced by R_jM_λ, C_jM_λ respectively.*

As far as the dichromate is concerned, we have the following results:

5.6. $TM = \prod_\eta T(TM_\eta)$.

This follows from Property 5.5 and the definition 2.5, 2.6 of M. Since the dichromate of a non-empty matroid contains no constant term, we see that:

5.7. *If M is separated into c components, $TM(x, y)$ contains no term of degree $< c$.*

We now require a result effectively due to Crapo [2], which is virtually a converse of Property 5.5.

Theorem 5.8. *Suppose that e_i is nondegenerate, R_iM is separated by a pair of sets*

(E_1^R, E_2^R) and C_iM by (E_1^C, E_2^C). Then M, R_iM and C_IM are all separated either by (E_1^R, E_2^R), or by (E_1^C, E_2^C), or by both.

Proof. Crapo gives a rather sophisticated proof. A proof based more directly on the matroid axioms is as follows.

For $S \subseteq E$ define

$$N_{\lambda\mu}(S) = |S \cap E_\lambda^R \cap E_\mu^C|, \tag{5.1}$$

and let $\mathsf{N}(S)$ be the 2×2 matrix with elements $N_{\lambda\mu}(S)$, $1 \leqslant (\lambda, \mu) \leqslant 2$. We wish to investigate what are the possible values of $\mathsf{N}(B_r)$ for bases B_r of R_iM. Choose one such base, say B_0, and set $\mathsf{N}(B_0) = \mathsf{n}$. Because R_iM is separated by (E_1^R, E_2^R),

$$N_{\lambda_1}(B_r) + N_{\lambda_2}(B_r) = |E_\lambda^R \cap B_r| = \rho_\lambda^R. \tag{5.2}$$

independently of B_r, i.e. the row sums of the matrices $\mathsf{N}(B_r)$ are independent of B_r. Hence in general $\mathsf{N}(B_r)$ can be written in the form

$$\mathsf{N}(B_r) = \mathsf{N}(B_0) + \begin{bmatrix} -P_r & P_r \\ -Q_r & Q_r \end{bmatrix} = \begin{bmatrix} (n_{11} - P_r) & (n_{12} + P_r) \\ (n_{21} - Q_r) & (n_{22} + Q_r) \end{bmatrix}$$

for some integers P_r, Q_r. However, by Property 5.3 there is another base, B_{rs}, consisting of the part of B_r in E_1^R and the part of B_s in E_2^R. Hence

$$\mathsf{N}(B_{rs}) = \begin{bmatrix} (n_{11} - P_r) & (n_{12} + P_r) \\ (n_{21} - Q_s) & (n_{22} + Q_s) \end{bmatrix} \tag{5.3}$$

is also a possible value of N(base of R_iM).

Now if b_u is a base of C_iR, be definition $b_u \cup e_i$ is a base of M which includes e_i. Hence by Properties 2.3 and 2.4, given B_r there exists at least one element e_h and one base b_r of C_iM such that

$$B_r - b_r \cup e_h, \tag{5.4}$$

and conversely for any base b_r of C_iM we can find at least one e_h and B_r, a base of R_iM satisfying eq. (5.4). Thus, each $\mathsf{N}(b_r)$ is obtained from some $\mathsf{N}(B_r)$ by subtracting 1 from exactly one of the 4 elements $N_{\lambda\mu}(B_r)$, i.e. for some λ, μ,

$$N_{\lambda\mu}(B_r) = N_{\lambda\mu}(b_r) + 1, \tag{5.5}$$

while for the other 3 elements $N_{\alpha\beta}(B_r) = N_{\alpha\beta}(b_r)$; and $\mathsf{N}(B_r)$ is similarly related to some $\mathsf{N}(b_r)$. In particular we will assume that

$$\mathsf{N}(b_0) = \begin{bmatrix} (n_{11} - 1) & n_{12} \\ n_{21} & n_{22} \end{bmatrix} \tag{5.6}$$

(as can always be arranged by relabeling rows and columns if need be). Since $b_u = a$ typical base of C_iM is separated by (E_1^C, E_2^C), the column sums of $\mathsf{N}(b_u)$ are independent of b_u, and in particular, the sum of the elements in the second column is always $n_{12} + n_{22}$. Hence, for every $\mathsf{N}(B_{rs})$ given by eq. (5.3) it must be possible to subtract 1 from just one element to make the second column sum to $n_{12}+n_{22}$. Remembering that by definition $P_0=0=Q_0$, this implies that the only other possible value for P_r or for Q_s would be 1, but we cannot have both $P_r=1=Q_s$. A similar argument shows that the general form of $\mathsf{N}(b_{uv})=\mathsf{N}(\text{base of } C_iM)$ is

$$\mathsf{N}(b_{uv})=\begin{bmatrix}(n_{11}-1+p_u) & (n_{12}+q_v)\\ (n_{12}-p_u) & (n_{22}-q_v)\end{bmatrix},$$

and that $p_u=0$ or 1, $q_v=0$ or 1, but not $P_u=1=q_v$. Now suppose if possible that we could have both $P_r=1=p_u$; then we would have the possible values for the matrices $\mathsf{N}(\text{base of } R_iM)$, $\mathsf{N}(\text{base of } C_iM)$, namely

$$\mathsf{N}(B_0)=\begin{bmatrix}n_{11} & n_{12}\\ n_{21} & n_{22}\end{bmatrix},\qquad \mathsf{N}(b_0)=\begin{bmatrix}(n_{11}-1) & n_{12}\\ n_{21} & n_{22}\end{bmatrix},$$

$$\mathsf{N}(B_r)=\begin{bmatrix}(n_{11}-1) & (n_{12}+1)\\ n_{21} & n_{22}\end{bmatrix},\qquad \mathsf{N}(b_u)=\begin{bmatrix}n_{11} & n_{12}\\ (n_{12}-1) & n_{22}\end{bmatrix}. \tag{5.7}$$

Recall that the bases of M take the forms B_t and $(b_w \cup e_i)$. Now the $(1,2)$ element of $\mathsf{N}(B_r)$, namely $N_{12}(B_r)=n_{12}+1$, is greater than the N_{12} element of matrices of all other forms in (5.7); hence B_r interects $E_1^R \cup e_i$ in a maximal independent set containing $(n_{11}+n_{12})$ elements. But $(b_u \cup e_i)$ intersects $E_1^R \cup e_i$ in an independent set of $n_{11}+n_{12}+1$ elements, thus contradicting the matroid axioms. Hence we cannot have $P_r=1=p_u$.

Similarly from a consideration of the intersections of B_0 and b_v with $E_1^R \cup e_i$, we see that we cannot have $P_r=1=q_v$; and by considering the intersections of B_5 and b_0 with $E_2^R \cup e_i$, we cannot have $Q_s=1=p_u$ or q_v. Thus, at most one of P_r, Q_s, p_u, q_v, can differ from 0. But $P_r=0=Q_s$ for all r, s implies that M, R_iM and C_iM are all separated by (E_1^C, E_2^C), while $p_u=0=q_v$ implies that they are all separated by (E_1^R, E_2^R), establishing Theorem 5.8.

The result 5.7 is completed by the following theorem due to Crapo [2, 3]:

Theorem 5.9. *If a matroid M has at least 2 elements, the first degree terms in the polynomial $TM(x, y)$ are of the form $(\beta M)(x+y)$, where βM, the Crapo invariant of M, is a nonnegative integer. Also $\beta M>0$ if and only if M is nonseparable.*

Proof. For $|E|=2$, true by direct verification. Hence proceed by induction, assuming that $|E|>2$, and that Theorem 5.9 holds for the matroids R_jM, C_jM. If M is nonseparable, then by Theorem 5.8, at least one of R_jM, C_jM are nonseparable. Hence 2.5 and 5.7 show that Theorem 5.9 holds also for M, completing the induction.

Corollary. *If $|E|\geqslant 2$, there are exactly $2\beta M$ one-element fragments of M in any one fragmentation, and of these, βM are isthmuses and βM are loops.*

References

[1] R.L. Brooks, C.A.B Smith, A.H. Stone and W.T. Tutte, The dissection of rectangles into squares, Duke Math. J. 7, (1940) 312–340.

[2] H.H. Crapo, A higher invariant for matroids, J. Combinatorial Theory 2 (1967) 406–417.

[3] H.H. Crapo, The Tutte Polynomial, Aequationes Math. 3 (1969) 211–219.

[4] A.P. Heron Matroid Polynomials, In: D.J.A. Welsh and D.R. Woodall, eds., Combinatorics (Institute of Mathematics and its Applications, Southend-on-Sea, 1972) 164–202.

[5] G. Kirchoff, Ueber die Auflösung der Gleichungen, auf welche man bei der Untersuchung der linearen Vertheilung galvanischer Ströme geführt wird, Annalen der Physik und Chemie 72 (1847) 497–508.

[6] C.A.B. Smith, Map Colourings and linear mappings, In: D.J.A. Welsh, ed., Combinatorial Mathematics and its Applications (Academic Press, London, 1969) 259–283.

[7] C.A.B. Smith, Electric currents in regular matroids, In: D.J.A. Welsh and D.R. Woodall, eds., Combinatorics (Institute of Mathematics and its Applications, Southend-on-Sea, 1972) 262–284.

[8] C.A.B. Smith, Patroids, J. Combinatorial Theory, 16 (1974) 64–76.

[9] W.T. Tutte, A ring in graph theory, Proc. Cambridge Philos. Soc. 43 (1967) 26–40.

[10] W.T. Tutte, A contribution to the theory of polynomials, Can. J. Math. 6 (1953) 80–91.

[11] D.J.A. Welsh, Matroid Theory (Academic Press, London, 1976).

Annals of Discrete Mathematics 3 (1978) 259–268.

HAMILTONIAN CYCLES AND UNIQUELY EDGE COLOURABLE GRAPHS

A. G. THOMASON

Department of Pure Mathematics and Mathematical Statistics, University of Cambridge, Cambridge CB2 1SB, England

0. Introduction

A theorem of Smith (see Tutte [8]) states that in any cubic graph the number of hamiltonian cycles containing a given edge is even. If the graph is cubic and bipartite, a theorem of Kotzig (see Bosák [2]) tells us that the total number of hamiltonian cycles in the graph is even too. These two theorems are in fact consequences of a more general result, which we prove in Section 1 below. We also look at sets of edge-disjoint hamiltonian cycles in multigraphs (loops are allowed). Let $m \geq 2$ and for two edges x and y of a multigraph G (with at least three vertices) let $P(x, y)$ be the set of all collections of m edge-disjoint hamiltonian cycles in G. The main result of Section 2 states that $|P(x, y)|$ is even.

These results were discovered whilst investigating uniquely edge colourable graphs. We denote by $\chi'(G)$ the edge chromatic number of a graph G. (We adopt the terminology of [1].) If G has no isolated vertices, and if all edge colourings of G induce the same partition of the edges into independent sets, we say that G is *uniquely k-edge colourable* (where $k = \chi'(G)$); this is sometimes abbreviated to *uniquely edge colourable*. Let α and β be two of the colours used to colour a uniquely k-edge colourable graph, and let $C_{\alpha\beta}$ be the subgraph induced by the edges of colour α and the edges of colour β. We may swap the colours α and β in any component of $C_{\alpha\beta}$ and get another edge colouring of G; hence $C_{\alpha\beta}$ is connected, and is a path or an (even) cycle. If G is k-regular then $C_{\alpha\beta}$ is a hamiltonian cycle, since there is an edge of colour α (and one of colour β) at each vertex.

Obviously any uniquely 2-edge colourable graph is a path or an even cycle; it is clear also that the star $K_{1,k}$ is uniquely k-edge colourable ($K_{1,k}$ has vertex set $\{u\} \cup \{v_1, \ldots, v_k\}$ and edge set $\{uv_1, \ldots, uv_k\}$). Suppose now that G is uniquely 3-edge colourable. If G contains a triangle we may contract the triangle to a single vertex and get another uniquely 3-edge colourable multigraph; conversely we may replace any vertex of degree 3 by a triangle to get a larger uniquely 3-edge colourable graph. This fact led Greenwell and Kronk [4] to conjecture that every uniquely 3-edge colourable graph other than $K_{1,3}$ contains a triangle; they

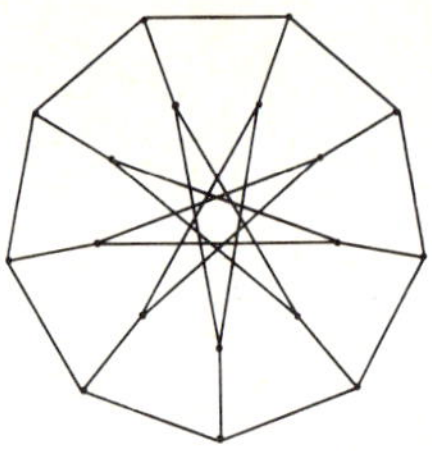

Fig. 1. Tutte's counterexample.

also conjectured that every cubic graph with exactly three hamiltonian circuits is uniquely edge colourable. A counterexample to the first conjecture was found by Tutte [9]; see Fig. 1.

A conjecture of Cantoni (see [9]) states that every cubic planar graph with exactly three hamiltonian cycles contains a triangle. This leads naturally to the conjecture stated by Fiorini [3], that every uniquely 3-edge colourable planar graph other than $K_{1,3}$ contains a triangle.

For $\chi'(G) \geqslant 4$ the stars are the only uniquely edge colourable graphs; we prove this in Section 3. It was first stated by Wilson [10] as a conjecture.

1. Hamiltonian cycles

Throughout this section we shall be concerned with hamiltonian paths in a multigraph $G = (V, E)$ which begin with a certain sequence of edges. (Paths and cycles are always considered as sequences or sets of edges, rather than as sequences of vertices.) We select a path $s = e_1, \ldots, e_m$ in G, where the endvertices of the edge e_i are v_i and v_{i+1}, $1 \leqslant i \leqslant m$. The path s is called a *stick*. The definitions to follow, and the statement of Theorem 1.1, depend on our choice of s; we obtain corollaries to Theorem 1.1 by making suitable specific choices of s.

Let $|V| = n$, and for a vertex $v \in V$ let $d(v)$ be the degree of v in G. Further let $\varepsilon(v)$ be the number of edges between v and the set of vertices $\{v_1, \ldots, v_m\}$, that is, all the vertices of the stick except the last. Let $h = e_1, \ldots, e_{n-1}$ be a hamiltonian path beginning with the stick s, where the edge e_i has endvertices v_i and v_{i+1}, $1 \leqslant i \leqslant n-1$. Let e_n be another edge with endvertices v_n and v_k, $k \geqslant m+1$, where $e_n \neq e_{n-1}$. Then the set ϙ $= \{e_1, \ldots, e_n\}$ is called a *lollipop*.[1] It contains two hamiltonian paths beginning with the stick s, namely $h = e_1, \ldots, e_n$ and $h' = e_1, \ldots, e_{k-1}, e_n, e_{n-1}, \ldots, e_{k+1}$. Note that if e_n is a loop then $h = h'$; we regard ϙ as then containing two copies of h.

We now define the *lollipop graph* ϙ(G, s) to be a multigraph whose vertex set is the set of hamiltonian paths of G beginning with the stick s. ϙ(G, s) has an edge e for each lollipop ϙ of G, the endvertices of e being the vertices h and h' of ϙ(G, s). Again, note that if $h = h'$ then e will be a loop of ϙ(G, s).

[1] The letter ϙ (koppa) is an episemon, originally coming between π and ρ in the Greek alphabet.

Suppose h is a hamiltonian path in G beginning with the stick s and ending in a vertex v_n. Then the degree of h in ⚲ (G, s) is exactly the number of copies of h contained in the lollipops, namely $d(v_n)-\varepsilon(v_n)-1$; this holds even if there are loops in G at v_n.

Theorem 1.1. *The number of hamiltonian paths in G beginning with the stick s and ending in a vertex of the set $W=\{w \in V\colon d(w)-\varepsilon(w)$ is even$\}$ is even.*

Proof. These paths are exactly the vertices of odd degree in ⚲(G, s).

Corollary 1.2. *Let G be a multigraph, let $u, v \in V$, and suppose that $d(w)$ is odd for each vertex $w \in V-\{u, v\} \neq \emptyset$. Then the number of hamiltonian paths in G from u to v is even.*

Proof. We may assume that u and v are adjacent vertices (if they are not we may add an edge between them); let e be an edge between u and v. We choose the stick s to be the edge e with $u=v_1$ and $v=v_2$; if $w \in V$ then $\varepsilon(w)$ is the number of edges from u to w. Consequently a hamiltonian path h beginning with s and ending in w gives rise to exactly $\varepsilon(w)$ hamiltonian paths from u to v. But by Theorem 1.1 the number of such paths ending in the set $W=\{w \in V\colon \varepsilon(w)$ is odd$\}$ is even.

Note that the case of Corollary 1.2 in which G is cubic and u is adjacent to v is precisely Smith's theorem.

Corollary 1.3. *Let G be a multigraph with n vertices, $n \geq 4$. Let $u, v, w \in V$ and suppose that $d(x)$ is odd if $x \in V-\{u, v, w\}$. Suppose that every path of length $n-2$ from v to w passes through the vertex u. Then the number of paths of length $n-2$ from u to v which do not contain w is even.*

We prove Corollary 1.3 in the following equivalent form.

Corollary 1.4. *Let G be a multigraph with n vertices, $n \geq 4$. Let $u, v, w \in V$, with $uw, wv \in E$, and let $d(x)$ be odd if $x \in V-\{u, v, w\}$. Suppose that every $(n-1)$-cycle in G passes through the vertex u. Then the number of hamiltonian cycles containing both the edges uw and wv is even.*

Proof. We take our stick to be $s=e_1, e_2$ where $e_1=uw$, $e_2=wv$, $v_1=u$, $v_2=w$ and $v_3=v$. Let h be a hamiltonian path starting with s and ending in a vertex v_n. Then v_n cannot be joined to w since there is no $(n-1)$-cycle in G which doesn't pass through the vertex u. Thus v_n is joined to u by $\varepsilon(v_n)$ edges and so h gives rise to $\varepsilon(v_n)$ hamiltonian cycles containing the edges e_1 and e_2. By Theorem 1.1, the number of such paths ending in the set $W=\{x \in V\colon \varepsilon(x)$ is odd$\}$ is even, and the result then follows.

In the particular case when G is cubic and bipartite, let $w \in V$, and let w have neighbours u_1, u_2 and u_3. By Corollary 1.4 the number of hamiltonian cycles containing the edges u_1w and wu_2 is even; similarly for u_1w and wu_3 and for u_2w and wu_3. Thus the total number of hamiltonian cycles in G is even, and we obtain Kotzig's theorem.

If we restrict ourselves to cubic graphs we can obtain the following stronger result.

Corollary 1.5. *Let G be a cubic graph, and let H be the number of hamiltonian cycles in G. For any vertex $v \in V$, let $g(v)$ be the number of $(n-1)$-cycles not containing v, and for any two incident edges e and f let $h(e, f)$ be the number of hamiltonian cycles containing both e and f. Then*

$$g(v) \equiv h(e, f) \equiv H (\text{mod } 2).$$

Proof. Let $s = e_1, e_2$ be a stick in G. Let a be the number of hamiltonian paths beginning with s and ending in a vertex adjacent to v_1 but not v_2. Let b be the number of hamiltonian paths beginning with s and ending in a vertex adjacent to v_2 but not v_1. Let c be the number of hamiltonian paths beginning with s and ending in a vertex adjacent to both v_1 and v_2. Then $h(e_1, e_2) = a + c$, and since G is cubic, $g(v_0) = b + c$. By Theorem 1.1, $a + b$ is even, and so $h(e_1, e_2) \equiv g(v_0)$ (mod 2). Let now f_1, f_2 and f_3 be the edges incident with a vertex w. The number of hamiltonian cycles not containing the edge f_1 is $h(f_2, f_3)$, so by Smith's theorem $H \equiv h(f_2, f_3)$ (mod 2), and the proof is complete.

Corollary 1.6. *Let G be a graph in which every vertex has even degree. Let u be a vertex of G, and let e be an edge incident to u. Then the number of hamiltonian paths in G which begin at u, contain e, and end in a vertex not adjacent to u, is even.*

Given a multigraph G and a hamiltonian path h beginning with a stick s we can always construct the lollipops which contain h and thus find the vertices adjacent to h in the lollipop graph $♀(G, s)$; thus we have an algorithm for constructing the component of $♀(G, s)$ which contains h. This is particularly simple in the case when G is cubic, since then the components of $♀(G, s)$ are paths and cycles. This algorithm is illustrated in Fig. 2, where given one hamiltonian cycle containing the two dark edges we may find another, since there is no 9-cycle which doesn't contain the vertex x. (This algorithm, applied to cubic planar graphs, was discovered independently by Price [6].)

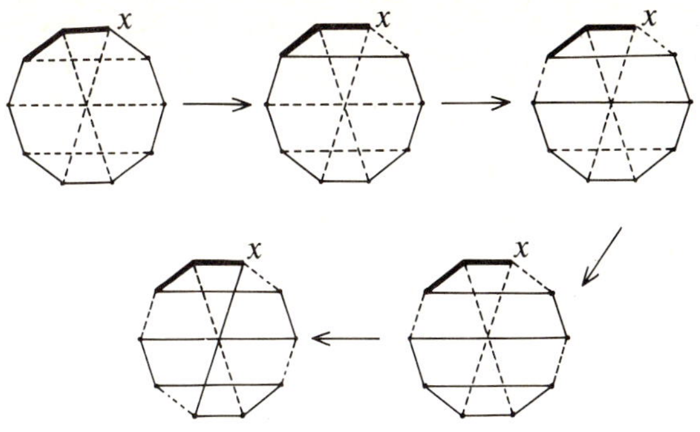

Fig. 2. An algorithm illustrated.

2. Hamiltonian decompositions

Given a multigraph $G=(V, E)$, a partition of E into edge-disjoint hamiltonian cycles is called a *hamiltonian decomposition* of G. A pair $\{h, \bar{h}\}$ of edge-disjoint hamiltonian cycles is called a *hamiltonian pair*. Let now G be 4-regular, that is, $d(v)=4$ for each $v \in V$, and let P be the set of all hamiltonian pairs. Since G is 4-regular a hamiltonian pair is a hamiltonian decomposition of G. For $x, y \in E$, let $P(x, y)$ be the set of hamiltonian pairs in which x and y lie in the same cycle, and let $Q(x, y)$ be the set of hamiltonian pairs in which x and y lie in different cycles; thus $Q(x, y)=P-P(x, y)$. Note that if x, y_1, y_2 and y_3 are the edges incident to a vertex $v \in V$, then $P=\bigcup_{i=1}^{3} P(x, y_i)$ and so $|P|=\sum_{i=1}^{3} |P(x, y_i)|$; in particular if each $|P(x, y_i)|$ is even then so is $|P|$.

I would like to express here my thanks to Mr. Richard Pinch, of Trinity College, Cambridge, whose computing work helped guide me towards the next theorem.

Theorem 2.1. *Let G be a 4-regular multigraph with at least three vertices, and let x and y be any two edges of G. Then the number of hamiltonian pairs in which x and y lie in the same cycle is even.*

Proof. Suppose that the theorem is false, and let G be a counter-example with fewest vertices. Then $|P|>0$, so G is connected and has no loops. Since the only loopless 4-regular multigraph on 3 vertices is the fat triangle (Fig. 3) it follows that $|V| \geqslant 4$.

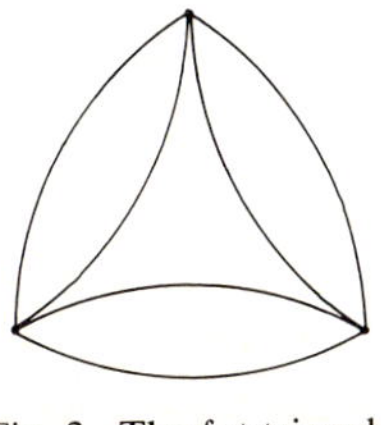

Fig. 3. The fat triangle.

Let z_1 and z_2 be edges with a common endvertex v; say v is joined to vertices u_1 and u_2 by z_1 and z_2 respectively and to vertices $\bar{u}_1$ and $\bar{u}_2$ by edges $\bar{z}_1$ and $\bar{z}_2$ respectively. The multigraph G' is constructed from G by removing v, z_1, z_2, $\bar{z}_1$ and $\bar{z}_2$, and by then adding the edge z between u_1 and u_2 and the edge $\bar{z}$ between $\bar{u}_1$ and $\bar{u}_2$. Given $\{h, \bar{h}\} \in P(z_1, z_2)$ with $z_1, z_2 \in h$, say, then $\bar{z}_1, \bar{z}_2 \in \bar{h}$, and there is a corresponding hamiltonian pair $\{h', \bar{h}'\}$ in G' with $z \in h'$ and $\bar{z} \in \bar{h}'$. Similarly it is clear that to each pair $\{k', \bar{k}'\} \in Q(z, \bar{z})$ there corresponds a pair $\{k, \bar{k}\} \in P(z_1, z_2)$, and so $|P(z_1, z_2)| = |Q(z, \bar{z})|$. But since G' is not a counterexample to the theorem it follows by the remarks made earlier that G' contains evenly many hamiltonian pairs, and so $|Q(z, \bar{z})|$ is even. Hence in G, $|P(z_1, z_2)|$ is even for any two incident edges z_1 and z_2, and in particular $|P|$ is even.

Let now x and y be any two edges of G, and let $x, y_1, y_2, \ldots, y_{r-1}, y_r = y$ be a sequence of edges forming a path whose end edges are x and y. Now for any edge z, the identity

$$Q(x, y) = P(x, z) \,\Delta\, P(z, y)$$

holds (where the triangle denotes symmetric difference) since z is in either the cycle containing x or that containing y. Hence we have for $1 \leqslant i \leqslant r-1$,

$$\begin{aligned} |P(x, y_{i+1})| &= |P| - |Q(x, y_{i+1})| \equiv |Q(x, y_{i+1})| \\ &= |P(x, y_i) \,\Delta\, P(y_i, y_{i+1})| \equiv |P(x, y_i)| + |P(y_i, y_{i+1})| \\ &\equiv |P(x, y_i)| \pmod 2, \end{aligned}$$

since y_i and y_{i+1} have a common endvertex. Thus

$$|P(x, y)| = |P(x, y_r)| \equiv |P(x, y_{r-1})| \equiv \cdots \equiv |P(x, y_1)| \equiv 0 \pmod 2,$$

contradicting our choice of G as a counterexample.

Theorem 2.1 answers a question of Sloane [7], who asked whether the existence of a hamiltonian pair in a graph G implied the existence of another such pair. Sloane showed that if G contains a hamiltonian pair then it contains a third hamiltonian cycle; Sloane's result was improved somewhat by Ninčák [5] who showed that G must contain at least six hamiltonian cycles. Corollary 2.2 includes a further improvement on the estimate of the number of hamiltonian cycles in G.

Corollary 2.2. *Let G be a $2m$-regular multigraph with at least three vertices, where $m \geqslant 1$. If G has a hamiltonian decomposition, then*

(i) *each edge of G is in at least $3m-2$ hamiltonian cycles,*

(ii) *G contains at least $m(3m-2)$ hamiltonian cycles, and*

(iii) *G has at least $(3m-2)(3m-5)\cdots 7.4 \geqslant 3^{m-1}(m-1)!$ hamiltonian decompositions.*

In particular if G has a unique hamiltonian decomposition then G is a cycle.

Proof. We prove statements (i), (ii) and (iii) by induction on m; they are obvious if $m=1$. Suppose $m=2$. By Theorem 2.1 the number $|P|$ of hamiltonian decompositions of G is even. Suppose $e \in E$ and $\{h_1, \bar{h}_1\}, \{h_2, \bar{h}_2\} \in P$ with $e \in h_i$, $i=1, 2$. Then there is an edge $f \in h_1\text{-}h_2$, so $\{h_1, \bar{h}_1\} \in P(e, f)$, and since $|P(e, f)|$ is even it follows that there is a third hamiltonian pair in G. Thus $|P| \geq 4$, G has at least 8 hamiltonian cycles and each edge is in at least 4 hamiltonian cycles.

Now suppose $k > 2$ and the statements are true for all values of $m \leq k-1$. Let $e \in E$ and let $\{h_1, \ldots, h_k\}$ be the given hamiltonian decomposition, with $e \in h_1$, say. Let G_i be the 4-regular subgraph induced by $h_1 \cup h_i$, $2 \leq i \leq k$. G_i has a hamiltonian decomposition, and there are at least three further hamiltonian decompositions $\{h_{il}, \bar{h}_{il}\}$, $1 \leq l \leq 3$, where $e \in h_{il}$. Now if $i \neq j$ then $h_{il} \cap h_{jl'} \subset h_1$ and so $h_{il} \neq h_{jl'}$. Let $H = \{h_1\} \cup \{h_{il}: 2 \leq i \leq k, 1 \leq l \leq 3\}$; then $|H| = 3k-2$ and so statement (i) is proved. Since each hamiltonian cycle contains $n=|V|$ edges it follows that G contains at least $kn \cdot (3k-2)/n$ hamiltonian cycles, and so statement (ii) is proved. Further, if $h \in H$ let $G_h = (V, E\text{-}h)$. Then G_h is $2(k-1)$-regular and has a hamiltonian decomposition, namely $\{h_2, \ldots, h_k\}$ if $h = h_1$ and $\{h_2, \ldots, h_{i-1}, \bar{h}_{il}, h_{i+1}, \ldots, h_k\}$ if $h = h_{il}$, $2 \leq i \leq k$, $1 \leq l \leq 3$. Thus G_h has at least $(3k-5) \cdots 7 \cdot 4$ hamiltonian decompositions, and so G has at least $(3k-2)\ (3k-5) \cdots 7 \cdot 4$, proving statement (iii).

An examination of a few arbitrarily chosen 4-regular graphs with fewer than 20 vertices suggested that the number of hamiltonian pairs in a 4-regular graph with n vertices increases rapidly with n. However, for every $n \geq 10$ there is a graph on n vertices with exactly 32 hamiltonian pairs. Consider first the 4-regular graph T_n, $n \geq 5$, with vertex set $\{0, 1, \ldots, n-1\}$ and with the vertex j joined to the vertices $j \pm 1$ and $j \pm 2$ (addition mod n). T_{12} is illustrated in Fig. 4.

For $0 \leq k \leq n-1$, the sequence of vertices $0, 1, \ldots, k-1, k+1, k, k+2, k+3, \ldots, n-1$ gives rise to a hamiltonian cycle, and the remaining edges also form a hamiltonian cycle; thus T_n has at least n hamiltonian pairs. If n is odd the cycle $0, 1, 2, \ldots, n-1$ also yields a hamiltonian pair. Suppose now that $\{h, \bar{h}\}$ is a hamiltonian pair. It is easily shown that if neither h nor $\bar{h}$ is given by $0, 1, \ldots, n-1$ then h, say, must contain a path of the form $j, j+2, j+1, j+3$, say the path 0, 2, 1, 3. Since 3, 2, 4 is a path in $\bar{h}$ the edge $(3, 4)$ must be in h, so $(3, 5) \in \bar{h}$, so $(4, 5) \in h$ etc., and we see that $\{h, \bar{h}\}$ is one of the pairs described above, and that

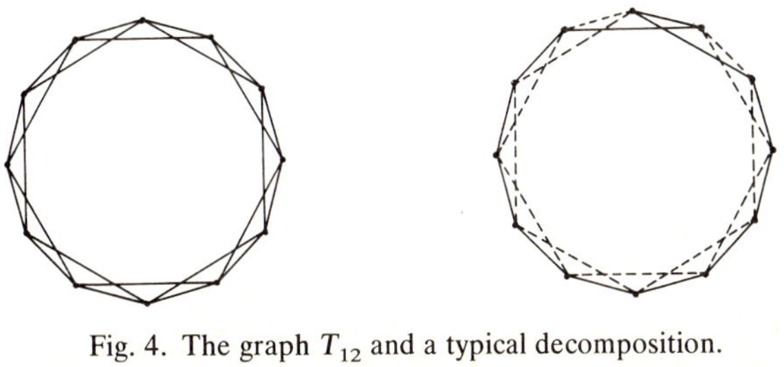

Fig. 4. The graph T_{12} and a typical decomposition.

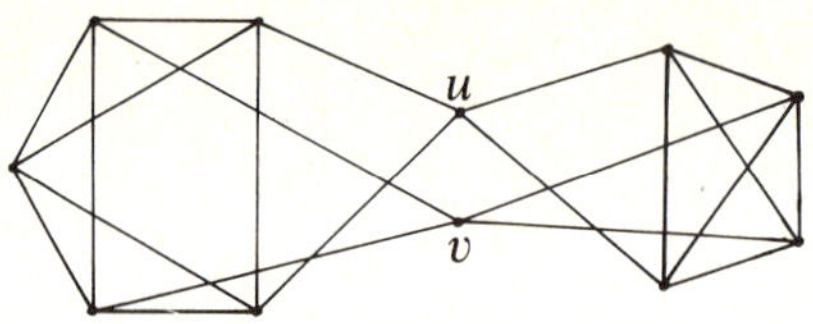

Fig. 5. A graph with 11 vertices and 32 hamiltonian pairs.

T_n has exactly $2\{\frac{1}{2}n\}$ hamiltonian pairs, $\{r\}$ denoting the least integer greater than or equal to the real number r.

Now let $n \geq 10$, and let $n_1 + n_2 = n$, with $n_i \geq 5$, $i = 1, 2$. Let G_i, $i = 1, 2$, be formed from T_{n_i} by removing the vertex 0 and its incident edges and adding vertices u_i and v_i; u_i is joined to 1 and $n_i - 1$ in T_{n_i} and v_i is joined to 2 and $n_i - 2$. Form G by identifying u_1 with u_2 and v_1 with v_2 (see Fig. 5). Then the number of hamiltonian pairs in G is $2p_1p_2$, where p_i is the number of pairs in T_{n_i} in which the edges $(0, 1)$ and $(0, n-1)$ are in different cycles. But by the above remarks $p_i = 4$ and so G has exactly 32 hamiltonian pairs.

3. Uniquely edge colourable graphs

Let G be a graph with $\chi'(G) = 4$, and suppose that G is edge coloured with the colours b, g, r and y. We denote by $u(b)$, say, a vertex u of degree 3 none of whose incident edges are coloured b, and by $v(g, r)$, say, a vertex v of degree 2 whose incident edges are coloured neither g nor r; that is, they are coloured b and y.

If G is uniquely edge colourable, then the subgraph induced by the edges of two given colours is connected, and so is a path or a cycle. We call these colour paths and colour cycles.

Lemma 3.1. *Suppose that $K_{1,4}$ is not the only uniquely 4-edge colourable graph. Then there is a uniquely 4-edge colourable graph G satisfying one of the following two properties:*

(i) *G is 4-regular, or*

(ii) *There are two vertices $u, v \in V$ such that $d(w) = 4$ for each $w \in V - \{u, v\}$; furthermore u and v both have degree 2 and their incident edges are coloured with the same two colours.*

Proof. Let H be a uniquely 4-edge colourable graph. We saw earlier that the subgraph induced by the edges of any two given colours is connected. In particular if H is a tree this means that H has no path of length three: thus $H = K_{1,4}$. Suppose now $H \neq K_{1,4}$. If v is a vertex of degree 1, then the removal of v and its incident edge gives a graph H' which is also uniquely 4-edge colourable; since then H is not a tree we may assume that each vertex of H has degree at least 2. We set about adding edges and vertices to H to obtain uniquely edge colourable graphs with fewer vertices of degree less than 4. If at some stage our graph were to have two vertices of degree 3, u and v say, then either $u = u(b)$ and $v = v(b)$ or $u = u(b)$ and $v = v(g)$. In the first case we add the b-coloured edge uv, and in the second we add the vertex w with a b-coloured edge uw and a g-coloured edge vw. This shows that we may assume H has at most one vertex of degree 3; since H cannot have just one vertex of odd degree, it has none at all.

Let now H have q vertices of degree 2, all other vertices having degree 4. If $q = 0$ then H is regular and we may take $G = H$, so we assume $q \geqslant 1$. Let H have p colour paths; then $p \leqslant \binom{4}{2} = 6$. Furthermore each vertex of degree 2 is an endvertex of exactly 4 colour paths (for instance, $u(b, g)$ is an endvertex of the b–r, b–y, g–r and g–y colour paths), and so $2p = 4q$; that is, $p = 2q$. Since $q \geqslant 1$ we have $p \geqslant 2$, and since each path has two ends we must then have $q \geqslant 2$; thus $q = 3$ or $q = 2$.

Suppose that $q = 3$ (and so $p = 6$) and that u, v, w are the vertices of degree 2. If $u = u(b, g)$ and $v = v(b, g)$, say, then neither u nor v is an endvertex of the b–g colour path, which is impossible since the b–g colour path has two ends. Thus we may assume that $u = u(b, g)$ and $v = v(g, r)$. Then we may add a g-coloured edge uv. We now have two vertices of degree 3 and by the remarks above this reduces to the case $q = 2$.

In the final case $q = 2$ let u and v be the vertices of degree 2, and let $u = u(b, g)$. Then the colour paths are coloured b–r, b–y, g–r and g–y, and so either $v = v(b, g)$ or $v = v(r, y)$, since v is the other endvertex of each of these paths. If $v = v(b, g)$ we may take $G = H$. If $v = v(r, y)$ we may identify u and v to get a 4-regular uniquely edge colourable graph.

Theorem 3.2. *The only uniquely k-edge colourable graph for $k \geqslant 4$ is the star, $K_{1,k}$.*

Proof. If G is uniquely k-edge colourable and G' is the subgraph induced by the edges of k' of the colours, $k' \leqslant k$, then G' is uniquely k'-edge colourable, so we need prove Theorem 3.2 only in the case $k = 4$.

Suppose then that $G \neq K_{1,4}$ is a uniquely 4-edge colourable graph. We may assume that G satisfies property (i) or property (ii) of Lemma 3.1. If G satisfies property (i) then any colour cycle of G is a hamiltonian cycle which is contained in a hamiltonian pair, hence G has at least 3 hamiltonian pairs. But given any hamiltonian pair we may colour one cycle b–g and the other r–y to get an edge colouring of G: this means that G has exactly 3 hamiltonian pairs. But this is impossible by Theorem 2.1 and so G must satisfy property (ii).

Suppose then G has property (ii), and so has two vertices $u(b, g)$ and $v(b, g)$, say. Then the (b–g)-coloured subgraph of G is an $(n-2)$-cycle C_1 (recall that G has n vertices) and the (r–y)-coloured subgraph is a hamiltonian cycle C_2. Let the neighbours of u and v be u_1, u_2 and v_1, v_2 respectively. Construct the multigraph G' from G by removing u and v and their incident edges and adding the edges

$x = u_1u_2$ and $y = v_1v_2$. Then C_1 and C_2 give rise to a hamiltonian pair $\{C_1', C_2'\}$ in G' such that $\{x, y\} \subseteq C_2'$. By Theorem 2.1 there is another hamiltonian pair $\{D_1', D_2'\}$ in G' such that $\{x, y\} \subseteq D_2'$. Hence there is an $(n-2)$-cycle D_1 in G and an edge-disjoint hamiltonian cycle D_2 such that $\{C_1, C_2\} \neq \{D_1, D_2\}$. By colouring D_1 with b and g and colouring D_2 with r and y we get a new edge colouring of G. This contradiction completes the proof of the theorem.

Acknowledgement

I am grateful to Dr. Bollobás of Cambridge for his encouragement and advice whilst this work was being done. Thanks are due also to the Science Research Council for financial help.

References

[1] B. Bollobás, Extremal Graph Theory (Academic Press, London, 1978).
[2] J. Bosák, Hamiltonian Lines in Cubic Graphs, in: P. Rosenstiehl, ed., Théorie des Graphes ICC (Dunod, Paris, 1967) 35–46.
[3] S. Fiorini, On the chromatic index of a graph, III. Uniquely edge-colourable graphs, Quart. J. Math. Oxford Ser. 26 (3) (1975) 129–140.
[4] D. Greenwell and H. Kronk, Uniquely line-colorable graphs, Can. Math. Bull. 16 (1973) 525–529.
[5] J. Ninčák, An estimate of the number of Hamiltonian cycles in a multigraph, in: M. Fiedler, ed., Recent Advances in Graph Theory, Proc. Second Czech. Symp. (Academia, Prague, 1975) 431–438.
[6] W. L. Price, A topological transformation algorithm which relates the hamiltonian circuits of a cubic planar map, J. London Math. Soc. 15 (2) (1977) 193–196.
[7] N. J. A. Sloane, Hamiltonian cycles in a graph of degree 4, J. Combinatorial Theory 6 (1969) 311–312.
[8] W. T. Tutte, On Hamiltonian circuits, J. London Math. Soc. 21 (1946) 98–101.
[9] W. T. Tutt, Hamiltonian circuits, Colloquio Internazionale sulle Teorie Combinatorie, Atti Convegni Lincei 17, Accad. Naz. Lincei, Roma I (1976) 193–199.
[10] R. J. Wilson, Problem 2, in: C. St. J. A. Nash-Williams and J. Sheehan, eds., Proc. Fifth British Comb. Conf. 1975, (Utilitas Mathematica, Winnipeg, 1976) 696.

Annals of Discrete Mathematics 3 (1978) 269–277.

HAMILTONIAN PATHS IN SQUARES OF INFINITE LOCALLY FINITE BLOCKS

Carsten THOMASSEN

Matematisk Institut, Ny Munkegade, 8000 Aarhus, Denmark

1. Introduction

In 1971 Fleischner [2] proved the Plummer and Nash-Williams conjecture asserting that the square of every finite block is Hamiltonian. This result has subsequently been extended in various directions. For example, Fleischner [3] has shown that for any vertex x of a block G, G^2 contains a Hamiltonian cycle C such that the two edges of C incident with x are edges of G.

Nash-Williams [7] proposed to study the analogue of Fleischner's theorem for countable graphs. As demonstrated by the graph of [7, Fig. 2], it is not so that the square of every infinite block has a Hamiltonian path. An obvious necessary condition for a graph to possess a one-way infinite Hamiltonian path is that the deletion of any finite vertex set results in graph with only one infinite component. Nash-Williams [7] asked if this condition is also sufficient when the graph in question is the square of a countable block. The purpose of this paper is to answer this question in the affirmative when restricted to locally finite graphs.

2. Terminology and preliminaries

We adopt the notation and terminology of Harary [5] except that we say *vertex* and *edge* instead of point and line, respectively. The set of vertices and edges of the graph G are denoted by $V(G)$ and $E(G)$, respectively. A *graph* contains no loops or multiple edges, a *multigraph* may contain multiple edges but no loops, and *pseudograph* may contain loops and multiple edges. An edge or multiple edge joining vertices x and y is denoted xy and is called an *x-y edge* (or multiple edge). An *x-loop* is a loop incident with x. An *x-y path* or *trail* is a path or a trail with x and y as end vertices (we only consider open trails, i.e. $x \neq y$). An x-y trail is denoted $x_1x_2 \cdots x_k$ where $x_1 = x$, $x_k = y$ and $x_2, \ldots, x_{k-1}$ are the intermediate vertices. A 1-∞ (one-way infinite) *path* or *trail* is a path or trail of the form $x_1x_2 \cdots$. A 2-∞ (two-way infinite) path or trail is defined analogously.

A *multipath* is a multigraph obtained from a path by replacing some edges with double edges. A *Euler trail* of a multigraph G is a trail containing all edges of G, and a *Hamiltonian path* or, for short, an *H-path*, is a path containing all vertices of G. A *Euler cover* of a finite multigraph is a collection $J_1, J_2, \ldots, J_k$ of mutually edge disjoint trails such that each trail joins two vertices of odd degree in G and such that each edge of G is in one of the trails. A *Hamiltonian cover* of the graph G' is a collection of mutually disjoint paths $P_1, P_2, \ldots, P_m$ such that each vertex of G' is in one of the paths. The Hamiltonian cover of G' is *compatible* to the Euler cover of G if $V(G) = V(G')$, $k = m$, and J_i has the same end vertices as P_i for $i = 1, 2, \ldots, k$. It is an easy consequence of Euler's theorem that every finite multigraph (with at least two vertices of odd degree in every component) has an Euler cover.

If G is a multigraph and H is a subgraph of G, then an *exterior path* in G with respect to H is a path which has precisely its ends in common with H, and an *exterior cycle* with respect to H is a cycle having precisely one vertex in common with H.

If G is a multigraph and $A \subseteq E(G)$, then $(G, A)^2$ is the graph with the same vertex set as G such that two vertices are adjacent if and only if they are adjacent in G, or if they are joined by a path of length 2 using only edges of A, or both. Thus $G^2 = (G, E(G))^2$. If xyz is a path of length 2 using only edges of A, then we may consider the graph G' obtained from G by deleting the edges xy and yz and adding the edge xz (if it is not already present), and we put $A' = A \setminus \{xy, yz\}$. We call the new edge xz a *short-cut* at y and observe that if $(G', A')^2$ has a 1-∞ H-path, then also $(G, A)^2$ has.

An infinite graph G is *d-indivisible* ($d \geq 2$) if the deletion of any finite vertex set from G results in a graph with fewer than d infinite components. It is an easy exercise to show that if G is *locally finite* (i.e. every vertex of G has finite degree), then G is d-indivisible if and only if G^2 is d-indivisible.

We shall use the following results.

Theorem 2.1. (Erdös *et al.* [1]). *A countably infinite connected multigraph G has a 1-∞ Euler trail starting at x_1 if and only if x_1 has odd or infinite degree, no other vertex has odd degree, and the deletion of a finite edge set from G results in a graph with only one infinite component.*

Theorem 2.2. *Let G be a finite connected graph and S a subset of $V(G)$ with at least two elements such that each endblock of G contains a vertex of S (distinct from the cutvertex of the block). Then G^2 has a Hamiltonian cover each path of which connects two vertices of S.*

Proof (due to Fleischner). Let G' be the graph obtained from G by adding a new vertex x and joining x by edges to each vertex of S. Then G' is a block and by a result of Fleischner [3, Theorem 3], G' has a Hamiltonian cycle such that the two edges of C incident with x are edges of G'. But then $C \cap G^2$ is a Hamiltonian cover of G^2 with the desired property.

3. Characterization of infinite, locally finite, connected, d-indivisible graphs

The theorem of this paper is based on the following characterization of infinite, locally finite, 2-indivisible connected graphs.

Proposition 3.1. *Let G be an infinite, locally finite, connected graph. Then G is 2-indivisible if and only if for every vertex x_1 of G, G contains a 1-∞ path $P: x_1 x_2 \cdots$ such that every component of $G - V(P)$ is finite.*

Proof. The "if" part is easy to prove so we proceed to the "only if" part. Since G is connected and locally finite, G is countable so we can enumerate the vertices of G: $y_1, y_2, \ldots$ where $y_1 = x_1$. We now define recursively a sequence $P_1, P_2, \ldots$ of finite paths starting at x_1 such that P_i is a segment of P_{i+1} for $i = 1, 2, \ldots$, and the end vertex of P_i other than x_1 is adjacent to the infinite component of $G - V(P_i)$ (by assumption, there is only one infinite component in this graph). We let P_1 consist of $y_1 = x_1$ only. Having already defined P_i, we define P_{i+1} as follows: Let z_i denote the end vertex of P_i other than x_1 and let $k(i)$ be the smallest integer j such that y_j belongs to the infinite component of $G - V(P_i)$. Then $G - [V(P_i) \setminus \{z_i\}]$ contains a $z_i - y_{k(i)}$ path $z_i u_1 u_2 \cdots u_r$ $(u_r = y_{k(i)})$. Now let $l(i)$ be the largest integer j such that u_j is adjacent to the infinite component of $G - [V(P_i) \cup \{u_1, u_2, \ldots, u_j\}]$. Then $l(i) \leq r$ and it is easy to see that $l(i) \geq 1$. Let P_{i+1} be the union of P_i and the path $z_i u_1 u_2 \cdots u_{l(i)}$. Then either $y_{k(i)}$ is in P_{i+1} or it belongs to a finite component of $G - V(P_{i+1})$. Now the 1-∞ path $P = \bigcup_{i=1}^{\infty} P_i$ has the desired properties since for each i, the vertex y_i is either in P_i or in a finite component of $G - V(P_i)$ and hence it is also in a finite component of $G - V(P)$.

Proposition 3.1 can also be derived from a result of Jung [6, Satz 3] since a 2-indivisible graph has only one "Ende" in the sense of Halin [4] and thus a $1-\infty$ path of the type described in Proposition 3.1 is a "Hauptweg" in the sense of Jung [6].

From Proposition 3.1 we easily get a characterization of d-indivisible locally finite graphs.

Proposition 3.2. *The infinite, locally finite graph G is d-indivisible if and only if it contains a set of k 1-∞ paths $P_1, P_2, \ldots, P_k$, where $k < d$, such that every component of $G - \bigcup_{i=1}^{k} V(P_i)$ is finite.*

Proof. Again, the "if" part is easy so we now assume that G is d-indivisible and we let k be the smallest integer such that G is $(k+1)$-indivisible. Then $k < d$ and G contains a finite vertex set S such that G-S has k infinite components $G_1, G_2, \ldots, G_k$. By assumption, each G_i $(i = 1, 2, \ldots, k)$ is 2-indivisible and contains therefore, by Proposition 3.1, a 1-∞ path P_i such that each component of $G_i - V(P_i)$ is finite. Then also each component of $G - \bigcup_{i=1}^{k} V(P_i)$ is finite because S is finite and each vertex of S has finite degree.

4. Euler trails versus Hamiltonian paths in a certain infinite graphs

In this section we show how Hamiltonian paths can be obtained from Euler trails in certain infinite graphs. We need the following lemma about finite graphs.

Lemma 4.1. *Let $P: x_1x_2 \cdots x_n$ ($n \geq 2$) be a multipath and let G be a graph obtained from P by adding a (possibly empty) system of pairwise disjoint exterior cycles and a new (possibly empty) set S of vertices and joining each vertex of S to precisely one vertex of P (by a single edge) such that G satisfies the conditions below.*

(i) *No vertex of P is joined to more than two vertices of S and at most one vertex of P is joined to two vertices of S.*

(ii) *No vertex of P is at the same time joined to a vertex of S and contained in an exterior cycle.*

(iii) *Every vertex not in $S \cup \{x_1, x_n\}$ has even degree in G.*

(iv) *x_n has not degree 3 in G, and if P contains a vertex which has degree 6 and which is adjacent to two vertices of S, then both of x_1, x_n has even degree (note that* (i) *and* (ii) *imply that no vertex of G has degree greater than 6 and that x_1 and x_n have degree at most 4).*

Let $J_1, J_2, \ldots, J_k$ be a Euler cover of G such that at most one trail starts at x_1. Then $(G, A)^2$ has a compatible Hamiltonian cover $P_1, P_2, \ldots, P_k$, where A is the set of edges incident with or contained in P.

Proof. We prove the lemma by induction on the number of edges of G. The lemma is easily verified for $n = 2$ so we proceed to the induction step and assume $n \geq 3$.

First we consider the case where P has a vertex x_j ($2 \leq j \leq n-1$) such that x_j is joined to each of x_{j-1} and x_{j+1} by a double edge and to two vertices u_1 and u_2 of S. If the path $u_1x_ju_2$ is one of the trails J_i (say J_1), then we put P_1 equal to the path u_1u_2 (in $(G, A)^2$) and use induction on $G - \{u_1, u_2\}$. So assume w.l.g. that the trail J_1 starts with the path $u_1x_jx_{j+\delta}$ where $\delta = \pm 1$. Let J_s be the trail starting with u_2. Then J_s contains a segment of the form $u_2x_jx_{j+\sigma}$ where $\sigma = \pm 1$. If $\sigma = \delta$, then some trail, say J_m, distinct from J_1 and J_s contains the segment $x_{j-\delta}x_jx_{j-\delta}$. We delete the double edge $x_{j-\delta}x_j$ from G (and from J_m) and apply the induction hypothesis on the components of the resulting multigraph. So we can assume without loss of generality that $\delta = -1$, $\sigma = 1$.

Now let J_m be the trail containing the segment $x_{j-1}x_jx_{j+1}$ (m may equal 1 or s or both). Vertex x_j partitions J_m into two trails J_m^1 (containing an edge $x_{j-1}x_j$) and J_m^2 (containing an edge x_jx_{j+1}). Let G_1 be the subgraph of G induced by $x_1, x_2, \ldots, x_{j-1}$ and all their neighbours and the exterior cycles containing an x_i ($1 \leq i < j$) and the vertex u_1. Let G_2 be the graph consisting of the edges not in G_1 and their ends. Then those of the trails $J_1, J_2, \ldots, J_{m-1}, J_m^1, J_m^2, \ldots, J_k$ which are contained in G_1 (resp. G_2) form an Euler cover of G_1 (resp. G_2). By the induction hypothesis there is a compatible Hamiltonian cover of $(G_i, [A \cap E(G_i)])^2$ for $i = 1, 2$. Let P^i_m be the path corresponding to J^i_m $(i = 1, 2)$ and let P_s be the path corresponding to J_s $(s = 1, 2, \ldots, k;\ s \neq m)$. Then $P_1, P_2, \ldots, P^1_m \cup P^2_m$, $P_{m+1}, \ldots, P_k$ is a Hamiltonian cover of $(G, A)^2$ compatible to $J_1, J_2, \ldots, J_k$.

Next we consider the case where there is a j $(2 \leq j \leq n-1)$ such that x_j and x_{j-1} are joined only by a single edge. Suppose J_1 is the trail containing this edge. As in the previous case, we write G as the union $G_1 \cup G_2$ where $G_1 \cap G_2 = \{x_j\}$ and if x_j is adjacent to a vertex of S or contained in an exterior cycle, then we let this vertex (resp. cycle) be contained in G_2. Vertex x_j partitions J_1 into two trails J^1_1, J^2_1 such that J^i_1 is contained in G_i for $i = 1, 2$. By applying the induction hypothesis on each G_i $(i = 1, 2)$, we get a Hamiltonian cover of G (one of the paths being the union of the paths corresponding to J^1_1 and J^2_1, respectively).

If x_{n-1} is joined to x_n by a single edge, then we argue as in the previous case. The only difference is that now $G_1 \cap G_2 = \{x_{n-1}\}$ (instead of $\{x_n\}$) and if x_{n-1} is adjacent to a vertex of S or contained in an exterior cycle, then we let this vertex (resp. cycle) belong to G_2.

So we can assume that each edge of P is a double edge and that no vertex of $\{x_2, x_3, \ldots, x_{n-1}\}$ is adjacent to a vertex of S, for if one of these vertices were adjacent to a vertex of S, then it would be adjacent to two vertices of S and this situation we have already considered. This implies that k, the number of trails in the Euler cover, equals 1 or 2. Now the structure of G and its Euler cover is so simple that it is easy to find a compatible Hamiltonian cover. Let us here only consider the case where x_1 is joined to precisely one vertex u_1 of S and x_n is joined to precisely two vertices, say u_2 and u_3, of S, and J_1 connects u_1 with u_2, and J_2 connects x_1 with u_3. Then $(G, A)^2$ contains the two paths $P_1: u_1x_2x_4 \cdots u_2$ and $P_2: x_1x_3x_5 \cdots u_3$. If $x_jy_1y_2 \cdots y_rx_j$ is an exterior cycle, then we replace the edge $x_{j-1}x_{j+1}$ (which is present in either P_1 or P_2) by the path $x_{j-1}y_1y_2 \cdots y_rx_{j+1}$. By doing this for every exterior cycle, we obtain a Hamiltonian cover of $(G, A)^2$ compatible to J_1, J_2. The other cases are treated analogously.

This completes the proof of the lemma.

Using this lemma we can prove the main result of this section.

Proposition 4.2. *Let G be a locally finite graph obtained from a* 1-∞ *path $P: x_1x_2x_3 \cdots$ by adding exterior paths and cycles to P in such a way that no two exterior paths or cycles intersect outside P. Let A be the set of edges contained in or incident with P. Then $(G, A)^2$ has a* 1-∞ *H-path starting at x_1.*

Proof. Throughout the proof, the terms exterior path and exterior cycle mean exterior path (resp. cycle) with respect to P. An *exterior z-cycle* is an exterior cycle having z in common with P. In order to describe the system of exterior paths and cycles, we introduce a pseudograph $\Gamma_l(G)$ with vertex set $V(P)$ such that for every exterior x_i-x_j path (resp. exterior x_i-cycle) there is in $\Gamma_l(G)$ an x_i-x_j edge (resp. x_i loop). The multigraph obtained from $\Gamma_l(G)$ by deleting all loops is denoted $\Gamma(G)$.

We first show that it is no loss of generality that $\Gamma(G)$ is a forest. For otherwise, we consider a maximal system of mutually edge disjoint cycles (by Zorn's lemma, such a system exists). Let $C_1, C_2, C_3, \ldots$ by the cycles in this system and let z_i be a vertex of C_i $(i = 1, 2, \ldots)$. By making short-cuts at the vertices of C_i (other than z_i), we transform the exterior paths (in G) corresponding to the edges of C_i into an exterior z_i-cycle. By doing this for every cycle C_i, we transform G into a graph G' such that $\Gamma(G')$ is a forest. (Note that G' (and hence also $\Gamma_l(G')$) is locally finite because $\{i : z_i = z_j\}$ is finite for every j).

So we assume that $\Gamma(G)$ is a forest and show next that we can even assume without loss of generality that every component of $\Gamma(G)$ is a path. For this, let Π be any component of $\Gamma(G)$ and let y_0 be a (fixed) vertex of Π. For any vertex z of Π, let $z_1, z_2, \ldots, z_{2k+\delta}$ $(\delta = 0$ or $1)$ be the vertices of Π adjacent to z and not in the y_0-z path in Π. By making k short-cuts at z, we transform G into a graph G' such that in $\Gamma(G')$, z is adjacent to precisely δ of the vertices $z_1, z_2, \ldots, z_{2k+\delta}$. We do this for every vertex z of Π and every component Π of $\Gamma(G)$. In this way G is transformed into a graph G'' such that no vertex of $\Gamma(G'')$ has degree greater than 2. It is easy to see that no component of $\Gamma(G'')$ is a cycle, so every component is a path.

We can therefore assume that each component of $\Gamma(G)$ is a path. Now, let Π be such a path (of length ≥ 1). If a vertex z of Π is incident with k loops of $\Gamma_l(G)$, then by making $k+1$ short-cuts at z (or k short-cuts if z is an end vertex of Π), we can get rid of these loops and if Π is finite (say of length r), we can transform Π into a path of length 1 by making $r-1$ short-cuts. Similarly, k loops at a vertex can be transformed into just one loop. Combining these observations, we can assume without loss of generality:

(1) Every component of Γ_l is either a 2-∞ path or a 1-∞ path or a path of length 1 or a cycle of length 1 (i.e. a loop together with its incident vertex).

We now denote by G_M the multigraph obtained from G by replacing some edges of P with double edges such that in G_M, x_1 has odd degree and every other vertex has even degree. It is easy to see that this can be done and that G_M is uniquely determined. Let P_M denote the multigraph in G_M corresponding to P. It is easily verified that G_M satisfies the condition of Theorem 2.1 for a multigraph to contain a 1-∞ Euler trail. For if S is any finite edge-set of G_M and n is an integer such that no edge of S is incident with a vertex x_{n+k} $(k \geq 0)$, then all but finitely many vertices of $G_M - S$ belong to the same component as the vertices $x_n, x_{n+1}, \ldots$.

Separately we consider the cases where P_M has finitely, respectively infinitely, many double edges. Consider first the case where there exists an integer n_0 such that no edge between x_i and x_{i-1} is a double edge in G_M for $i \geq n_0$. We can assume that in $\Gamma(G)$ every x_j with $j \leq n_0$ has degree one or zero. For otherwise we achieve this by making a short-cut at every x_j ($j \leq n_0$) which has degree 2 in $\Gamma(G)$.

Now let $J: x_1 z_2 z_3 \cdots$ be an Euler trail of G_M. We shall modify J so as to obtain an H-path of $(G, A)^2$. Let L be the subgraph of G_M induced by the vertices and exterior cycles adjacent to (resp. containing) some of the vertices $x_1, x_2, \ldots, x_{n_0-1}$ (in particular, $x_{n_0} \in V(L)$). By Lemma 4.1, we can modify that part J' of J which is in L so as to obtain a system of disjoint paths in L^2 such that these paths together with the part of J outside L form a trail J_1 of $(G, A)^2$ which includes every vertex of L and every edge outside L. We now transform J_1 into an H-path of $(G, A)^2$ by considering each segment $z_{i-1} z_i z_{i+1}$ of J_1 such that $z_i \in \{x_{n_0}, x_{n_0+1}, \ldots\}$ and $z_{i-1} \in V(G) \setminus V(P)$, and by replacing it with a short-cut at z_i, when $z_{i+1} \in V(G) \setminus V(P)$ and when z_{i+1} is the successor of z_i in P.

Consider next the case where infinitely many edges of P_M are double edges. We consider first the subcase where no component of $\Gamma_l(G)$ is an infinite path, i.e., every component of $\Gamma_l(G)$ is a path or cycle of length 1. If x_i and x_{i+1} are joined by a double edge in G_M then $\Gamma(G)$ contains an edge of the form $x_s x_t$ where $s \leq i < t$ because G_M has a Euler trail starting at x_1. So we can select integers $i_1, i_2, \ldots$ and $s_1, s_2, \ldots$ and $t_1, t_2, \ldots$ such that $s_j \leq i_j < t_j$ and the vertices x_{i_j}, x_{i_j+1} are joined by a double edge in G_M and x_{s_j}, x_{t_j} are adjacent in $\Gamma(G)$. By choosing i_{j+1} sufficiently large compared to i_j, we can assume that $t_j < s_{j+1}$ for each j. Let $\bar{G}_M$ denote the graph obtained from G_M by deleting all the double edges $x_{i_j} x_{i_j+1}$ ($j = 1, 2, \ldots$). Then $\bar{G}_M$ is connected, x_1 has odd degree in $\bar{G}_M$, and every other vertex has even degree. Moreover, the deletion from $\bar{G}_M$ of any finite edge set results in a graph with only one infinite component. So $\bar{G}_M$ has a Euler trail J. By using Lemma 4.1, we modify for each $j = 0, 1, 2, \ldots$ the part of J which is in the graph of $\bar{G}_M$ induced by the vertices in $\{x_r : i_j + 1 \leq r \leq i_{j+1}\}$ and the exterior cycles containing one of these vertices and the neighbouring vertices (here $i_0 = 0$), and we thereby obtain an H-path of $(G, A)^2$ starting at x_1.

Finally, we have to consider the case where P_M has infinitely many double edges and $\Gamma(G)$ has at least one 1-∞ path, say Π_0. $\Gamma(G)$ has only countably many 1-∞ paths, so we can write a sequence $\Pi_1, \Pi_2, \ldots$ of 1-∞ paths of $\Gamma(G)$ such that every 1-∞ path in $\Gamma(G)$ occurs infinitely often in this sequence. We define recursively sequences $i_1, i_2, \ldots$ and $s_1, s_2, \ldots$ and $t_1, t_2, \ldots$ of natural numbers as follows: We let s_1 be any vertex of Π_0. Suppose we have already defined $i_1, \ldots, i_{j-1}$, $s_1, s_2, \ldots, s_j$, $t_1, t_2, \ldots, t_{j-1}$. Then we let i_j be the smallest integer greater than or equal to s_j such that x_{i_j} and x_{i_j+1} are joined by a double edge and let t_j be any integer greater than i_j such that x_{t_j} is a vertex of Π_j and we let s_{j+1} be any integer greater than t_j such that $x_{s_{j+1}}$ is a vertex of Π_0.

We can assume that the vertices x_{s_j}, x_{t_j} ($j = 1, 2, \ldots$) are the only vertices of degree 2 in $\Gamma(G)$, for otherwise this can be achieved by making short-cuts at all other vertices of degree 2 in $\Gamma(G)$. By doing this, we still have a graph of the same type as described in the proposition because every 1-∞ path of $\Gamma(G)$ occurs infinitely often in the sequence $\Pi_1, \Pi_2, \ldots$. We now denote by $\bar{G}_M$ the multigraph obtained from G_M by deleting all double edges $x_{i_j} x_{i_j+1}$ ($j = 1, 2, \ldots$). Then $\bar{G}_M$ is connected (because each edge $x_{i_j}x_{i_j+1}$ is compensated for by the exterior path corresponding to the edge $x_{s_j}x_{s_{j+1}}$ of $\Gamma(G)$).

Furthermore, x_1 is the only vertex of odd degree in $\bar{G}_M$ and the deletion of a finite set of edges of $\bar{G}_M$ results in a graph with only one infinite component. So $\bar{G}_M$ has an Euler trail J. By the choice of the integers i_j, each vertex x_{s_j} has degree 4 in $\bar{G}_M$. So for every j $(j=0, 1, 2, \ldots)$ we modify, using Lemma 4.1, the part of J in the subgraph of $\bar{G}_M$ induced by the vertices $x_{i_j+1}, x_{i_j+2}, \ldots, x_{i_{j+1}}$ and the neighbouring vertices and the exterior cycles containing one of these vertices so as to obtain an H-path of $(G, A)^2$ starting at x_1.

The proof is complete.

5. Hamiltonian paths in infinite 2-indivisible locally finite blocks

By combining Theorem 2.2 and Propositions 3.1, 4.2, we easily prove the theorem of this paper.

Theorem 5.1. *Let G be an infinite, locally finite, 2-indivisible block, and let x_1 be any vertex of G. Then G^2 has a 1-∞ Hamiltonian path starting at x_1.*

Proof. By Proposition 3.1, G has a 1-∞ path $P: x_1x_2x_3\cdots$ such that each component of $G-V(P)$ is finite. Consider such a component, say K. Let S be those vertices of K which are joined to P by edges. Then the pair K, S satisfies the assumption of Theorem 2.2 because G is a block so K^2 contains a Hamiltonian cover $P_1, P_2, \ldots, P_k$ such that each P_i can be extended to an exterior cycle or path with respect to P by adding two edges of G. We form these exterior paths and cycles for every component K of $G-V(P)$ and obtain a graph G' satisfying the assumption of Proposition 4.2. By this proposition, $(G', A)^2$ has a 1-∞ Hamiltonian path starting at x_1 where A is the set of edges incident with or contained in P. Since $A \subseteq E(G)$, this Hamiltonian path is also a Hamiltonian path of G^2 and the proof is complete.

6. Concluding remarks

If G satisfies the assumption of Proposition 4.2, then also $(G, A)^2$ has a two-way infinite Hamiltonian path. In order to prove this, we define in the proof of Proposition 4.2 the multigraph G_M in such a way that every vertex has even degree and we replace "1-∞ Euler trail" with "2-∞ Euler trail" throughout the proof and we use the conditions [1] for the existence of such a trail. Using this result, we can modify the theorem to obtain the result that the square of every infinite locally finite 2-indivisible block has a two-way infinite Hamiltonian path. Moreover, by using Proposition 3.2 (with $d=3$), we can also prove by the methods of this paper that the square of every infinite, locally finite 3-indivisible block has a two-way infinite Hamiltonian path.

Acknowledgement

The author is indebted to Dr. H. Fleischner for several stimulating discussions on the subject and, in particular, for the proof of Theorem 2.2.

References

[1] P. Erdös, T. Grünwald and E. Vàzsonyi, *Über Euler-Linien unendlicher* Graphen, J. Math. Phys. Mass. Inst. Technology 17 (1938) 59–75.

[2] H. Fleischner, *The square of every two-connected graph is Hamiltonian*, J. Combinatorial Theory, 16(B) (1974) 29–34.

[3] H. Fleischner, *In the square of graphs, Hamiltonicity and pancyclicity, Hamiltonian connectedness and panconnectedness are equivalent concepts*, Monatsh. Math. 82 (1976), 125–149.

[4] R. Halin, Über unendliche Wege in Graphen, Math. Ann. 157 (1964) 125–137.

[5] F. Harary, *Graph Theory*, (Addison Wesley, Reading, MA, 1969).

[6] H.A. Jung, Normale Wurzelbäume und unendliche Weg in Graphen, Math. Nachr. 41 (1969) 1–22.

[7] C.St.J.A. Nash-Williams, *Unexplored and semi-explored territories in graph theory*, in: (F. Harary, ed.,) New Directions in the Theory of Graphs (Academic Press, New York, 1973) 149–186.

Annals of Discrete Mathematics 3 (1978) 279–287.

AN INVESTIGATION OF COLOUR-CRITICAL GRAPHS WITH COMPLEMENTS OF LOW CONNECTIVITY

Bjarne TOFT

Department of Mathematics, University of Odense, DK-5230 Odense M, Denmark

A k-chromatic graph Γ is called v-critical if all vertices x of the graph are critical, i.e. $\chi(\Gamma - x) < \chi(\Gamma) = k$. It is called e-critical if it is connected and all edges e of the graph are critical, i.e. $\chi(\Gamma - e) < \chi(\Gamma)$. It follows from these definitions that an e-critical graph is also v-critical. The critical 3-chromatic graphs are the odd circuits. It seems hopeless to try to determine the structure of all critical graphs with chromatic number 4 or more.

G.A. Dirac observed that a graph $\bar{\Gamma}$ is the complement of a v-critical (e-critical) graph Γ if and only if each connected component of $\bar{\Gamma}$ is the complement of a v-critical (e-critical) graph. This result gives a method for constructing new critical graphs from known ones. Let us for example take two disjoint odd circuits and join them completely by edges. The result is an e-critical 6-chromatic graph Γ, and if the two circuits are of equal length then the number of edges of Γ is $> \frac{1}{4}n^2$, where n denotes the number of vertices of Γ. This shows the existence of e-critical 6-chromatic graphs with "many edges", i.e. with more than $c \cdot n^2$ edges, where c is a positive constant.

The aim of this note is to give an account of an investigation of those critical graphs Γ, whose complements $\bar{\Gamma}$ have low vertex-connectivity, thus continuing the line of study initiated by Dirac's result. The investigation has been carried through for $\bar{\Gamma}$ of connectivity ≤ 2, and the critical graphs of this structure have been characterized.

The main-tool of the investigation is

Lemma 1. *Let Γ be a graph, and let C be a subset of the set of vertices of Γ such that $\bar{\Gamma} - C$ is disconnected. Let Γ_1 and Γ_2 be subgraphs of Γ such that* (*cf. Fig.* 1)

(1) *$\Gamma_1 \cap \Gamma_2$ is the subgraph of Γ spanned by C,*
(2) *$\Gamma_1 - C \neq \emptyset$ and $\Gamma_2 - C \neq \emptyset$,*
(3) *Γ consists of $\Gamma_1 \cup \Gamma_2$ together with all edges from $\Gamma_1 - C$ to $\Gamma_2 - C$.*

Let S denote the set of all subsets of C including the empty set and the whole set. Then

$$\chi(\Gamma) = \min_{\Lambda \in S} \{\chi(\Gamma_1 - (C - \Lambda)) + \chi(\Gamma_2 - \Lambda)\}.$$

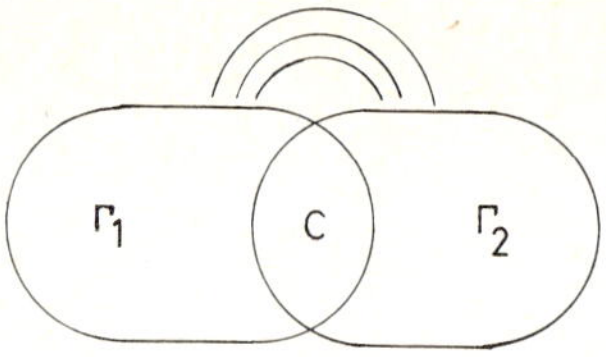

Fig. 1

Proof. Let $\Lambda \in S$. Colour $\Gamma_1-(C-\Lambda)$ with $\chi(\Gamma_1-(C-\Lambda)$ colours and $\Gamma_2-\Lambda$ with $\chi(\Gamma_2-\Lambda)$ other colours. Each vertex of Γ has then got a colour. The result is a $(\chi(\Gamma_1-(C-\Lambda))+\chi(\Gamma_2-\Lambda))$-colouring of Γ. Hence

$$\chi(\Gamma)\leq \min_{\Lambda\in S}\{\chi(\Gamma_1-(C-\Lambda))+\chi(\Gamma_2-\Lambda)\}.$$

On the other hand, let K be a $\chi(\Gamma)$-colouring of Γ. A set $\Lambda\in S$ is defined as follows. A vertex x of C is contained in Λ if and only if x has the same colour in K as at least one vertex of Γ_1-C. Then in K the colours of the vertices of $\Gamma_1-(C-\Lambda)$ are all different from the colours of the vertices of $\Gamma_2-\Lambda$. Hence

$$\chi(\Gamma)\geq\chi(\Gamma_1-(C-\Lambda))+\chi(\Gamma_2-\Lambda)\geq \min_{\Lambda\in S}\{\chi(\Gamma_1-(C-\Lambda))+\chi(\Gamma_2-\Lambda)\}$$

and the Lemma follows.

Note that if in the Lemma $\Gamma-C$ has ≥ 3 connected components, then the two graphs Γ_1 and Γ_2 satisfying (1), (2) and (3) are not uniquely determined from the structure of Γ.

For $\bar{\Gamma}$ of connectivity ≤ 1 the results of the investigation are:

Theorem 2. *A graph $\bar{\Gamma}$ is the complement of a v-critical graph Γ if and only if each block of $\bar{\Gamma}$ is the complement of a v-critical graph. Moreover, if Γ is v-critical then*

$$\chi(\Gamma)=\left(\sum_{j=1}^{b}\chi(\Gamma_j)\right)-(b-c),$$

where $\Gamma_1, \Gamma_2, \ldots, \Gamma_b$ are the blocks of $\bar{\Gamma}$ and c is the number of connected components of $\bar{\Gamma}$.

Theorem 3. *If Γ is e-critical then $\bar{\Gamma}$ has no cutvertices, i.e. each connected component of $\bar{\Gamma}$ consists of precisely one block.*

Proof of Theorem 2. The proof is by induction over the number b of blocks of $\bar{\Gamma}$. If $b=1$ then the theorem is obviously true. Suppose that $b\geq 2$ and that the theorem is true for graphs containing $<b$ blocks.

If $\bar{\Gamma}$ is disconnected then by the induction-hypothesis the theorem holds for each connected component of $\bar{\Gamma}$. By the above result of Dirac the theorem then holds for $\bar{\Gamma}$. We may, therefore, assume that $\bar{\Gamma}$ is connected.

Since $\bar{\Gamma}$ is connected and has $\geqslant 2$ blocks, Γ has the structure described in Lemma 1 with C consisting of a single vertex x. Then by Lemma 1

(1) $\quad \chi(\Gamma)=\min\{\chi(\Gamma_1-x)+\chi(\Gamma_2),\ \chi(\Gamma_1)+\chi(\Gamma_2-x)\}.$

Suppose first that Γ is v-critical. Then

(2) $\quad \chi(\Gamma)=\chi(\Gamma-x)+1=\chi(\Gamma_1-x)+\chi(\Gamma_2-x)+1.$

From this and (1) it follows that we may choose the notation such that $\chi(\Gamma_1-x)=\chi(\Gamma_1)-1$. Then by (1)

(3) $\quad \chi(\Gamma)=\chi(\Gamma_1)+\chi(\Gamma_2)-1.$

By (2) and (3) x is critical vertex of both Γ_1 and Γ_2.

Let z be a vertex of Γ_i-x. Then by (1) used on $\Gamma-z$, and by $\chi(\Gamma-z)=\chi(\Gamma)-1$ and (3), it follows that

(4) $\quad \chi(\Gamma-z)=\min\{\chi(\Gamma_i-z-x)+\chi(\Gamma_j),\ \chi(\Gamma_i-z)+\chi(\Gamma_j-x)\}$
$\quad\quad =\chi(\Gamma_1)+\chi(\Gamma_2)-2.$

By (4) either $\chi(\Gamma_i-z-x)=\chi(\Gamma_i)-2$ or $\chi(\Gamma_i-z)=\chi(\Gamma_i)-1$. The first alternative also implies $\chi(\Gamma_i-z)=\chi(\Gamma_i)-1$. Hence z is a critical vertex of Γ_i.

We have then proved that both Γ_1 and Γ_2 are v-critical. Both $\bar{\Gamma}_1$ and $\bar{\Gamma}_2$ have $<b$ blocks, hence by the induction-hypothesis all blocks of $\bar{\Gamma}_1$ and $\bar{\Gamma}_2$, and therefore of $\bar{\Gamma}$, are complements of v-critical graphs. The formula for $\chi(\Gamma)$ follows from the corresponding formulas for $\chi(\Gamma_1)$ and $\chi(\Gamma_2)$, which are true by the induction-hypothesis, and (3).

Suppose on the other hand that each block of Γ is the complement of a v-critical graph. Then by the induction-hypothesis Γ_1 and Γ_2 are v-critical. By (1) we get again (3) and, therefore,

$$\chi(\Gamma-x)=\chi(\Gamma_1-x)+\chi(\Gamma_2-x)=\chi(\Gamma_1)-1+\chi(\Gamma_2)-1=\chi(\Gamma)-1,$$

i.e. x is a critical vertex of Γ. Again by (1) used on $\Gamma-z$, and by Γ_1 and Γ_2 v-critical, (4) follows. Then by (3) and (4) $\chi(\Gamma-z)=\chi(\Gamma)-1$, thus z is a critical vertex of Γ. Hence Γ is v-critical. This completes the proof of Theorem 2.

Proof of Theorem 3. Suppose that the theorem is false. We shall obtain a contradiction. Let Γ be e-critical and let x be a cutvertex of $\bar{\Gamma}$. The vertex x is a cutvertex of the connected component of $\bar{\Gamma}$ in which it is contained, hence by the result of Dirac we may assume that $\bar{\Gamma}$ is connected.

Let again Γ_1 and Γ_2 be as in Lemma 1 with $C=\{x\}$. Since both $\bar{\Gamma}_1$ and $\bar{\Gamma}_2$ are connected and with $\geqslant 2$ vertices, there exists for $i=1, 2$ a vertex z_i in Γ_i-x such that (x, z_i) is an edge of $\bar{\Gamma}_i$.

By (3) $\chi(\Gamma)=\chi(\Gamma_1)+\chi(\Gamma_2)-1$, and hence, since $e=(z_1, z_2)$ is a critical edge of

Γ, $\chi(\Gamma - e) = \chi(\Gamma_1) + \chi(\Gamma_2) - 2$. In a $(\chi(\Gamma_1) + \chi(\Gamma_2) - 2)$-colouring of $\Gamma - e$ the vertices z_1 and z_2 necessarily have the same colour, moreover the only further vertex that can possibly have this colour is x. Hence

$$\chi(\Gamma - z_1 - z_2 - x) \leq \chi(\Gamma_1) + \chi(\Gamma_2) - 3.$$

On the other hand

$$\chi(\Gamma - z_1 - z_2 - x) = \chi(\Gamma_1 - z_1 - x) + \chi(\Gamma_2 - z_2 - x)$$
$$\leq \chi(\Gamma_1) - 1 + \chi(\Gamma_2) - 1.$$

The last inequality follows since z_i and x are not joined by an edge of Γ_i for $i = 1$, 2.

A contradiction has then been obtained. This completes the proof of Theorem 3.

Like Dirac's result the above Theorem 2 gives a method for constructing new v-critical graphs from known ones. As an example, let Γ be a graph such that each block of $\bar{\Gamma}$ is the complement of an odd circuit. By Theorem 2 Γ is v-critical and $\chi(\Gamma) = 2b + c$. If $b = 2$ and $c = 1$ then $\chi(\Gamma) = 5$, and if the two odd circuits are of equal length then the number of edges of Γ is $> \frac{1}{4}n^2$.

Simonovits has described a more general construction-method [2, p. 71, proof of Theorem 2], that he obtained while studying large independent sets of vertices in critical graphs.

It is also easy to prove that Theorem 2 implies:

Corollary 4. *If $\bar{\Gamma}$ is a connected complement of a v-critical k-chromatic graph Γ ($k \geq 5$) and x is a vertex of Γ, then $\bar{\Gamma} - x$ may have anything up to $(k-1)/2$ connected components. Moreover, $\bar{\Gamma} - x$ has precisely $(k-1)/2$ connected components if and only if $\bar{\Gamma}$ consists of the complements of odd circuits of lengths ≥ 5, each pair of which has just x in common.*

The result of Corollary 4 is in sharp contrast to Theorem 3.

Since K_2 is not the complement of a v-critical graph, Theorem 2 also has the consequence that if Γ is v-critical then $\bar{\Gamma}$ has no bridges. This generalizes the result of Dirac [1, p. 463], that no vertex of a v-critical graph is joined to all other vertices except one.

For $\bar{\Gamma}$ of connectivity ≤ 2 the results of the investigation are more complicated, and perhaps rather unattractive. For v-critical graphs the characterization splits into four possible cases, and for e-critical graphs into two possible cases. Each of these six cases gives rise to one or more methods for constructions of critical graphs, and a variety of new examples of critical graphs can be obtained.

We shall describe the four v-critical cases in constructive terms.

Case 1. Let Γ_1 and Γ_2 be two graphs, each having ≥ 3 vertices, and $\Gamma_1 \cap \Gamma_2$ consisting of two vertices x_1 and x_2, these two vertices being independent in both

Γ_1 and Γ_2. Let Γ be obtained from $\Gamma_1 \cup \Gamma_2$ by joining all vertices of $\Gamma_1 - x_1 - x_2$ to all vertices of $\Gamma_2 - x_1 - x_2$ by edges. If Γ_1 and Γ_2 satisfy

(1) Γ_1 is v-critical,

(2) $\chi(\Gamma_2 - x_1) = \chi(\Gamma_2 - x_2) = \chi(\Gamma_2)$,

$\chi(\Gamma_2 - x_1 - x_2) = \chi(\Gamma_2) - 1$,

$\forall z \in V(\Gamma_2 - x_1 - x_2)$: either

$\chi(\Gamma_2 - x_1 - z) = \chi(\Gamma_2) - 1$

or

$\chi(\Gamma_2 - x_2 - z) = \chi(\Gamma_2) - 1$,

then Γ is v-critical and $\chi(\Gamma) = \chi(\Gamma_1) + \chi(\Gamma_2) - 1$.

Case 2. Let Γ_1 and Γ_2 be two graphs, each having $\geqslant 3$ vertices, and $\Gamma_1 \cap \Gamma_2$ consisting of two vertices x_1 and x_2 and the edge (x_1, x_2). Let Γ be obtained from $\Gamma_1 \cup \Gamma_2$ as in Case 1. If Γ_1 and Γ_2 satisfy

(1) $\chi(\Gamma_1 - x_1) = \chi(\Gamma_1 - x_2) = \chi(\Gamma_1 - x_1 - x_2) = \chi(\Gamma_1) - 1$,

$\forall z \in V(\Gamma_1 - x_1 - x_2)$: either

$\chi(\Gamma_1 - z) = \chi(\Gamma_1) - 1$

or

$\chi(\Gamma_1 - x_1 - x_2 - z) = x(\Gamma_1) - 2$,

(2) $\chi(\Gamma_2 - x_1) = \chi(\Gamma_2 - x_2) = \chi(\Gamma_2)$,

$\chi(\Gamma_2 \;\; x_1 \;\; x_2) - \chi(\Gamma_2) - 1$,

$\forall z \in V(\Gamma_2 - x_1 - x_2)$: either

$\chi(\Gamma_2 - x_1 - z) = \chi(\Gamma_2) - 1$

or

$\chi(\Gamma_2 - x_2 - z) = \chi(\Gamma_2) - 1$,

then Γ is v-critical and $\chi(\Gamma) = \chi(\Gamma_1) + \chi(\Gamma_2) - 1$.

Case 3. Let Γ_1, Γ_2 and Γ be as in Case 2. If Γ_1 and Γ_2 satisfy

Γ_i is v-critical for $i = 1, 2$, and

$\chi(\Gamma_i - x_1 - x_2) = \chi(\Gamma_i) - 2$ for $i = 1, 2$,

then Γ is v-critical and $\chi(\Gamma) = \chi(\Gamma_1) + \chi(\Gamma_2) - 2$.

Case 4. Let Γ_1, Γ_2, and Γ be as in Case 2. If Γ_1 and Γ_2 satisfy

(1) $\chi(\Gamma_1 - x_1 - x_2) = \chi(\Gamma_1) - 2$,

$\forall z \in V(\Gamma_1 - x_1 - x_2)$: either

$\chi(\Gamma_1 - x_1 - z) = \chi(\Gamma_1) - 2$

or

$\chi(\Gamma_1 - x_2 - z) = \chi(\Gamma_1) - 2$,

(in particular these conditions imply that Γ_1 is v-critical),

(2) Γ_2 is v-critical and $\chi(\Gamma_2 - x_1 - x_2) = \chi(\Gamma_2) - 1$,

then Γ is v-critical and $\chi(\Gamma) = \chi(\Gamma_1) + \chi(\Gamma_2) - 2$.

The verification of the four cases is based on Lemma 1. It is not difficult, and we leave it to the reader. Also based on Lemma 1 one can conversely prove:

Theorem 5. *Let Γ be v-critical and $\bar{\Gamma}$ 2-fold-connected with $C = \{x_1, x_2\}$ a cutset of Γ of size 2. Let Γ_1 and Γ_2 be as in Lemma 1.*

Then the notation may be chosen so that Γ_1 and Γ_2 satisfy one of the above four cases. In particular Γ is obtainable by one of the four constructions.

As said above: for each of the four cases a variety of examples can be obtained. We shall not describe these examples here, however a few of them are indicated in the following corollary.

Corollary 6. *Let Γ be v-critical and $\bar{\Gamma}$ 2-fold-connected with $\{x_1, x_2\}$ a cutset of $\bar{\Gamma}$ of size 2.*

(a) *For $\chi(\Gamma) \geq 4$ and (x_1, x_2) not an edge of Γ, the number of connected components of $\bar{\Gamma} - x_1 - x_2$ may be anything from 2 up to $\chi(\Gamma) - 2$. It is equal to $\chi(\Gamma) - 2$ if and only if $\bar{\Gamma}$ consists of the complement of an odd circuit of length ≥ 5 and $\chi(\Gamma) - 3$ circuits of length 4, each of these $\chi(\Gamma) - 2$ graphs having pairwise just x_1, x_2 and (x_1, x_2) in common.*

(b) *For $\chi(\Gamma) \geq 3$ and (x_1, x_2) an edge of Γ, the number of connected components of $\bar{\Gamma} - x_1 - x_2$ may be anything from 2 up to $\chi(\Gamma) - 1$. It is equal to $\chi(\Gamma) - 1$ if and only if $\bar{\Gamma}$ consists of a path of length 2 and $\chi(\Gamma) - 2$ paths of length 3, each of these $\chi(\Gamma) - 1$ paths having x_1 and x_2 as their end-vertices, and each pair of them having only x_1 and x_2 in common.*

We shall now turn to the e-critical case. Let thus Γ be an e-critical graph with $\bar{\Gamma}$ 2-fold-connected and $\{x_1, x_2\}$ a cutset of $\bar{\Gamma}$. Since an e-critical graph is also v-critical it follows from Theorem 5 that Γ is obtainable by one of the Cases 1–4. However, by an argument similar to the one used in the proof of Theorem 3, the

Cases 3 and 4 can be excluded. Moreover, of course, the conditions for Γ_1 and Γ_2 stated in the Cases 1 and 2 are only necessary, but not sufficient, for Γ to be e-critical. Some further necessary conditions can be obtained.

Theorem 7. *Let Γ be e-critical and $\bar{\Gamma}$ 2-fold-connected with $C=\{x_1, x_2\}$ a cutset of $\bar{\Gamma}$ of size 2. Let Γ_1 and Γ_2 be as in Lemma 1. Then the notation may be chosen so that Γ_1 and Γ_2 satisfy one of the above Cases 1 and 2. In particular Γ is obtainable by one of the constructions described in the Cases 1 and 2.*

Moreover, if the notation is as in Case 1 then Γ_1 and Γ_2 satisfy the following additional necessary conditions:

(1) *Γ_1 is e-critical, and $\Gamma_1-x_1-x_2$ is v-critical,*
(2) *Γ_2-x_1 and Γ_2-x_2 are both v-critical.*

If the notation is as in Case 2 then Γ_1 satisfies the following additional necessary condition:

(1) *$\Gamma_1-x_1-x_2$ is v-critical.*

It is possible to further extend the necessary conditions of Theorem 7 to necessary and sufficient conditions for Γ to be e-critical; however, the additional conditions are very unattractive (some of them are "mixed" conditions for Γ_1 and Γ_2), and we shall not discuss them here.

Theorem 7 has the following corollary:

Corollary 8. *Let Γ be e-critical and $\bar{\Gamma}$ 2-fold-connected with $\{x_1, x_2\}$ a cutset of $\bar{\Gamma}$ of size 2.*

(a) *If (x_1, x_2) is not an edge of Γ then $\chi(\Gamma)\geq 6$, and if $\chi(\Gamma)=6$ then the number of vertices of Γ is odd and ≥ 11 (and there are precisely two non-isomorphic examples with 11 vertices, see Fig. 2).*

(b) *If $\chi(\Gamma)\leq 5$, then $\bar{\Gamma}$ contains a vertex of valency 2, i.e. Γ contains a vertex joined to all other vertices except 2.*

(c) *$\bar{\Gamma}-x_1-x_2$ has precisely two connected components.*

Only the proof of (c) presents some difficulties. We shall outline a proof of (c).

Let Γ_1 and Γ_2 be as described in Theorem 7. If $\bar{\Gamma}_1-x_1-x_2$ has 2 or more than 2 connected components, then by Theorem 7 each of these connected components is the complement of a v-critical graph. Let t_1 be a vertex of a connected component $\bar{\Delta}_1$ of $\bar{\Gamma}-x_1-x_2$ such that (t_1, x_1) is an edge of $\bar{\Gamma}$. Let t_2 be a vertex of $\bar{\Delta}_2=\bar{\Gamma}-x_1-x_2-\Delta_1$ such that (t_2, x_2) is an edge of $\bar{\Gamma}$. Such vertices t_1 and t_2 exist, since x_1, x_2 is a minimal cutset of $\bar{\Gamma}$. By giving x_1 and t_1 the same colour, and x_2 and t_2 the same other colour, and using that Δ_1 and Δ_2 are both v-critical, it

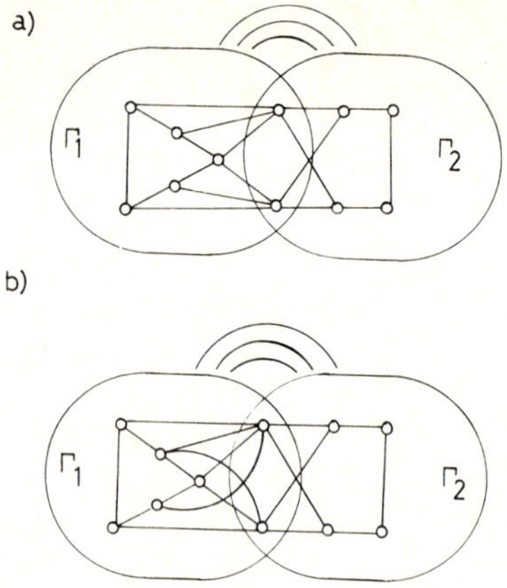

Fig. 2

follows that

$$\begin{aligned}\chi(\Gamma_1) &\leq \chi(\Delta_1 - t) + \chi(\Delta_2 - t_2) + 2\\ &= \chi(\Delta_1) - 1 + \chi(\Delta_2) - 1 + 2\\ &= \chi(\Gamma_1 - x_1 - x_2).\end{aligned}$$

This is a contradiction, since by Theorem 7 $\chi(\Gamma_1 - x_1 - x_2) = \chi(\Gamma_1) - 1$.

If $\bar{\Gamma} - x_1 - x_2$ has 3 or more than 3 connected components then, as noted after the proof of Lemma 1, there is a freedom in the choice of Γ_1 and Γ_2. Using this freedom, the above fact that $\bar{\Gamma}_1 - x_1 - x_2$ only has one connected component, and this component always being the complement of a v-critical graph, it follows that all connected components of $\bar{\Gamma} - x_1 - x_2$ are complements of v-critical graphs.

From this it follows as above that if $\bar{\Gamma}_2 - x_1 - x_2$ has 2 or more than 2 connected components, then

$$\chi(\Gamma_2) \leq \chi(\Gamma_2 - x_1 - x_2),$$

contradicting that $\chi(\Gamma_2 - x_1 - x_2) = \chi(\Gamma_2) - 1$ by Theorem 7. This proves Corollary 8(c).

By a result in [5], Corollary 8(b) implies that all e-critical 4-chromatic graphs whose complements are not 3-fold-connected are known.

The investigation described above was started in 1968 in the hope that it would produce e-critical 4- and 5-chromatic graphs with many edges. However, it only succeeded for v-critical 5-chromatic graphs, cf. the remarks to Theorems 2 and 3. This was somewhat disappointing, also because the obtained graphs with many edges were already known. In fact, I learned later that Zeidl in 1957 had obtained examples of v-critical 4- and 5-chromatic graphs with more than $\frac{1}{6}n^2$ edges (see [6]). The existence of e-critical 4- and 5-chromatic graphs with many edges was established in 1969 using other methods. Thus in [3] e-critical 4-chromatic graphs with more than $\frac{1}{16}n^2$ edges were obtained.

The v-critical 4-chromatic graphs of Zeidl (or the e-critical 4-chromatic graphs of [3]) and odd circuits can be combined as indicated in Theorem 2 to provide examples of v-critical 6-chromatic graphs with more than $\frac{3}{10}n^2$ edges (more than $\frac{8}{30}n^2$ edges). This shows that the graphs consisting of two disjoint odd circuits of equal length and completely joined by edges are not the v-critical 6-chromatic graphs with a maximum number of edges. Whether they are the e-critical ones is still an unsolved problem, however, in view of the above I suspect this not to be so.

Because of the "failure" of the investigation to produce interesting critical graphs with many edges, perhaps the most striking feature of it is the exhibition of qualitative differences between v-critical and e-critical graphs (in terms of their complements), cf. Corollary 4 and Theorem 3, and Corollary 6 and Corollary 8(c).

The proof-methods of the investigation are simple and straightforward, but somewhat tedious. In principle the investigation can be continued to cover critical graphs, whose complements contain cutsets of 3, 4, 5, ... and so on vertices. Maybe a continued investigation could produce more light, in particular this would be desirable in the e-critical case.

A rather detailed account of the investigation can be found in [4]. An improved and slightly extended version can be obtained on request from the author.

References

[1] G.A. Dirac, Map colour theorems related to the Heawood colour formula, J. London Math. Soc. 31 (1956) 460–471.

[2] M. Simonovits, On colour-critical graphs, Studia Sci. Math. Hung. 7 (1972) 67–81.

[3] B. Toft, On the maximal number of edges of critical k-chromatic graphs, Studia Sci. Math. Hung. 5 (1970) 461–470.

[4] B. Toft, Some contributions to the theory of colour-critical graphs, Ph.D. thesis, University of London (1970), published as No. 14 in Various Publications Series, Matematisk Institut, Aarhus Universitet.

[5] B. Toft, On critical subgraphs of colour-critical graphs, Discrete Math. 7 (1974) 377–392.

[6] B. Zeidl, Uber 4- und 5-chrome Graphen, Monatsh. Math. 62 (1958) 212–218.

Annals of Discrete Mathematics 3 (1978) 289–295.

THE SUBGRAPH PROBLEM

W.T. TUTTE

Department of Combinatorics and Optimization, University of Waterloo, Waterloo, Ontario N2L 3GI Canada

1. The 1-Factor Theorem

A 1-*factor* of a finite graph G can be defined as a regular spanning subgraph of G of valency 1. Petersen's Theorem [2, Chapter 10; 4] asserts that if a cubic graph is without a 1-factor then it has at least three isthmuses, not all in one arc.

Let G be any finite graph, with vertex-set $V(G)$. If S is any subset of $V(G)$ let G_S be the graph obtained from G by deleting the vertices of S and their incident edges. Let a component of G_S be called *odd* or *even* according as the number of its vertices is odd or even. Let us write $|S|$ for the number of elements of S, and $h(S)$ for the number of odd components of G_S. The following theorem is proved in [5].

Theorem 1.1. (1-Factor Theorem). *G is without a 1-factor if and only if there is a subset S of $V(G)$ such that $h(S)>|S|$.*

Now if the number of vertices of G is odd there can be no 1-factor. And we then have $h(\Omega)>|\Omega|$, where Ω is the null subset of $V(G)$. So in applications we can arrange, by excluding trivialities, that $|V(G)|$ is even. It then follows that

$$h(S)\equiv|S| \mod 2, \tag{1}$$

for each subset S of $V(G)$.

Let us see how the 1-Factor Theorem can be used to prove Petersen's Theorem. Let G be cubic and without a 1-factor. Then $|V(G)|$ is even and, by the 1-factor Theorem, there is a subset S of $V(G)$ such that $|S|<h(S)$. Using (1) we deduce that $|S|\leqslant h(S)-2$. The number of edges joining a given odd component of G_S to S must be odd. Let it be 1 in m cases and 3 or more in n cases. Then $|S|$ is at least $\frac{1}{3}(m+3n)$. We now have

$$\frac{1}{3}(m+3n)\leqslant|S|\leqslant h(S)-2=m+n-2,$$
$$m\leqslant 3m-6,$$
$$m\geqslant 3.$$

Accordingly G has at least 3 isthmuses and its isthmuses are not all in one arc, as Petersen's Theorem asserts.

2. The Subgraph Theorem

In this Section we generalize the notion of a 1-factor.

Let us suppose given a mapping f of $V(G)$ into the set of non-negative integers. We define an *f-factor* of G as a spanning subgraph of G in which the valency of x is $f(x)$, for each x in $V(G)$. By the "subgraph problem" we mean the problem of finding a usable necessary and sufficient condition for the general finite graph G to have an f-factor. When $f(x)=1$ for each vertex x an f-factor reduces to a 1-factor.

We define a *graph-triple* as an ordered triple (S, T, U), where S, T and U are disjoint subsets of $V(G)$ having $V(G)$ as their union. The components of the subgraph of G induced by U are called simply the *components of U*.

If $x \in V(G)$ and $X \subseteq V(G)$ let us write $\lambda(X, x)$ for the number of edges having x as one end and with the other end in X. If $x \in X$ then loops on x are to be counted twice in the evaluation of $\lambda(X, x)$. If X and Y are disjoint subsets of $V(G)$ we write $\lambda(X, Y)$ for the number of edges joining X to Y.

Consider a graph-triple $B=(S, T, U)$. If C is any component of U we write

$$J(B; C)=\sum_{b \in V(C)} \{f(b)+\lambda(T, b)\}. \tag{2}$$

We then say that C is an *odd* or an *even* component of U, with respect to B, according as $J(B; C)$ is odd or even. We write $h(B)$ for the number of odd components of U, with respect to B.

We define the *deficiency* $\delta(B)$ of B as follows.

$$\delta(B)=h(B)-\sum_{a \in S} f(a)+\sum_{c \in T} \{f(c)-\lambda(T, c)-\lambda(U, c)\}. \tag{3}$$

We call B an *f-barrier* of G if $\delta(B)>0$. The main result of [6] can be stated as follows.

Theorem 2.1. (Subgraph Theorem). *For a given f the graph G has either an f-factor or an f-barrier, but not both.*

This theorem is proved in [6] by the method of alternating paths. It is assumed in the proof that $f(x)$ is never zero, and that G has no loops. But these restrictions are removed in [7].

Let us check the Subgraph Theorem in some trivial cases. If G is to have a 1-factor it is clear that the sum of the "weights" $f(x)$ over all the vertices of G must be even. And if the weight-sum is odd we have the f-barrier $(\Omega, \Omega, V(G))$, of deficiency 1.

Let the valency of a vertex x of G be denoted by $\operatorname{val}(G, x)$. If G is to have an f-factor it is necessary that $f(x) \leqslant \operatorname{val}(G, x)$ for each x. But if $f(x) > \operatorname{val}(G, x)$ for some x then G has the f-barrier $(\Omega, V(G)-\{x\}, \{x\})$.

Neglecting the trivial cases just considered we may suppose that the weight-sum is even and that $f(x) \leq \text{val}(G, x)$ for each vertex x. We can then define a mapping f' of $V(G)$ into the set of non-negative integers such that

$$f'(x) = \text{val}(G, x) - f(x) \tag{4}$$

for each vertex x. If F is any f-factor of G then the edges of G not belonging to F determine an f'-factor F' of G. Evidently G has an f'-factor if and only if it has an f-factor.

We return to the graph-triple $B = (S, T, U)$ and the numbers $h(B)$, $\delta(B)$ and $J(B; C)$ calculated for it in terms of f. When f is replaced by f' let these numbers be replaced by $h'(B)$, $\delta'(B)$ and $J'(B; C)$ respectively. Let us also write $B' = (T, S, U)$. We note that

$$\begin{aligned} J(B; C) + J'(B', C) &= \sum_{b \in V(C)} \{\text{val}(G, b) + \lambda(S, b) + \lambda(T, b)\} \\ &\equiv \sum_{b \in V(C)} \lambda(U, b) \mod 2 \\ &\equiv 0 \mod 2 \end{aligned}$$

for any component C of U.

The numbers $J(B; C)$ and $J'(B'; C)$ being of the same parity for each component C of U we deduce that

$$h'(B') = h(B). \tag{5}$$

We can write $\lambda(S, c) - \text{val}(G, c)$ for $-\lambda(T, c) - \lambda(U, c)$ on the right of (3). We can therefore rewrite that equation as

$$\delta(B) = h(B) + \lambda(S, T) - \sum_{a \in S} f(a) - \sum_{c \in T} f'(c). \tag{6}$$

Using (5) we can then deduce that

$$\delta'(B') = \delta(B). \tag{7}$$

These results lead to the following theorem.

Theorem 2.2. (Interchange Theorem). *If B is a maximal f-barrier of G then B' is a maximal f'-barrier of G, and conversely.*

3. The Transfer Theorem

The Subgraph Theorem (2.1) is difficult to apply. Suppose for example that we wish to deduce the 1-Factor Theorem (1.1) from it. We put $f(x) = 1$ for each x, and assert that if no 1-factor of G exists then G has an f-barrier $B = (S, T, U)$. If T happens to be null then the 1-Factor Theorem follows immediately, $h(B)$ being the number $h(S)$ of Section 1. But why should T be null, and what can we do if it is not?

Some of the difficulties are overcome in [7]. If G has an f-barrier, for a general f, then it has a *maximal* f-barrier, that is one with the greatest possible deficiency. In [7] we are recommended to use the Subgraph Theorem in the following form: *G has either an f-factor or a maximal f-barrier, but not both.* It is found that when the f-barrier $B=(S, T, U)$ is maximal the vertices of S and T have special properties, and these can be exploited in problems of graph-factorization. We arrive at them by considering the effect of transferring a vertex between S and U.

Consider the graph-triple $B=(S, T, U)$ and let x be a vertex of S. We put $B_x=(S-\{x\}, T, U\cup\{x\})$ and consider the relation between $\delta(B)$ and $\delta(B_x)$.

Write $\mu(x)$ for the number of odd components of U, with respect to B, that are joined to x by edges of G. In B_x all such odd components are incorporated in a single component K of $U\cup\{x\}$, and K includes also the vertex x. The other odd components of U remain as the odd components, other than K, of $U\cup\{x\}$. Some even components of U, with respect to B, may also be contained in K.

It is easy to verify that

$$J(B_x, K)\equiv\mu(x)+f(x)+\lambda(T, x) \mod 2. \tag{8}$$

Let us define $\eta(x)$ as the number 0 or 1 having the same parity as $\mu(x)+f(x)+\lambda(T, x)$. Then we can deduce from the foregoing results that

$$h(B)-h(B_x)=\mu(x)-\eta(x). \tag{9}$$

It now follows from (3) that

$$\delta(B)-\delta(B_x)=\mu(x)-\eta(x)-f(x)+\lambda(T, x). \tag{10}$$

It should be emphasized that the expression on the right of (10) is always even, by the definition of $\eta(x)$.

If B is a maximal f-barrier the expression on the right cannot be negative; if B_x is a maximal f-barrier the expression cannot be positive. If the expression is zero and one of the graph-triples B and B_x is a maximal f-barrier then so is the other.

We take account also of the case in which x belongs to U instead of S. We then define $\mu(x)$ and $\eta(x)$ with respect to the graph-triple $B_1=(S\{x\}, T, U-\{x\})$, and use (10) with B_1 replacing B and B replacing B_x.

We state the consequences of (10) in the following theorem.

Theorem 3.1. (Transfer Theorem). *Let $B=(S, T, U)$ be a maximal f-barrier. If x is in S then*

$$f(x)\leqslant\mu(x)+\lambda(T, x)-\eta(x).$$

If x is in U then

$$f(x)\geqslant\mu(x)+\lambda(T, x)-\eta(x).$$

if

$$f(x) = \mu(x) + \lambda(T, x) - \eta(x)$$

then x can be transferred between S and U without affecting the maximality of B.

In the case of equality x is called in [7] a "left-neutral vertex".

In [7] we find some analogous results concerned with the transfer of a vertex between T and U. However, it is not necessary to state these separately; we can obtain them by applying the Transfer Theorem with f' replacing f and B' replacing B.

One consequence of the Transfer Theorem derived in [7], is that if G has no f-factor it has a maximal f-barrier $B = (S, T, U)$ such that $f(x) > 1$ for each x in T. Thus if $f(x) = 1$ for each vertex x we can arrange that T is null. This observation removes the difficulty we have encountered in proving the 1-factor Theorem as a consequence of the Subgraph Theorem.

Starting with the Subgraph and Transfer Theorems we can construct short proofs of Berge's extension of the 1-Factor Theorem (see [1, p. 154]) and of the Erdös-Gallai Theorem on valency-sequences [3]. In the next section we give a new example, a solution of a problem brought to the author's attention by Erdös.

4. A theorem on regular graphs

Our purpose in this Section is to give an example of the application of the Subgraph and Transfer Theorems (2.1 and 3.1). We use them to establish the following result.

Theorem 4.1. *Let G be a regular graph of valency k. Let r be an integer satisfying* $0 \leq r \leq k$. *Then there exists a spanning subgraph H of G such that* $\operatorname{val}(H, x) = r$ *or* $r+1$ *for each vertex x of G.*

Neglecting trivial cases we may assume $0 < r < k$. Suppose first that the number of vertices of G is an even number $2q$.

We introduce a new vertex w and join it to each vertex of G by a single new edge. We then attach q loops to w. Let the resulting graph be denoted by G_1. Then the valency of w in G_1 is $4q$, and the valency of each other vertex of G_1 is $k+1$.

Let f be the mapping of $V(G_1)$ into the set of non-negative integers defined as follows. $f(w) = 2q$, and $f(x) = r+1$ for each vertex x of G. Correspondingly we can write $f'(w) = 2q$ and $f'(x) = k - r$.

Let H be a spanning subgraph of G satisfying the conditions of Theorem 4.1. Then it has $2q$ vertices in all, and an even number of vertices of odd valency. It thus has an even number $2p$ of vertices of valency r. We can derive an f-factor F

of G_1 from H by adding to it w and the edges joining w to vertices of valency r in H, and then adjoining $q-p$ loops on w. Conversely if F is any f-factor of G_1 its intersection with G is a spanning subgraph H of G satisfying the conditions of the theorem.

Assume that Theorem 4.1 fails. Then by the above reasoning G_1 can have no f-factor. Hence, by the Subgraph Theorem, G_1 has a maximal f-barrier $B=(S, T, U)$.

Suppose w in S. Then, by the Transfer Theorem,

$$2q \leq \mu(w)+\lambda(T, w)-\eta(w). \tag{11}$$

But $\mu(w)$ cannot exceed $\lambda(U, w)$. Hence the expression on the right of (11) cannot exceed the number $2q$ of links incident with w. If (11) holds at all it must do so with strict equality. But then, by the Transfer Theorem, we can transfer w to U without destroying the maximality of B.

If instead we suppose w in T we can apply similar reasoning with f' replacing f and B' replacing B. We get an equation identical with (11) except that T is replaced by S. Using the Interchange Theorem (2.2) we deduce that if w is in T it can be transferred to U.

It is now permissible to assume that w is in U. But then U has only one component, since w is joined to every other vertex of G_1. Hence $h(B)$ is at most 1. Since $\delta(B)$ is at least 1 it follows from (6) that

$$(r+1)\,|S|+(k-r)\,|T| \leq \lambda(S, T). \tag{12}$$

Suppose first that $\lambda(S, T)$ is not zero. Then S and T are both non-null. Let R be one of them, the smaller of the two if they have different sizes. Then, by (12),

$$(k+1)\,|R| \leq \lambda(S, T). \tag{13}$$

We deduce from (13) that some vertex of R is incident with $k+1$ or more edges joining S to T. But this is impossible since the valency of each vertex of G is only k(in G).

In the remaining case $\lambda(S, T)=0$. Hence S and T are both null, by (12). Since B is an f-barrier we must suppose G_1 to be the only odd component of U. This means that the sum of the weights $f(x)$ over all vertices x of G_1 must be odd. But in fact the sum is $2q(r+1)+2q$. The foregoing contradictions establish the theorem in the special case for which G has an even number of vertices.

Let us now suppose the number of vertices of G to be odd. Let G_2 be the union of two disjoint copies of G. The theorem holds for G_2 by the result already proved. It therefore holds for G.

References

[1] C. Berge, Graphes et Hypergraphes (Dunod, Paris, 1970).
[2] N.L. Biggs, E.K. Lloyd and R.J. Wilson, Graph Theory 1736–1936 (Oxford, 1976).

[3] P. Erdös and T. Gallai, Graphs with prescribed degrees of vertices, Mat. Lapok 11 (1960) 264–273 (in Hungarian).
[4] J. Petersen, Die Theorie der regularen Graphs, Acta Math. 15 (1891) 193–220.
[5] W.T. Tutte, The factorization of linear graphs, J. London Math. Soc. 22 (1947) 107–111.
[6] W.T. Tutte The factors of graphs, Can. J. Math. 4 (1952) 314–328.
[7] W.T. Tutte, Spaning subgraphs with specified valencies, Discrete Math. 9 (1974) 97–108.